Language Evolution

# Studies in the Evolution of Language

*General Editors*
James R. Hurford, *University of Edinburgh*
Frederick J. Newmeyer, *University of Washington*

This series provides a forum for work in linguistics and related fields on the origins and evolution of language. It places a premium on scholarship, readability, and rational thought.

### Published

*The Origins of Vowel Systems*
Bart de Boer

*The Transition to Language*
Edited by Alison Wray

*Language Evolution*
Edited by Morten H. Christiansen and Simon Kirby

### Published in Association with the Series

*Language Diversity*
Daniel Nettle

*Function, Selection, and Innateness:*
*The Emergence of Language Universals*
Simon Kirby

*The Origins of Complex Language:*
*An Inquiry into the Evolutionary Beginnings of Sentences, Syllables, and Truth*
Andrew Carstairs-McCarthy

# Language Evolution

*Edited by*
Morten H. Christiansen
Simon Kirby

OXFORD
UNIVERSITY PRESS

# OXFORD

UNIVERSITY PRESS

Great Clarendon Street, Oxford OX2 6DP

Oxford University Press is a department of the University of Oxford.
It furthers the University's objective of excellence in research, scholarship,
and education by publishing worldwide in

Oxford New York

Auckland Bangkok Buenos Aires Cape Town Chennai
Dar es Salaam Delhi Hong Kong Istanbul Karachi Kolkata
Kuala Lumpur Madrid Melbourne Mexico City Mumbai
Nairobi São Paulo Shanghai Taipei Tokyo Toronto

Oxford is a registered trade mark of Oxford University Press
in the UK and in certain other countries

Published in the United States
by Oxford University Press Inc., New York

British Library Cataloguing in Publication Data
Data available

Library of Congress Cataloging in Publication Data
Data applied for
ISBN 0 19 924483 9 (hbk.)
ISBN 0 19 924484 7 (pbk.)

10 9 8 7 6 5 4 3 2

Typeset in Minion
by Peter Kahrel Ltd.
Printed in Great Britain
on acid-free paper by
Biddles Ltd., King's Lynn

# Contents

# Preface

In January 2000, we were invited by Janet Wiles and her colleagues to speak at the Workshop on Evolutionary Computation and Cognitive Science arranged in connection with the Fifth Australasian Cognitive Science Conference in Melbourne. It was one of those trips that remind you why you became an academic. The weather was wonderful; Melbourne is a great city, and Janet, as anyone who knows her will attest, is a fantastic host. The workshop was fascinating, and led us to the realization that language evolution really was a research topic whose time had come.

On one of our free days, we set off for a hike in the Grampians National Park along with Jim Hurford. In between enjoying the beautiful scenery and being awestruck by Eastern grey kangaroos, kookaburras, colourful parrots, and a blue-tongued lizard, we got talking about how quickly the interest in the origins and evolution of language was growing. It seemed that we were all used to fielding questions about the topic. These queries came from fellow academics, from students, and from interested friends. What were the key issues, the big questions, and the major controversies? Who studied the topic, and what backgrounds did they come from? Most importantly, was there any consensus on how, when, why, and in what manner language evolved?

We realized that we needed a definitive book on the subject: one that we could happily recommend to anyone interested in the area, one that gave the current states of the art, from the big names in every discipline that has a stake in answering these questions, and one that could form the foundation for courses on the evolution of language. This book is the result.

In a street café near St Kilda Beach, we began tackling our first challenge: compiling a list of contributors. This was very hard indeed, because we wished to give as broad a perspective as possible. There were many people we wished we could have included, and would have, had length not been an issue. That said, we think the resulting chapters, along with their further reading sections, should provide a good springboard into the wider, primary literature. The book spans an extensive range of different scientific disciplines, including: anthropology, archaeology, biology, cognitive science, computational linguistics, linguistics, neurophysiology, neuropsychology, neuroscience, philosophy, primatology, psycholinguistics, and psychology. Yet despite having their academic home in such different fields, our authors were overwhelmingly positive in their response to the idea of the book. It thus seems that our belief in the necessity of a collection like this was both timely and widely shared.

It was important for us that the book should be accessible to a wide audience of readers. The authors were therefore asked to provide their up-to-date perspectives on language evolution in as non-technical a way as possible without overly simplifying the issues. To fine-tune the book both in terms of coverage and readability, we gave it a test-run in a combined advanced undergraduate and graduate class on the evolution of language that Morten taught at Cornell University. Every chapter in the book was debated and critiqued in class. As part of the course, students were asked to submit electronic questions regarding each chapter. These e-questions were then passed on to the authors who used them to revise their chapters. Our authors did a great job of incorporating the students' comments and suggestions, improving their already very well written chapters even further. The final result is a very readable volume that also makes an excellent textbook for courses on the origin and evolution of language.

There have been many people without whom this book would have been no more than an idea that we discussed in Australia. In particular, we are indebted to the students in Morten's class. So, many thanks to Mike Brantley, Chris Conway, Rick Dale, Erin Hannon, Ben Hiles, Gary Lupyan, Janice Ng, Makeba Parramore, and Karen Tsui for their very helpful feedback. Others who have helped out along the way include: Jelle Zuidema, Richard Dawkins, and John Maynard Smith. Thanks are also due to John Davey, Jacqueline Smith, and Sarah Barrett at OUP, and to Jim Hurford, who helped get this project off the ground in the first place, and made sure it kept flying.

Very special thanks go out to our loved ones, Anita Govindjee, Sunita Christiansen, and Anna Claybourne, for their patience, encouragement, and support throughout the editorial work on this book.

Finally, of course, we are grateful to our contributing authors. Thank you for your patience with inevitable delays, your readiness to make changes, your enthusiasm for the project, but most of all for your chapters. We think that together, they will stimulate further interest in and understanding of the origins and evolution of what makes us human: language.

Morten H. Christiansen and Simon Kirby

*Ithaca and Edinburgh 2003*

# List of Figures

# List of Tables

# Notes on the Contributors

MICHAEL A. ARBIB's first book, *Brains, Machines and Mathematics* (McGraw-Hill, 1964), set the main theme of his career: the brain is not a computer in the current technological sense, but we can learn much about machines from studying brains, and much about brains from studying machines. He currently focuses on brain mechanisms underlying the co-ordination of perception and action, working closely with the experimental findings of neuroscientists on mechanisms for eye-hand co-ordination in humans and monkeys. As in this article, he is now using his insights into the monkey brain to develop a new theory of the evolution of human language.

DEREK BICKERTON is Emeritus Professor of Linguistics at the University of Hawaii. He originally specialized in creole languages, and developed the controversial 'bioprogram theory' which claimed that these languages are originated by children in a single generation from unstructured input, and represent Universal Grammar in a purer form than do older languages. His major goal at present is to integrate the empirical findings (not necessarily the formalisms) of generative grammar with current knowledge of biological evolution and neurological structure, thereby producing a realistic model of language and its origins.

TED BRISCOE is Reader in Computational Linguistics at the Computer Laboratory, University of Cambridge where he has been a member of staff since 1989. He works on statistical and constraint-based approaches to natural language processing as well as evolutionary modelling and simulation of language development and change.

MORTEN H. CHRISTIANSEN is an Assistant Professor of Psychology at Cornell University. His research integrates connectionist modeling, statistical analyses, behavioural experimentation, and event-related potential (ERP) methods in the study of the learning and processing of complex sequential structure, in particular as related to the acquisition, processing, and evolution of language. He received his Ph.D. in Cognitive Science from the University of Edinburgh, Scotland, and has held several interdisciplinary positions before joining Cornell University. He is the editor of *Connectionist Psycholinguistics* (with Nick Chater, Ablex, 2001).

MICHAEL C. CORBALLIS was born in New Zealand and completed his Ph.D. at McGill University in Montreal, Canada, where he taught from 1968 until 1977. He then returned to his present position as Professor of Psychology at the University of Auckland. He has carried out research in a number of areas of cognitive and evolutionary neuroscience, including human laterality, visual perception, the split brain, and the evolution of language. His books include *The Psychology of Left and Right* (with I. L. Beale), *The Lopsided Ape*, *The Descent of Mind* (with S. E. G. Lea), and *From Hand to Mouth*.

IAIN DAVIDSON is Professor of Archaeology and Paleo-anthropology at the University of New England in Armidale, NSW, Australia. Since 1988 he has published a book (*Human evolution, language and mind*, CUP 1996) and more than 35 other publications on the evolutionary emergence of language, mostly jointly with psychologist William Noble. His other research work has been concerned with the Upper Palaeolithic, particularly of Spain, and with the archaeology of Australia including its stone artefacts and rock art.

TERRENCE W. DEACON received his Ph.D. from Harvard University in 1984 with research tracing the primate homologues to cortical language circuits. Since then he has held faculty positions at Harvard University (1984–92), Boston University (1992–2002), Harvard Medical School (1992–2000), and University of California at Berkeley (2002). Professor Deacon's research combines human evolutionary biology and neuroscience, with the aim of investigating the evolution of human cognition. His work extends from laboratory-based cellular-molecular neurobiology to the study of semiotic processes underlying animal and human communication, especially language. Many of these interests are explored in his book *The Symbolic Species: The Coevolution of Language and the Brain* (Norton, 1997). His new book, *Homunculus* (Norton, in preparation) explores the relationship between self-organization, evolution, and semiotic processes.

ROBIN I. M. DUNBAR graduated from the University of Oxford with a B.Sc. in Psychology and Philosophy, and then gained a Ph.D. in Psychology from the University of Bristol. He has subsequently held research and academic posts at the University of Cambridge, University of Stockholm (Sweden) and University College London. He is currently Professor of Evolutionary Psychology at the University of Liverpool, where he leads a research group of 3–5 staff and 12–15 postgraduate students whose research focuses on the behavioural ecology and evolutionary psychology of large mammals (mainly primates and ungulates) and humans.

W. TECUMSEH FITCH studies the evolution of cognition in animals and man focusing on the evolution of communication. Originally trained in ethology and evolutionary biology, he has more recently applied his graduate training in speech science to animal vocal communication. He is interested in all aspects of vocal communication in terrestrial vertebrates, particularly aspects of vocal production that bear on questions of meaning and honesty in animal communication systems, including human language. Fitch received his B.A. in Biology and Medicine and his Ph.D. in Cognitive and Linguistic Sciences, both from Brown University. He joined the faculty at Harvard as a Lecturer in Biology and Psychology in 1999. He conducts research on humans and various vertebrates (including alligators, birds, and monkeys). In 2002 he was a Fellow at the European Institute for Advanced Studies, in Berlin.

LOUIS GOLDSTEIN is Professor of Linguistics and Psychology at Yale University and Senior Research Scientist at Haskins Laboratories. He has developed, along with Catherine Browman, a gesture-based theory of phonology called 'articulatory phonology'. His research includes experimental work that tests the role of gestures in phonological encoding and development modeling work on the dynamics of inter-gestural coordination and the different modes of coordination that can be employed in human languages and simulation work on the self-organization of phonology through mutual attunement of computational agents.

MARC D. HAUSER's research sits at the interface between evolutionary biology and cognitive neuroscience and is aimed at understanding the processes and consequences of cognitive evolution. Observations and experiments focus on captive and wild primates, incorporating methodological procedures from ethology, infant cognitive development, cognitive neuroscience and neurobiology. Current foci include studies of numerical abilities, the role of inhibitory control in problem solving, cortical physiology of acoustic processing in primates, the nature of conceptual acquisition in a non-linguistic species, the shared and unique computational mechanisms subserving the faculty of language, and the mechanisms underlying the production and perception of vocal signals in primates. Hauser is a Professor at Harvard University in the Department of Psychology and the Program in Neurosciences. He is the author of three books including, most recently, *Wild Minds: What Animals Really Think* (Holt, 2000).

JAMES R. HURFORD has a broad interest in reconciling various traditions in Linguistics which have tended to conflict. In particular, he has worked on articulating a framework in which formal representation of grammars in individual minds interacts with statistical properties of language as used in communities. The framework emphasizes the interaction of evolution, learning and communication. He is perhaps best known for his computer simulations of various aspects of the evolution of language.

SIMON KIRBY is a research fellow in the Language Evolution and Computation Research Unit at the University of Edinburgh. At the LEC he has pioneered a computational approach to understanding the origins and evolution of language which treats human language as a complex adaptive system. His previous book–Function, Selection and Innateness: The Emergence of Language Universals—is also published by Oxford University Press.

NATALIA L. KOMAROVA studied Theoretical Physics in Moscow State University. Her Master's thesis was with Alexander Loskutov on Chaos Control in 2D maps. She received her Ph.D. in Applied Mathematics in 1998 from the University of Arizona, where she studied Non-linear Waves and Natural Pattern Formation under the supervision of Alan Newell. She came to the Institute for Advanced Study (IAS), Princeton, in 1999, as a member at the School of Mathematics, and then she joined the Program in Theoretical Biology (headed by Martin Nowak) at the IAS. She is interested in applying mathematical tools to describe natural phenomena, of which evolution of language is one of the most fascinating and challenging problems.

PHILIP LIEBERMAN received degrees in Electrical Engineering in 1958 and a Ph.D. in Linguistics in 1966 at the Massachusetts Institute of Technology. His interests have included the prosody of language, voice analysis of laryngeal pathologies, and psychological stress. His primary focus has been on the nature and evolution of the biological bases of human language. This includes studies on the evolution of human speech producing anatomy and the human brain, with special attention to the role of subcortical basal ganglia. His research complements many independent studies that indicate that the neural bases of motor control (particularly speech), syntactic competence, and cognitive ability are interrelated.

FREDERICK J. NEWMEYER specializes in syntax and the history of linguistics and has as his current research program the attempt to synthesize the results of formal and functional linguistics. He is the author or editor of twelve books, including *Linguistic Theory in America* (Academic Press, 1980) and *Language Form and Language Function* (MIT Press, 1998). Newmeyer gained his Ph.D. from the University of Illinois in 1969 and since then has taught in the Department of Linguistics at the University of Washington in Seattle. He served as President of the Linguistic Society of America in 2002.

MARTIN A. NOWAK studied biochemistry and mathematics at the University of Vienna, where he received his Ph.D. in 1989. Subsequently, he went to the University of Oxford to work with Robert May. In 1992, Nowak became a Wellcome Trust Senior Research Fellow and in 1997 Professor of Mathematical Biology. In 1998 he moved from Oxford to Princeton to establish the first program in Theoretical Biology at the Institute for Advanced Study. In 2002, he moved to Harvard University as Professor of Mathematics and Biology. Nowak is interested in all aspects of applying mathematical thinking to biology. In particular, he works on the dynamics of infectious diseases, cancer genetics, the evolution of cooperation and human language.

STEVEN PINKER received his BA from McGill University in 1976 and his Ph.D. in Psychology from Harvard in 1979. After serving on the faculties of Harvard and Stanford Universities for a year each, he moved to MIT in 1982, where he is currently Peter de Florez Professor in the Department of Brain and Cognitive Sciences and a MacVicar Faculty Fellow. Pinker's research has focused on visual cognition and the psychology of language. In addition to his research papers, he has written two technical books on language acquisition *Language Learnability and Language Development* (Harvard University Press, 1984), and *Learnability and Cognition* (MIT Press, 1989), and four books for a wider audience *The Language Instinct* (HarperCollins/Morrow. 1994), *How the Mind Works* (Norton, 1997), *Words and Rules: The Ingredients of Language* (HarperCollins, 1999), and *The Blank Slate* (Viking, 2002).

MICHAEL STUDDERT-KENNEDY has a BA in Classics from Cambridge University and a Ph.D. in Experimental Psychology from Columbia University. He is Professor Emeritus of Communications at the City University of New York, Professor Emeritus of Psychology at the University of Connecticut and former President of Haskins Laboratories. He has been a research scientist at Haskins Laboratories since 1961, with particular interests in speech perception, hemispheric specialization for speech perception and, most recently, in the ontogeny and evolution of speech.

MICHAEL TOMASELLO is co-director of the Max Planck Institute for Evolutionary Anthropology, Leipzig, Germany. His research interests focus on processes of social cognition, social learning, and communication in human children and great apes. Books include *Primate Cognition* (with J. Call, Oxford University Press, 1997), The *New Psychology of Language: Cognitive and Functional Approaches to Language Structure* (edited, Lawrence Erlbaum, 1998), and *The Cultural Origins of Human Cognition* (Harvard University Press, 1999).

# 1

---

# Language Evolution:
# The Hardest Problem in Science?

*Morten H. Christiansen and Simon Kirby*

What is it that makes us human? If we look at the impact that we have had on our environment, it is hard not to think that we are in some way 'special'—a qualitatively different species from any of the ten million others. Perhaps we only feel that way because it is hard to be objective when thinking about ourselves. After all, biology tells us that all species are exquisitely adapted to their respective ecological niches. Nevertheless, there is something odd about humans. We participate in hugely complex and diverse types of social systems. There are humans living in almost every environment on earth. We mould the world around us in unprecedented ways, creating structures that can be seen from space, and then going into space to see them.

One of our achievements, especially over the previous century, has been a staggering growth in our scientific understanding of the universe we live in. We are closing in on a complete unitary theory of its building blocks, and we know much about how it started. Yet despite this, our understanding of our place in this universe is far from complete. We still have only a hazy understanding of what exactly it is that makes us human.

Advances are being made, however. The cognitive neurosciences are bringing our view of the brain into focus, and the recent success of human genome sequencing gives us a recipe book for how we are built. However, these approaches to humanity mostly show us how similar we are to other forms of life. The essence of human uniqueness remains elusive.

In this book, we contend that the feature of humanity that leads to the strange properties listed above is language. To understand ourselves, we must understand language. To understand language, we need to know where it came from, why it works the way it does, and how it has changed.

To some it may be a surprise that, despite rapid advances in many areas of science, we still know relatively little about the origins and evolution of

this peculiarly human trait. Why might this be? We believe that at least pa
of the answer is that a deep understanding of language evolution can or
come from the concerted, joint effort of researchers from a huge range
disciplines. We must understand how our brains and minds work; how la
guage is structured and what it is used for; how early language and mode
language differ from each other and from other communication systems;
what ways the biology of hominids has changed; how we manage to acqui
language during development; and how learning, culture, and evolution
teract.

This book is intended to bring together, for the first time, all the maj
perspectives on language evolution, as represented by the various fields th
have a stake in language evolution research: psycholinguistics, linguisti
psychology, primatology, philosophy, anthropology, archaeology, biolo
neuroscience, neuropsychology, neurophysiology, cognitive science, a
computational linguistics. The chapters are written by the key authoriti
in each area, and together they cast the brightest light yet on questions s
rounding the origin and evolution of language.

## The Many Facets of Language Evolution

In 1859, when Charles Darwin published his book *The Origin of Spec.
there was already a great interest in the origin and evolution of language
plethora of ideas and conjectures flourished, but with few hard constrair
to limit the realm of possibility, the theorizing became plagued by outla
ish speculations. By 1866 this situation had deteriorated to such a degr
that the primary authority for the study of language at the time—the inf
ential Société de Linguistique de Paris—felt compelled to impose a ban
all discussions of the origin and evolution of language.

This ban effectively excluded all theorizing about language evoluti
from the scientific discourse for more than a century. The scientific int
est in language evolution was rekindled with the conference on 'Origins a
Evolution of Language and Speech', sponsored by the New York Acader
of Sciences in 1975. However, it took an additional decade and a half befo
the interest in language evolution resurged in full. Fuelled by theoretic
constraints derived from advances in the brain and cognitive sciences, t.
field finally emerged during the last decade of the twentieth century as a
gitimate area of scientific inquiry.

The landmark paper 'Natural Language and Natural Selection', published in 1990 by Steven Pinker and Paul Bloom in the respected journal *Behavioural and Brain Sciences*, is considered by many to be the catalyst that brought about the resurgence of interest in the evolution of language.[1] The paper proposed the theory that the human ability for language is a complex biological adaptation evolved by way of natural selection. In Chapter 2, Pinker updates the theory in the light of new empirical data and the theoretical alternatives that have emerged since the original paper (many of which are represented in subsequent chapters). He lists a number of properties of the language system that give the appearance of complex design. By analogy to the visual system, he argues that the only plausible explanation for the evolution of such complex adaptive design is one that involves natural selection. On this account, language has evolved as an innate specialization to code propositional information (such as who did what to whom, when, where, and why) for the purpose of social information-gathering and exchange within a humanly distinct 'cognitive niche'. In further support for his perspective, Pinker concludes his chapter with a discussion of recent evidence regarding the possible genetic bases of language and the application of mathematical game theory to language evolution (the latter described in more detail in Komarova and Nowak, Chapter 17). The work described in Briscoe (Chapter 16) on grammatical assimilation—an evolutionary genetic adaptation for language acquisition—from the viewpoint of computational linguistics also seems compatible with Pinker's approach.

In Chapter 3, James Hurford agrees with Pinker that humans have evolved a unique mental capacity for acquiring language, but disagrees with him over the role of cultural transmission (learning) in explaining language evolution. Hurford argues that language evolution needs to be understood as a combination of both biological pre-adaptations—that is, biological changes that may not be adaptive by themselves—and learning-based linguistic adaptations over generations. He points to several possible biological steps prior to the emergence of language: pre-adaptations for the production of speech sounds (phonetics), for organizing the sounds into

---

[1] According to the ISI Web of Knowledge index, the rate at which language evolution work appears in the literature increased tenfold in the decade following the Pinker and Bloom paper. Thus, when counting the papers that contain both 'language' and 'evolution' in title, keywords, or abstract, the publishing rate for 1981–1989 was 9 per year, whereas it was 86 per year for the period 1990–1999, and 134 per year between 2000 and 2002.

complex sequences (syntax), for forming basic and complex concepts and doing mental calculations with them (semantics), for complex social inter-action (pragmatics), and for an elementary ability to link sounds to con-cepts (symbolic capacity). Once humans were language-ready with these pre-adaptations in place, language systems would have grown increasingly complex due to the process of transmitting language across generations through the narrow filter of children's learning mechanisms. Hurford ex emplifies the processes of cultural transmission by reference to research on grammaticalization (see also Tomasello, Chapter 6). Grammaticalization refers to rapid historical processes by which loose and redundantly organ ized utterance combinations can become transformed into a more compact syntactic construction (e.g. *My dad . . . He plays tennis . . . He plays with his colleagues* may become *My dad plays tennis with his colleagues*). Additional work within the computational modelling of language evolution provides further illumination of the possible consequences of cultural transmis sion. Specifically, Hurford describes computer simulations in which simple but coordinated language systems emerge within populations of artificial agents through iterated learning across generations (this work is described in more detail in Kirby and Christiansen, Chapter 15).

This book is in many ways a testament to the many different disciplines that have become involved in the study of language evolution over the past decade. In Chapter 4, however, Frederick Newmeyer points to the surpris ing fact that researchers in linguistics—the study of language—have been slow to join the resurgence of interest in the evolution of language (see also Bickerton, Chapter 5). Part of the reason, he suggests, may be that linguists are not in agreement about how to characterize *what* evolved, and this com plicates uncovering *how* it may have evolved. Another possible stumbling block appears to be one of the key dogmas in linguistics: uniformitarian ism. Almost all linguists take it for granted that, in some important sense, all languages are equal. That is, there is no such thing as a 'primitive' language-the language of a nomadic tribe of hunter-gatherers is no less complex than the language spoken in an industrialized society. Newmeyer suggests that a more measured approach to uniformitarianism is needed because there may have been differences in the use of language across language evolution. For example, language may originally have been used as a tool for concept alization rather than communication. Newmeyer concludes that a less rigid view of uniformitarianism, combined with a better understanding of the biological bases for language and how languages change over time, is likely to lead to an increasing number of linguists raising their voices among the chorus of language evolution researchers.

Derek Bickerton strikes a more worried tone in Chapter 5 when discussing the odd fact that few linguists appear to be interested in language evolution. He is concerned that many non-linguists are proposing theories based on simplified 'toy' examples that may be inconsistent with the facts about language as seen from the viewpoint of linguistics. Against this backdrop, he suggests that when approaching language from an evolutionary perspective it is important to look at language not as a unitary phenomenon, but as the coming together of three things: modality, symbols, and structure. He argues that a largely cultural emergence of symbolic representation combined with a biological adaptation of brain circuitry capable of encoding syntactic structure were the two distinct evolutionary sources that gave rise to human language. Only later would a preference for the spoken modality have evolved, and then entirely contingent on the prior existence of the symbolic and structural components of language. From this perspective, the evolutionary dissociation of symbols and structure are reflected in ape language studies, where learning of symbolic relations approaches a near-human level of performance but where only a limited grasp of syntax has been demonstrated. Bickerton concludes that a capacity for structural manipulations of symbols may be the key adaptation that gives us, but no other species, language in all its intricate complexity.

Whereas Bickerton stresses the importance of linguistics in understanding the evolution of language, Michael Tomasello emphasizes the role of psychology. Nonetheless, Tomasello, in Chapter 6, also shares the view that it was the separate evolution of capacities for using symbols and grammar (that is, syntactic structure) that distinguishes human communication from the communication of other primates. In contrast to Pinker and Bickerton, he suggests that there was no specific biological adaptation for linguistic communication. Rather, Tomasello argues that there was an adaptation for a broader kind of complex social cognition that enabled human culture and, as a special case of that, human symbolic communication. A crucial part of this adaptation was an evolved ability to recognize other individuals as intentional agents whose attention and behaviour could be shared and manipulated. The capacity for grammar subsequently developed, and became refined through processes of grammaticalization occurring across generations (see also Hurford, Chapter 3)—but with no additional biological adaptations. In support for this perspective, Tomasello reviews psychological data from the study of language development in young children and from comparisons with the linguistic, social, and mental capacities of non-human primates (see also Hauser and Fitch, Chapter 9). More generally, Tomasello sees the origin and emergence of language as merely one part in the much larger process of the evolution of human culture.

In Chapter 7, Terrence Deacon also places the human ability for complex symbolic communication at the centre of the evolution of language. Contrary to Tomasello, however, Deacon does not find that the many sub-patterns of language structure that can be found across all the languages of the worlds—the so-called language universals—are products of cultural processes; neither does he think that they reflect a set of evolved innate constraints (a language-specific 'Universal Grammar') as proposed by Bickerton, Pinker, and others. Instead, drawing on research in philosophy and semiotics (the study of symbol systems), Deacon argues that they derive from a third kind of constraint originating from within the linguistic symbol system itself. Because of the complex relationships between words and what they refer to (as symbols), he suggests that semiotic constraints arise from within the symbol system when putting words together to form phrases and sentences. As an analogy, Deacon refers to mathematics. Although the mathematical concept of division has been around for millennia, it would seem incorrect to say that humans invented division. Rather, we would say that the concept was discovered. Indeed, we would expect that mathematical concepts, such as division, are so universal that they would be the same anywhere in the universe. As an example, Deacon points out that the SETI (Search for Extra-Terrestrial Intelligence) project transmits pulses counting out prime numbers into deep space with the idea that any alien beings would immediately recognize that these signals were generated by intelligent beings rather than by some natural astronomical source. Similarly, Deacon proposes that during the evolution of language humans have discovered the set of universal semiotic constraints. These constraints govern not only human language but also, by their very nature, any system of symbolic communication, terrestrial or otherwise.

Iain Davidson, too, focuses on the human use of symbols in Chapter 8, but this time illuminated from the viewpoint of archaeology. He argues that anatomical evidence from skeletal remains contributes little to the understanding of the evolution of language because of the difficulty in determining possible linguistic behaviours from fossilized bones (but see Lieberman, Chapter 14, for a different perspective). Instead Davidson points to the archeological record of artefacts because they may reveal something about the behaviour that produced them. In particular, analyses of ancient art objects provide evidence of symbol use dating back at least 70,000 years. To Davidson, these artefacts indicate sophisticated symbol use that incorporates two key features of language: open-ended productivity and the ability to use symbols to stand for things displaced in time and place. On the other hand, he notes that evidence of syntax has proved more elusive in the archaeological record. Like many of the other contributors, Davidson sees symbol use as the first crucial step toward modern human language, with syntax emerging through cultural learning processes that include grammaticalization and iterated learning across generations (see also Hurford, Tomasello, and Kirby and Christiansen, Chapters 3, 6, and 15).

The previous chapters have highlighted the use of symbols as a unique human ability. In Chapter 9, Marc Hauser and Tecumseh Fitch take a biologist's perspective on language evolution, advocating the use of a comparative method for exploring the various other components that make up the human language ability. They argue that studying animals, in particular non-human primates, is the only way to determine which components of language may be unique to humans and which may be shared with other species (see also Tomasello, Chapter 6, for a similar point). Hauser and Fitch review a wealth of data regarding the mechanisms underlying the production and perception of speech. When it comes to vocal production, they find very little that is unique to humans (but see Lieberman, Chapter 14, for a different perspective), except perhaps a much more powerful ability for combining individual sound units (phonemes and syllables) into larger ones (words and phrases). As for speech perception, the evidence suggests that the underlying mechanisms also are shared with other mammals. Moreover, Hauser and Fitch propose that the mechanisms underlying the production and perception of speech in modern humans did not evolve for their current purposes; rather, they evolved for other communicative or cognitive functions in a common ancestor to humans and chimpanzees. However, Hauser and Fitch share with Bickerton the suggestion that the fundamental difference between humans and non-human animals is the capacity to use recursive syntax—the ability to take units of language, such as words, and recombine them to produce an open-ended variety of meaningful expressions.

In Chapter 10, Michael Arbib outlines another language evolution perspective that is based on comparison with non-human primates, but with

a focus on brain anatomy. He suggests that biological evolution resulted
a number of pre-adaptations leading to a language-ready brain (see al
Hurford, in Chapter 3). One of the key pre-adaptations on this account
the evolution of a mirror system, providing a link between the producti
and perception of motor acts. The mirror system has been studied ext
sively in monkeys where it is found in an area of monkey cortex (F5) that
considered to be homologous to Broca's area in the human brain—an ar
that appears to play an important role in human language. Arbib sugge
that the mirror system forms the evolutionary basis for a link betweer
sender of a message and the perceiver of that message. The same subset
neurons appears to be active in the mirror system both when generating
particular motor act and when observing others producing the very sar
motor act. Following the evolution of a unique human ability for compl
imitation, Arbib proposes that language originated in a system of manu
gestures, and only later evolved into a primarily spoken form. Finally, Art
joins Hurford, Tomasello, and Davidson in arguing that syntax emerged
the result of subsequent cultural evolution.

Michael Corballis also sees language as originating with a system of ma
ual gestures, but comes to this conclusion from the viewpoint of cogniti
and evolutionary neuroscience. In Chapter 11 he reviews a broad range
data, including studies of language and communicative abilities in apes (s
also Tomasello, Chapter 6), the skeletal remains and artefacts in the arch
logical record (see also Davidson, Chapter 8), and the language abilities
hearing, deaf, and language-impaired human populations (see also Pink
Chapter 2). He argues that whereas non-human primates tend to gestu
only when others are looking, their vocalizations are not necessarily
rected at others—perhaps because of differences in voluntary control ov
gestures and vocalizations. Corballis suggests that one of the first steps
language evolution may have been the advent of bipedalism, which wou
have allowed the hands to be used for gestures instead of locomotion. I
follows Pinker in pointing to a gradual evolution of a capacity for gramm
though Corballis maintains that language remained primarily gestural u
til relatively late in our evolutionary history. The shift from visual gestur
to vocal ones would have been gradual, and he proposes that largely autc
omous vocal language arose following a genetic mutation between 100,0
and 50,000 years ago.

The gestural theories of language origin as outlined by Arbib and Corb
lis are not without their critics. Robin Dunbar argues in Chapter 12 that t

arguments in favour of a gestural origin of language are largely circumstantial. He moreover contends that gestural language suffers from two major disadvantages in comparison with spoken language: it requires direct line of sight, and it cannot be used at night. Instead, Dunbar proposes that language originated as a device for bonding in large social groups. He notes that grooming is the mechanism of choice among primates to bond social groups. However, human social groups tend to be too large for it to be possible for grooming to bond them effectively. Language, on this account, emerged as a form of grooming-at-a-distance, which is reflected in the large amount of time typically spent verbally 'servicing' social relationships. Dunbar sees the use of primate-like vocalizations in chorusing—a kind of communal singing—as a key intermediate step in the evolution of language. Once such cooperative use of vocalizations was in place, grammar could then emerge through processes of natural selection. Like Pinker, Dunbar also refers to recent mathematical game theory modelling of language evolution in support of this standard Darwinian perspective (for details see Komarova and Nowak, Chapter 17). In addition, he points to the extraordinary capacity of language to diversify into new dialects and distinct languages, suggesting that this property of language may have evolved to make it easier for members of social groups to identify each other. Thus, Dunbar's proposal about the social origin of language can explain both the origin and subsequent diversification of language.

In Chapter 13, Michael Studdert-Kennedy and Louis Goldstein also point to vocalization as the basis for language evolution, but focus on the mechanics involved in producing the sounds of human languages. They propose that a key pre-adaptation for language was the evolution of a system in which a limited set of discrete elements could be combined into an unlimited number of different larger units (see also Hauser and Fitch, Chapter 9, for a similar perspective). They suggest that the ability for vocal language draws on ancient mammalian oral capacities for sucking, licking, chewing, and swallowing. Subsequent evolutionary pressures for more intelligible information exchanges through vocalizations would then have led to a further differentiation of the vocal tract. On their account, this resulted in the evolution of six different brain-controlled motor systems to modify the configuration of the vocal tract, comprising the lips, tongue tip, tongue body, tongue root, velum (the soft part in the back of the roof of the mouth), and the larynx (the 'voice box' containing the vocal cords). Different configurations of these discrete systems result in different phonetic gestures (not to be confused with the manual gestures mentioned by Arbib and Corballis, Chapters 10 and 11). Studdert-Kennedy and Goldstein argue that subsequent expansion, elaboration, and combination of phonetic gestures into larger complex structures would have occurred through processes of cultural evolution involving attunement among speakers through vocal mimicry.

In Chapter 14 Philip Lieberman, too, emphasizes the importance of speech production in language evolution. He reviews a wide range of neuropsychological and neurophysiological data relevant to explaining the evolution of language. Like Corballis, he points to the advent of bipedalism as the first step toward the evolution of language. However, Lieberman argues that upright walking would have resulted in biological adaptations of basal ganglia—a collection of subcortical brain structures—for the learning and sequencing of more complex movements. These changes to basal ganglia formed the key adaptation en route to language. In support for this connection between language and basal ganglia, Lieberman discusses a range of different language impairments, including impairments following strokes (aphasia), Parkinson's disease, and disordered language development—all of which appear to involve damage to basal ganglia. A consequence of this view is that language has a rather long evolutionary history, with simple symbol use (in the form of naming) and rudimentary syntax dating back to some of the earliest hominids. Lieberman notes, however, that modern speech would have emerged considerably later in human evolution, given his interpretation of the fossil record and comparisons with the vocalization abilities of extant apes. He argues that speech production may thus be the crucial factor that differentiates human and non-human primate communication (but see Hauser and Fitch, Chapter 9, for a different perspective).

Lieberman and Corballis both point to the evolution of more complex sequential learning and processing abilities as forming part of the foundation for the origin of language. In the first part of Chapter 15, Simon Kirby and Morten Christiansen similarly relate general properties of sequential learning to the structure of language. Specifically, they propose that many language universals—that is, invariant sub-patterns of language—may derive from underlying constraints on the way we learn and process sequential structure, rather than from an innate biological adaptation for grammar (see Pinker, Bickerton, Dunbar, and Briscoe, Chapters 2, 5, 12, and 16). Kirby and Christiansen present evidence from computational simulations and psychological experiments involving the learning of simple artificial languages, indicating that specific language universals can be explained by sequential learning constraints. This perspective further suggests that languages themselves can be viewed as evolving systems, adapting to the innate constraints of the human learning and processing mechanisms. Kirby and Christiansen report on computational modelling work in which coordinated communication systems emerge among groups of artificial agents through a process of iterated learning over many generations (also described by Hurford, Chapter 3). They argue that the broader consequences of this work are that language evolution must be understood through processes that work on three different, but partially overlapping timescales: the individual timescale (through learning in development), the cultural timescale (through iterated learning across generations), and the biological timescale (through natural selection of the species).

Whereas Kirby and Christiansen approach language evolution by investigating how the properties of cultural transmission across generations may affect language structure, Ted Briscoe focuses on the possible emergence of biological adaptations for grammar. In Chapter 16, he suggests that we may be able to understand how language-specific learning biases could have arisen in our evolutionary history by exploring how learning itself may impact on the ability to procreate. The assumption is that aspects of language, which were previously learned, would gradually become genetically encoded through 'genetic assimilation'—that is, through genetic adaptations for language selected to increase reproductive fitness (see also Pinker, Chapter 2, for a similar theoretical perspective). Based on a discussion of computational models of language acquisition, Briscoe contends that innate language-specific constraints are required in order to account for the full complexity of grammatical acquisition. Given this characterization of our current language ability, he argues that the only plausible way such innate constraints could have evolved in humans is through genetic assimilation. On this account, language started out relying on general-purpose learning mechanisms, but through biological adaptations learning gradually became language-specific. As support, Briscoe reviews a series of computational simulations in which grammatical assimilation emerges in populations of language-learning agents. In contrast, the simulations described in Kirby and Christiansen show how the task to be learned—in this case, language—may itself be shaped by the learner.

This volume concludes with Natalia Komarova and Martin Nowak who study language evolution from the viewpoint of mathematical game theory.[2] In Chapter 17 they first argue, on the basis of evidence from formal language theory, that innate constraints on language acquisition are a logical necessity. Komarova and Nowak note, however, that these results do not determine whether such innate constraints must be linguistic in nature—in fact, they could equally well derive from more general cognitive constraints— they only demonstrate that innate constraints on learning are needed. The formal language work is combined with an evolutionary approach based on game theory in order to provide a general mathematical framework for exploring the evolution of language. Within this framework language evolution can be studied in terms of populations of language-learning agents whose survival and ability to procreate depend on their capacity for language. The results indicate that natural selection would tend to favour systematic mechanisms for encoding grammatical knowledge. Such a system, for example, could be instantiated in terms of recursive rules, though any system capable of generating an infinite number of sentences would suffice. Although this research does directly address the question of whether evolved constraints would have to be language-specific or not, others have taken it to support the idea of a biological adaptation for language (see e.g. Pinker and Dunbar, Chapters 2 and 12). Komarova and Nowak's modelling work, together with that of Kirby and Christiansen and of Briscoe, demonstrates how mathematical and computational modelling can be fruitfully applied to the study of language evolution.

## Consensus and Remaining Controversies

The chapters in this book provide a comprehensive survey of the state of the art in language evolution research. Many different disciplines are represented, and many different perspectives are expressed. Here, we seek to draw out the major points of consensus as well as the remaining controversies.

Possibly the strongest point of consensus is the notion that to fully understand language evolution, it must be approached simultaneously from many

---

[2] As editors, we realize that the chapter by Komarova and Nowak may appear daunting because of its mathematical content. However, we note that it is possible to gain a perfectly good grasp of the underlying ideas put forward in this chapter without necessarily understanding the maths behind them.

disciplines. This would certainly seem to be a necessary condition for language evolution research, in order to provide sufficient constraints on theorizing to make it a legitimate scientific inquiry. Nonetheless, most researchers in language evolution only cover parts of the relevant data, perhaps for the reason that it is nearly impossible to be a specialist in all the relevant fields. Still, as a whole, the field—as exemplified in this book—is definitely moving in the direction of becoming more interdisciplinary. Collaborations between researchers in different fields with a stake in language evolution may be a way in which this tendency could be strengthened even further.

Another area of consensus is the growing interest in using mathematical and computational modelling to explore issues relevant for understanding the origin and evolution of language. More than half of the chapters in this book were in some way informed by modelling results (though see Bickerton, Chapter 5, for cautionary remarks). Models are useful because they allow researchers to test particular theories about the mechanisms underlying the evolution of language. Given the number of different factors that may potentially influence language evolution, our intuitions about their complex interactions are often limited. It is exactly in these circumstances, when multiple processes have to be considered together, that modelling becomes a useful—and perhaps even necessary—tool. In this book, modelling work has been used to inform theories about biological adaptations for grammar (Pinker, Dunbar, Briscoe, Komarova and Nowak, Chapters 2, 12, 16, and 17), about the emergence of language structure through cultural transmission (Hurford, Deacon, Kirby and Christiansen, Chapters 3, 7, and 15), and about the evolution of phonetic gesture systems (Studdert-Kennedy and Goldstein, Chapter 13). We envisage that the interest in mathematical and computational modelling is likely to increase even further, especially as it becomes more sophisticated in terms of both psychological mechanisms and linguistic complexity.

There is a general consensus that to understand language evolution we need a good understanding of what language is. However, the field is divided over what the exact characterization of language should be, and in which terms it should be defined. Nonetheless, some agreement appears to be in sight regarding some of the necessary steps toward language. Specifically, there seems to be agreement that prior to the emergence of language some pre-adaptations occurred in the hominid lineage. There is less agreement about what these may have been, but one candidate that seems to be put forward by many is the ability for using symbols. Most also see grammatical structure as emerging during a later stage in language evolution, though opinions differ as to whether this was a consequence of an evolved innate grammar (Pinker, Bickerton, Dunbar, Briscoe, and Komarova and Nowak, Chapters 2, 5, 12, 16, and 17) or the emergence of grammar through cultural transmission (Hurford, Tomasello, Davidson, Arbib, and Kirby and Christiansen, Chapters 3, 6, 8, 10, and 15).

Of course, several major points of disagreement still remain. We have already touched upon the disagreement over whether constraints on language structure as reflected in human language are a consequence of biological adaptations for grammar or products of using and transmitting language across generations of learners with certain limited capacities. Another debated issue is whether language originated in manual gestures or evolved exclusively in the vocal domain. Although mathematical and computational modelling may help inform the discussions about how language came to have the structure it has today, it is less likely to be able to address issues related to language origin. However, evidence from other disciplines such as archaeology, comparative neuroanatomy, and cognitive neuroscience may provide clues.

One line of evidence that is likely to figure more prominently in future discussions of language evolution is results from the study of the human genome. A better understanding of the genetic bases of language and cognition, as well as its interaction with the environment, may provide strong constraints on language evolution theories, in particular with respect to issues related to the origin of language. Currently, however, the evidence appears to provide few constraints on such theorizing. In this book, Pinker Corballis, and Lieberman (Chapters 2, 11, and 14) each cite data regarding the newly discovered FOXP2 gene in support for their theories of language evolution—even though the theories differ substantially. However, they do seem to agree that the FOXP2 data suggest a late evolution of speech. Therefore, the genetic data may be particularly useful for our understanding of the timeline for language evolution.

## The Hardest Problem?

Understanding the evolution of language is a hard problem, but is it really the hardest problem in science, as we have provocatively suggested in the title of this chapter? This question is difficult to answer. Certainly, other scientific fields have their own intrinsic obstacles; those studying conscious-ness have already laid claim to the phrase 'the hard problem'. Nevertheless, it is worth considering the unique challenges that face language evolution researchers. Language itself is rather difficult to define, existing as it does both as transitory utterances that leave no trace and as patterns of neural connectivity in the natural world's most complex brains. It is never station-ary, changing over time and within populations which themselves are dy-namic. It is infinitely flexible and (almost) universally present. It is by far the most complex behaviour we know of—the mammoth efforts of twentieth-century language research across a multitude of disciplines only serve to re-mind us just how much about language we still have to discover.

There are good reasons to suppose that we will not be able to account for the evolution of language without taking into account all the various sys-tems that underlie it. This means that we can no longer afford to ignore the research on language in fields other than our own. Understanding the ori-gin of human uniqueness is a worthy goal for twenty-first-century science. It may not be the hardest problem; but we hope that this book will help us focus on the challenges ahead and go some way to showing what a complete theory of language evolution will look like.

## FURTHER READING

The article (and associated peer commentaries) that gave rise to the resurgence of interest in language evolution, Pinker and Bloom (1990), may be a good start-ing point—when combined with this book—for looking at how the field has pro-gressed over the last dozen years. The large volume of proceedings papers resulting from the 1975 conference sponsored by the New York Academy of Sciences on the origins and evolution of language and speech (Harnad et al. 1976) provides a good snapshot of the field a quarter of a century ago. It also contains an interesting paper by Hans Aarsleff on the history of language origin and evolution theories since the Renaissance.

A very good source of semi-technical papers covering a wide variety of topics and angles on language evolution can be found in the volumes based on selected pres-entations at the biennial conference on language evolution. So far, volumes have ap-peared from the 1996 conference in Edinburgh (Hurford et al. 1998), the 1998 con-ference in London (Knight 2000) and the 2000 conference in Paris (Wray 2002).

Cangelosi and Parisi (2002) provide a useful introduction to the modelling of language evolution—including chapters covering many different approaches to simulating the origin and evolution of language. For a competent and intelligible introduction to the issues relating to understanding the possible genetic bases for language, see Tomblin (in press).

2

# Language as an Adaptation
to the Cognitive Niche*

*Steven Pinker*

## Introduction

This chapter outlines the theory (first explicitly defended by Pinker and Bloom 1990), that the human language faculty is a complex biological adaptation that evolved by natural selection for communication in a knowledge-using, socially interdependent lifestyle. This claim might seem to be anyone's first guess about the evolutionary status of language, and the default prediction from a Darwinian perspective on human psychological abilities. But the theory has proved to be controversial, as shown by the commentaries in Pinker and Bloom (1990) and the numerous debates on language evolution since then (Fitch 2002; Hurford et al. 1998).

In the chapter I will discuss the design of the language faculty, the theory that language is an adaptation, alternatives to the theory, an examination of what language might be an adaptation for, and how the theory is being tested by new kinds of analyses and evidence.

## The Design of Human Language

The starting point in an analysis of the evolution of language must be an analysis of language itself (for other overviews, see Bickerton 1990; Jackendoff 2002; Miller 1991). The most remarkable aspect of language is its *expressive power*: its ability to convey an unlimited number of ideas from one person to another via a structured stream of sound. Language can communicate

*Supported by NIH grant HD-18381. I thank Morten Christiansen and the members of his seminar on the evolution of language for helpful comments on an earlier draft.

anything from soap opera plots to theories of the origin of the universe, from lectures to threats to promises to questions. Accordingly, the most significant aspects of the language faculty are those that make such information transfer possible (Pinker 1994; 1999). The first cut in dissecting the language faculty is to separate the two principles behind this remarkable talent.

### Words

The first principle underlies the mental lexicon, a finite memorized list of words. As Ferdinand de Saussure pointed out, a word is an arbitrary sign: a connection between a signal and a concept shared by the members of the community. The word *duck* does not look like a duck, walk like a duck, or quack like a duck, but I can use it to convey the idea of a duck because we all have learned the same connection between the sound and the meaning. I can therefore bring the idea to mind in a listener simply by making that noise. If instead I had to shape the signal to evoke the thought using some perceptible connection between its form and its content, every word would require the inefficient contortions of the game of charades.

The symbols underlying words are bidirectional. Generally, if I can use a word I can understand it when someone else uses it, and vice versa. When children learn words, their tongues are not moulded into the right shape by parents, and they do not need to be rewarded for successive approximations to the target sound for every word they hear. Instead, children have an ability, upon hearing somebody else use a word, to know that they in turn can use it to that person or to a third party and expect to be understood.

### Grammar

Of course, we do not just learn individual words; we combine them into larger words, phrases, and sentences. This involves the second trick behind language, grammar. The principle behind grammar was articulated by Wilhelm von Humboldt as 'the infinite use of finite media'. Inside every language user's head is a finite algorithm with the ability to generate an infinite number of potential sentences, each of which corresponds to a distinct thought. For example, our knowledge of English incorporates rules that say 'A sentence may be composed of a noun phrase (subject) and a verb phrase (object)' and 'A verb phrase may be composed of a verb, a noun phrase (object), and a sentence (complement)'. That pair of rules is *recursive*: a phrase is defined as a sequence of phrases, and one or more of those daughter phrases can be of the same kind as the mother phrase. This creates a loop that can generate sentences of any size, such as *I wonder whether she knows that I know that she knows that he thinks she is interested in him.* By means of generating an infinite number of sentences, we can convey an infinite number of distinct thoughts (see also Studdert-Kennedy and Goldstein, Chapter 13 below), since every sentence has a different meaning (most linguists believe that true synonymy is rare or nonexistent).

Grammar can express an astonishing range of thoughts because our knowledge of grammar is couched in abstract categories such as 'noun' and 'verb' rather than concrete concepts such as 'man' and 'dog' or 'eater' and 'eaten' (Pinker 1994; 1999). This gives us an ability to talk about new kinds of ideas. We can talk about a dog biting a man, or, as in the journalist's definition of 'news', a man biting a dog. We can talk about aliens landing in Roswell, or the universe beginning with a big bang, or Michael Jackson marrying Elvis's daughter. The abstractness of grammatical categories puts no restriction on the content of sentences; the recursive, combinatorial nature of grammar puts no limits on their complexity or number.

A grammar comprises many rules, which fall into subsystems. The most prominent is *syntax*, the component that combines words into phrases and sentences. One of the tools of syntax is linear order, which allows us to distinguish, say, *Man bites dog* from *Dog bites man.* Linear order is the most conspicuous property of syntax, but it is a relatively superficial one. Far more important is *constituency.* A sentence has a hierarchical structure, which allows us to convey complex propositions consisting of ideas embedded inside ideas. A simple demonstration comes from an ambiguous sentence such as *On tonight's program Dr Ruth will discuss sex with Dick Cavett.* It is composed of a single string of words in a particular order but with two different meanings, which depend on their constituent bracketings: [*discuss*] [*sex*] [*with Dick Cavett*] versus [*discuss*] [*sex with Dick Cavett*]. Of course, most sentences in context are not blatantly ambiguous, but ambiguity illustrates the essential place of constituency in interpreting meaning from sentences. As with other symbolic systems that encode logical information, such as arithmetic, logic, and computer programming, it is essential to get the parentheses right, and that's what phrase structure in grammar does.

Syntax also involves *predicate–argument* structure, the component of language that encodes the relationship among a set of participants (Pinker 1989). To understand a sentence one cannot merely pay attention to the order of words, or even the way they are grouped; one has to look up information associated with the predicate (usually the verb) which specifies how its arguments are placed in the sentence. For example, in the sentences *The man feared the dog* and *The man frightened the dog*, the word *man* is the subject in both cases, but its semantic role differs: in the first sentence the man experiences the fear; in the second he causes it. In understanding a sentence, one has to look up information stored with the mental dictionary entry of the verb and see whether it says (for instance) 'my subject is the one experiencing the fear' or 'my subject is the one causing the fear'.

A fourth trick of syntax is known as *transformations, movement,* or *binding traces*. Once one has specified a hierarchical tree structure into which the words of a sentence are plugged, a further set of operations can alter it in precise ways. For example, the sentence *Dog is bitten by man* contains the verb *bite*, which ordinarily requires a direct object. But here the object is missing from its customary location; it has been 'moved' to the front of the sentence. This gives us a way of shifting the emphasis and quantification of a given set of participants in an event or state. The sentences *Man bites dog* and *Dog is bitten by man* both express the same information about who did what to whom, but one of them is a comment about the man and the other is a comment about the dog. Similarly, sentences in which a phrase is replaced by a *wh*-word and moved to the front of a sentence, such as *Who did the dog bite?*, allow the speaker to seek the identity of one of the participants in a specified event or relationship. Transformations thus provide a layer of meaning beyond who did what to whom; that layer emphasizes or seeks information about one of the participants, while keeping constant the actual event being talked about.

Syntax, for all that complexity, is only one component of grammar. All languages have a second combinatorial system, *morphology*, in which simple words or parts of words (such as prefixes and suffixes) are assembled to produce complex words. The noun *duck*, for example, comes in two forms—*duck* and *ducks*—and the verb *quack* in four—*quack, quacks, quacked,* and *quacking*. In languages other than English morphology can play a much greater role. In Latin, for example, case suffixes on nouns convey information about who did what to whom, allowing one to scramble the left-to-right order of the words for emphasis or style. For example, *Canis hominem mordet* and *Hominem canis mordet* (different orders, same cases) have the same non-newsworthy meaning, and *Homo canem mordet* and *Canem homo mordet* have the same newsworthy meaning.

Language also embraces a third combinatorial system called *phonology*, which governs the sound pattern of a language. In no language do people form words by associating them directly with articulatory gestures like a movement of the tongue or lips. Instead, an inventory of gestures is combined into sequences, each defining a word. The combinations are governed by phonological rules and constraints that work in similar ways in all languages but whose specific content people have to acquire. English speakers, for example, sense that *bluck* is not a word but could be one, whereas *nguck* is not a word and could not be one (though it could be a word in other languages). All languages define templates for how words may be built out of hierarchically nested units such as feet, syllables, vowels and consonants, and features (articulatory gestures). Interestingly, whereas syntax and morphology are semantically compositional—one can predict the meaning of the whole by the meanings of the elements and the way they are combined—this is not true of phonology. One cannot predict the meaning of *duck* from the meaning of /d/, the meaning of /ʌ/, and the meaning of /k/. Phonology is a combinatorial system that allows us to have large vocabularies (e.g. 100,000 words is not atypical for an English speaker) without having to pair each word with a distinct noise. The presence of these two kinds of discrete combinatorial systems in language is sometimes called duality of patterning.

Phonology also contains a set of adjustment rules which, after the words are defined and combined into phrases, smooth out the sequence of articulatory gestures to make them easier to pronounce and comprehend. For instance, one set of rules in English causes us to pronounce the past-tense morpheme *-ed* in three different ways, depending on whether it is attached to *jogged*, *walked*, or *patted*. The adjustment for *walked* keeps the consonants at the end of a word either all voiced or all unvoiced, and the adjustment for *patted* inserts a vowel to separate two *d*-like sounds. These adjustments often function to make articulation easier or speech clearer in a way that is consistent across the language, but they are not merely products of a desire to be lazy or clear. These two goals are at cross purposes, and the rules of phonology impose shared conventions on the speakers of a language as to exactly when one is allowed to be lazy in which way.

*Interfaces of Language With Other Parts of the Mind*

Grammar is only one component of language, and it has to interface with at least four other systems of the mind: perception, articulation, conceptual knowledge (which provides the meanings of words and their relationships), and social knowledge (how language can be used and interpreted in a social context). While these systems also serve non-linguistic functions, and may have been carried over from earlier primate designs, at least some aspects of them may have evolved specifically to mesh with language. A likely example is the vocal tract: Darwin pointed to the fact that in humans every mouthful of food has to pass over the trachea, with some chance of getting lodged in it and causing death by choking. The human vocal tract has a low larynx compared to those of most other mammals, an arrangement that compromises a number of physiological functions but allows us to articulate a large range of vowel sounds. Lieberman (1984) has plausibly argued that physiological costs such as the risk of death by choking were outweighed in human evolution by the benefit of rapid, expressive communication.

## Is Language an Adaptation?

In the biologist's sense of the word, an 'adaptation' is a trait whose genetic basis was shaped by natural selection (as opposed to the everyday sense of a trait that is useful to the individual). What are the alternatives to the theory that language is an adaptation? And what are the reasons for believing it might be one?

*Is Language a Distinct Part of the Human Phenotype?*

One alternative is that language is not an adaptation itself, but a manifestation of more general cognitive abilities, such as 'general intelligence', 'a symbolic capacity', 'cultural learning', 'mimesis', or 'hierarchically organized behaviour' (see e.g. Bates et al. 1991; Deacon 1997; Tomasello 1999). If so, these more general cognitive capacities would be the adaptation.

These alternatives are difficult to evaluate, because no one has spelled out a mechanistic theory of 'general intelligence' or 'cultural learning' that is capable of acquiring human language. Intelligence, learning, symbol comprehension, and so on do not happen by magic but need particular mechanisms, and it is likely that different mechanisms are needed in different domains such as vision, motor control, understanding the physical and social worlds, and so on (Pinker 1997). The ability to acquire and use the cultural symbols called 'language' may require learning mechanisms adapted to that job. Attempts to model the acquisition of language using general-purpose algorithms such as those in traditional artificial intelligence or connectionist neural networks have failed to duplicate the complexity of human language (Pinker 1979; Pinker 1999; Pinker and Prince 1988).

Though it is hard to know exactly what is meant by terms like 'cultural learning' or 'general intelligence', one can see whether mastery of language in the human species resembles abilities that are unambiguously culturally acquired, like agricultural techniques, chess skill, knowledge of government, and mathematical expertise, or whether it looks more like a part of the standard human phenotype, like fear, humor, or sexual desire. Some very general properties of the natural history of language suggests that the latter is more accurate (see Jackendoff 2002; Lightfoot and Anderson 2002; Pinker 1994).

First, language is universal across societies and across neurological normal people within a society, unlike far simpler skills like farming techniques or chess. There may be technologically primitive peoples, but there are no primitive languages: the anthropologists who first documented the languages of technologically primitive societies a century ago were repeatedly astonished by their complexity and abstractness (Voegelin and Voegelin 1977). And despite stereotypes to the contrary, the language of uneducated, working-class, and rural speakers has been found to be systematic and rule-governed, though the rules may belong to dialects that differ from the standard one (Labov 1969; McWhorter 2002).

Second, languages conform to a universal design. A language is not just any conceivable code that maps efficiently from sound to meaning. The design specifications listed in the preceding section—and, indeed, far more subtle and complex properties of grammar—can be found in all human languages (Baker 2001; Comrie 1981; Greenberg et al. 1978; Hockett 1960).

A third kind of evidence is the ontogenetic development of language. Children the world over pass through a universal series of stages in acquiring a language (Brown 1973; Ingram 1989; Pinker 1994). That sequence culminates in mastery of the local tongue, despite the fact that learning a language requires solving the daunting problem of taking in a finite sample of sentences (speech from parents) and inducing a grammar capable of generating the infinite language from which they were drawn (Pinker 1979; 1984). Moreover, children's speech patterns, including their errors, are highly systematic, and can often be shown to conform to linguistic universals for which there was no direct evidence in parents' speech (Crain 1992; Gordon 1985; Kim et al. 1994).

A fourth kind of evidence also comes from the study of language acquisition. If children are thrown together without a pre-existing language that can be 'culturally transmitted' to them, they will develop one of their own. One example, studied by Bickerton, comes from the polyglot slave and servant plantations in which the only lingua franca among adults was a pidgin, a makeshift communicative system with little in the way of grammar. The children in those plantations did not passively have the pidgin culturally transmitted to them, but quickly developed creole languages, which differ substantially from the pidgins and which have all the basic features of established human languages (Bickerton 1981). Another example comes from deaf communities, where complex sign languages emerge quickly and spontaneously. A recent study in Nicaragua has tracked the emergence of a complex sign language in little more than a decade, and has shown that the most fluent and creative users of the language were the children (Senghas and Coppola 2001).

A fifth kind of evidence is that language and general intelligence, to the extent we can make sense of that term, seem to be doubly dissociable in neurological and genetic disorders. In aphasias and in the genetically caused developmental syndrome called Specific Language Impairment, intelligent people can have extreme difficulties speaking and understanding (Leonard 1998; Siegal et al. 2001; van der Lely and Christian 1998). Conversely, in a number of retardation syndromes, such as Williams syndrome and the sequelae of hydrocephalus, substantially retarded children may speak fluently and grammatically and do well on tests of grammatical comprehension and judgement (Clahsen and Almazen 1998; Curtiss 1989; Rossen et al. 1996). Few of these dissociations are absolute, with language or non-linguistic cognition completely spared or completely impaired. But the fact that the two kinds of abilities can dissociate quantitatively and along multiple dimensions shows that they are not manifestations of a single underlying ability.

## Did Language Evolve by Means Other Than Natural Selection?

A different alternative to the hypothesis that language is an adaptation is the possibility that it evolved by mechanisms other than natural selection, a hypothesis associated with Stephen Jay Gould and Noam Chomsky (Chomsky 1988; Gould 1997; see Piatelli-Palmarini 1989 and Pinker and Bloom 1990 for discussion). On this view, language may have evolved all at once as the product of a macromutation. Or the genes promoting language may have become fixed by random genetic drift or by genetic hitchhiking (i.e. genes that were near other genes that were the real target of selection). Or it may have arisen as a by-product of some other evolutionary development such as a large brain, perhaps because of physical constraints on how neurons can be packed into the skull.

It is hard to evaluate this theory (though, as we shall see, not impossible), because there have been no specific proposals fleshing out the theory (e.g. specifying the physical constraint that makes language a neurobiological necessity). So what is the appeal of the non-selectionist theories?

One is a general misconception, spread by Gould, that natural selection has become an obsolete or minor concept in evolutionary biology, and that explanations in terms of by-products (what he called 'spandrels') or physical constraints are to be preferred in principle (e.g. Piatelli-Palmarini 1989). This is a misconception because natural selection remains the only evolutionary force capable of generating complex adaptive design, in which a feature of an organism (such as the eye or heart) has a non-random organization that enables it to attain an improbable goal that fosters survival and reproduction (Dawkins 1986; Williams 1966). Moreover, natural selection is a rigorous concept which can be modelled mathematically or in computer simulations, measured in natural environments, and detected by statistical analyses of organisms' genomes (Kreitman 2000; Maynard Smith 1988; Przeworski et al. 2000; Weiner 1994).

A second appeal of non-selectionist theories comes from a scepticism that language could have provided enough reproductive benefits to have been selected for. According to one objection, popular among linguists, language has arbitrary features that do not obviously contribute to communication. However, *all* communication systems have arbitrary features (such as the particular sequences of dots and dashes making up Morse code), because arbitrary ways of linking messages to signals are useful as long as they are shared by sender and recipient. Moreover, since a feature that eases the task of the speaker (by omitting information or reducing the complexity of the signal) will complicate the task of the listener (by making the message more ambiguous or vulnerable to noise), a shared code must legislate arbitrary conventions that do not consistently favour any single desideratum (Pinker and Bloom 1990).

Another argument for non-selectionist theories is that grammar is more complicated than it needs to be to fulfil the communicative needs of a hunter-gatherer lifestyle. As one sceptic put it, 'How does recursion help in the hunt for mastodons?' But as Bloom and I pointed out, complex grammar is anything but a useless luxury: 'It makes a big difference whether a far-off region is reached by taking the trail that is in front of the large tree or the trail that the large tree is in front of. It makes a difference whether that region has animals that you can eat or animals that can eat you'. Since selection can proceed even with small reproductive advantages (say, one per cent), the evolution of complex grammar presents no paradox.

A third misconception is that if language is absent from chimpanzees, it must have evolved by a single macromutation. This is seen as an argument for a macromutational theory by those who believe that human language is qualitatively distinct from the communicative abilities of chimpanzees, and as an argument that human language cannot be qualitatively distinct from the communicative abilities of chimpanzees by those who believe that macromutations are improbable. But both arguments are based on a misunderstanding of how evolution works. Chimpanzees and bonobos are our closest living relatives, but that does not mean that we evolved from them. Rather, humans evolved from an extinct common ancestor that lived six to eight million years ago. There were many other (now-extinct) species in the lineage from the common ancestor to modern humans (australopithecines, *habilis, ergaster*, archaic *sapiens*, etc.) and, more important, many individuals making up the lineages that we group into species for convenience. Language could well have evolved gradually *after* the chimp/human split, in the 200,000–300,000 generations that make up the lineage leading to modern humans. Language, that is, could be an autapomorphy: a trait that evolved in one lineage but not its sister lineages.

The final appeal of the non-selectionist hypothesis is that language could only have been useful once it was completely in place: a language is useless if you are the only one to have evolved the ability to speak it. But this objection could be raised about the evolution of any communicative system, and we know that communication has evolved many times in the animal kingdom. The solution is that comprehension does not have to be in perfect synchrony with production. In the case of language, it is often possible to decode parts of an utterance in a language one has not completely mastered.

When some individuals are making important distinctions that can be c
coded by listeners only with cognitive effort, a pressure could thereby c
velop for the evolution of neural mechanisms that would make this deco
ing process become increasingly automatic and effortlessly learned (Pink
and Bloom 1990). The process whereby environmentally induced respons
set up selection pressures for such responses to become innate, triggerir
conventional Darwinian evolution that superficially mimics a Lamarckia
sequence, is known as the Baldwin Effect (Hinton and Nowlan 1987).

Opposing these spurious arguments for the non-selectionist hypothes
is a strong prima facie reason to favour the selectionist one: the standard a
gument in evolutionary biology that only natural selection can explain th
evolution of complex adaptive design (Dawkins 1986; Williams 1966). Th
information-processing circuitry necessary to produce, comprehend, ar
learn language requires considerable organization. Randomly organize
neural networks, or randomly selected subroutines from an artificial int
ligence library, do not give rise to a system that can learn and use a huma
language. As we saw, language is not just a set of symbolic labels for concep
not just the use of linear order, not just the use of hierarchical structure, ar
not just a blurting out of a sequence of sounds. It is an integrated syste
containing a lexicon, several components of grammar, and interfaces to i
put–output systems, possibly with language-specific modifications of the
own. And this complexity is not just there for show, but makes possible a r
markable ability: language's vast expressive power, rapid acquisition by ch
dren, and efficient use by adults.

As with other complex organs that accomplish improbable feats, the ne
essary circuitry for language is unlikely to have evolved by a process that
insensitive to the functionality of the end product, such as a single mu
tion, genetic drift, or arbitrary physical constraints. Natural selection is th
most plausible explanation of the evolution of language, because it is th
only physical process in which how well something works can explain ho
it came into existence.

## What Did Language Evolve For?

If language is an adaptation, what is it an adaptation for? Note that this
different from the question of what language is typically *used* for, especial

what it is used for at present. It is a question about the 'engineering design' of language and the extent to which it informs us about the selective pressures that shaped it.

What is the machinery of language trying to accomplish? The system appears to have been put together to encode propositional information—who did what to whom, what is true of what, when, where and why—into a signal that can be conveyed from one person to another. It is not hard to see why it might have been adaptive for a species with the rest of our characteristics to evolve such an ability. The structures of grammar are well suited to conveying information about technology, such as which two things can be put together to produce a third thing; about the local environment, such as where things are; about the social environment, such as who did what to whom, when where and why; and about one's own intentions, such as *If you do this, I will do that*, allowing people to convey the promises and threats that undergird relations of exchange and dominance.

### The Cognitive Niche

Gathering and exchanging information is, in turn, integral to the larger niche that modern *Homo sapiens* has filled, which John Tooby and Irven DeVore (1987) have called 'the cognitive niche' (it may also be called the 'informavore' niche, following a coinage by George Miller). Tooby and DeVore developed a unified explanation of the many human traits that are unusual in the rest of the living world. They include our extensive manufacture of and dependence on complex tools, our wide range of habitats and diets, our extended childhoods and long lives, our hypersociality, our complex patterns of mating and sexuality, and our division into groups or cultures with distinctive patterns of behaviour. Tooby and DeVore proposed that the human lifestyle is a consequence of a specialization for overcoming the evolutionary fixed defences of plants and animals (poisons, coverings, stealth, speed, and so on) by cause-and-effect reasoning. Such reasoning enables humans to invent and use new technologies (such as weapons, traps, co-ordinated driving of game, and ways of detoxifying plants) that exploit other living things before they can develop defensive countermeasures in evolutionary time. This cause-and-effect reasoning depends on intuitive theories about various domains of the world, such as objects, forces, paths, places, manners, states, substances, hidden biochemical essences, and other people's beliefs and desires.

The information captured in these intuitive theories is reminiscent of the information that the machinery of grammar is designed to convert into strings of sounds. It cannot be a coincidence that humans are special in their ability to outsmart other animals and plants by cause-and-effect rea soning, and that language is a way of converting information about cause and-effect and action into perceptible signals.

A distinctive and important feature of information is that it can be dupli cated without loss. If I give you a fish, I do not have the fish, as we know from sayings like *You can't have your cake and eat it*. But if I tell you how to fish it is not the case that I now lack the knowledge how to fish. Information is what economists call a non-rival good, a concept recently made famous by debates about intellectual property (such as musical recordings that can be shared without cost on the internet).

Tooby and DeVore have pointed out that a species that has evolved to rely on information should thus also evolve a means to *exchange* that informa tion. Language multiplies the benefit of knowledge, because a bit of know how is useful not only for its practical benefits to oneself but as a trade good with others. Using language, I can exchange knowledge with somebody else at a low cost to myself and hope to get something in return. It can also lower the original acquisition cost—I can learn about how to catch a rabbit from someone else's trial and error, without having to go through it myself.

A possible objection to this theory is that organisms are competitors, so that sharing information is costly because of the advantages it gives to one's competitors. If I teach someone to fish, I may still know how to fish, but they may now overfish the local lake, leaving no fish for me. But this is just the standard problem of the evolution of any form of cooperation or altruism and the solution in the case of language is the same. By sharing information with our kin, we help copies of our genes inside those kin, including genes that make language come naturally. As for non-kin, if we inform only those people who are likely to return the favour, both of us can gain the benefits of trade. It seems clear that we do use our faculties of social cognition to ration our conversation to those with whom we have established a non-exploita tive relationship; hence the expression 'to be on speaking terms'.

Language, therefore, meshes neatly with the other features of the cogni tive niche. The zoologically unusual features of *Homo sapiens* can be ex plained parsimoniously by the idea that humans have evolved an ability to encode information about the causal structure of the world and to share it among themselves. Our hypersociality comes about because information is a particularly good commodity of exchange that makes it worth people's while to hang out together. Our long childhood and extensive biparental investment are the ingredients of an apprenticeship: before we go out in the world, we spend a lot of time learning what the people around us have figured out. And because of the greater pay-off for investment in children, fathers, and not just mothers, have an incentive to invest in their children. This leads to changes in sexuality and to social arrangements (such as marriage and families) that connect men to their children and to the mothers of those children.

Humans depend on culture, and culture can be seen in part as a pool of local expertise. Many traditions are endemic to a people in an area because know-how and social conventions have spread via a local network of information sharing. Humans have evolved to have a long lifespan (one end of the evolutionarily ubiquitous trade-off between longevity and fecundity) because once you have had an expensive education you might as well make the most out of it by having a long period in which the expertise can be put to use. Finally, the reason that humans can inhabit such a wide range of habitats is that our minds are not adapted to a narrow, specialized domain of knowledge, such as how to catch a rabbit. Our knowledge is more abstract, such as how living things work and how objects collide with and stick to each other. That mindset for construing the world can be applied to many kinds of environment rather than confining us to a single ecosystem.

On this view, then, three key features of the distinctively human lifestyle—know-how, sociality, and language—co-evolved, each constituting a selection pressure for the others.

### Alternatives to the Cognitive Niche Theory

Several alternative hypotheses acknowledge that language is an adaptation but disagree on what it is an adaptation for. One possibility, inspired by an influential theory of the evolution of communication by Dawkins and Krebs (Dawkins 1982), is that language evolved not to inform others but to manipulate and deceive them. The problem with this theory is that, unlike signals with the physiological power to manipulate another organism directly, such as loud noises or chemicals, the signals of language are impotent unless the recipient actively applies complicated computations to decode them. It is impossible to use language to manipulate someone who does not understand the language, so hominids in the presence of the first linguistic

manipulators would have done best by refusing to allow their nascent lan
guage systems to evolve further, and language evolution would have bee
over before it began.

Another possibility is that language evolved to allow us to think rathe
than to communicate. According to one argument, it is impossible t
think at human levels of complexity without a representational mediur
for propositions, and language is that medium (Bickerton 1990). Accor
ing to another argument, we spend more time talking to ourselves tha
talking to other people, so if language has any function at all, it must b
thought rather than communication (Chomsky 2002). These theories hav
two problems. One is that they assume the strongest possible form of th
Whorfian hypothesis—that thought depends entirely on language—whic
is unliklely for a number of reasons (see Pinker 1994; 2002; Siegal et a
2001; Weiskrantz 1988). The other is that if language evolved to represer
information internally, much of the apparatus of grammar, which conver
logical relationships into perceptible signals, would be superfluous. La
guage would not need rules for defining word orders, case markers, phon
logical strings, adjustment rules, and so on, because the brain could mor
efficiently code the information to itself silently, using networks of va
ables and pointers.

Considerations of language design rule out other putative selection
pressures. Language is unlikely to have evolved as a direct substitute fc
grooming (Dunbar 1998), or as a courtship device to advertise the fitne
of our brains (Miller 2000), because such pressures would not have led t
an ability to code complex abstract propositions into signals. A fixed set c
greetings would suffice for the former; meaningless displays of virtuosi
as in scat singing, would suffice for the latter.


## New Tests of the Theory That Language is an Adaptation

Contrary to the common accusation that evolutionary hypotheses, esp
cially ones about language, are post hoc 'just so' stories, the hypothesis tha
language is an evolutionary adaptation can be made rigorous and put t
empirical test. I will conclude by reviewing two new areas of research o
the evolution of language that have blossomed since my 1990 paper wit
Bloom and which are beginning to support its major predictions.

*Language and Evolutionary Game Theory*

Good theories of adaptation can be distinguished from bad ones (Williams 1966). The bad ones try to explain one bit of our psychology (say, humor or music) by appealing to some other, equally mysterious bit (laughing makes you feel better; people like to make music with other people). The good ones use some *independently established* finding of engineering or mathematics to show that some mechanism can efficiently attain some goal in some environment. These engineering benchmarks can serve as predictions for how Darwinian organisms ought to work: the more uncannily the engineering specifications match the facts of the organism, the more confidently one infers that the organism was selected to carry out that function.

Evolutionary game theory has allowed biologists to predict how organisms ought to interact with other organisms co-evolving their own strategies (Maynard Smith 1982). Language, like sex, aggression, and cooperation, is a game it takes two to play, and game theory can provide the external criteria for utility enjoyed by the rest of evolutionary biology. Modellers assume only that the transmission of information between partners provides them with an advantage (say, by exchanging information or coordinating their behaviour), and that the advantage translates into more offspring, with similar communicative skills. The question then is how a stable communication system might evolve from repeated pairwise interactions and, crucially, whether such systems have the major design features of human language.

The first such attempt was a set of simulations by Hurford (1989) showing that one of the defining properties of human language, the arbitrary, bidirectional sign, will drive out other schemes over evolutionary time (Hurford 1989). More recently, Nowak and his collaborators have now done the same for two of the other central design features of language (Nowak and Krakauer 1999; Nowak et al. 1999a; Nowak 2000).

Nowak and his colleagues pointed out that in all communication systems, errors in signalling or perception are inevitable, especially when signals are physically similar. Imagine organisms that use a different sound (say, a vowel) for every concept they wish to communicate. As they communicate more concepts, they will need additional sounds, which will be physically closer and hence harder to discriminate. At some point adding new signals just makes the whole repertoire more confusable and fails to increase its

net communicative power. Nowak and colleagues showed that this limitation can be overcome by capping the number of signals and stringing them
together into sequences, one sequence per concept. The sequences are what
we call words, and as I mentioned earlier, the combination of meaningless
vowels and consonants into meaningful words by rules of phonology is a
universal property of language, half of the trait called 'duality of patterning'. Nowak and his colleagues have shown how its evolution is likely among
communicators with a large number of messages to convey, a precondition
that plausibly characterizes occupants of the cognitive niche.

Nowak and his colleagues have recently motivated another hallmark of
language. Imagine a language in which each message was conveyed by a
single word. For any word to survive in a community, it must be used frequently enough to be heard and remembered by all the learners. As new
words are added to the vocabularies of speakers, old words must be used
less often, and they are liable to fade, leaving the language no more expressive than before. Nowak et al. point out that this limitation can be overcome
by communicators who use compositional syntax: rather than pairing each
word with an entire event, they pair each word with a *component* of an event
(a participant, an action, a relationship), and string the words together in
an order that reflects their roles (e.g. *Dog bites man*). Such communicators
need not memorize a word for every event, reducing the word-learning
burden and allowing them to talk about events that lack words. Syntax and
semantics, the other half of the duality of patterning, will evolve.

Nowak et al. note that syntax has a cost: the requirement to attend to
the order of words. Its benefits exceed the costs only when the number of
events worth communicating exceeds a threshold. This 'syntax threshold'
is most likely to be crossed when the environment, as conceptualized by
the communicators, has a combinatorial structure: for example, when any
of a number of actors (dogs, cats, men, women, children) can engage in
any of a number of actions (walking, running, sleeping, biting). In such a
world, the number of words that have to be learned by a syntactic com-
municator equals the sum of the number of actors, actions, places, and so
on, whereas the number that must be learned by a nonsyntactic commu-
nicator equals their *product*, a potentially unlearnable number. Nowak et
al. thus proved the theoretical soundness of the conjecture of Pinker and
Bloom (1990) that syntax is invaluable to an analytical mind in a combinatorial world.

### Language and Molecular Evolution

Mathematical models and computer simulations can show that the advantages claimed for some features of language really can evolve by known mechanisms of natural selection. These models cannot, of course, show that language *in fact* evolved according to the proposed scenario. But recent advances in molecular and population genetics may provide ways of testing whether selection in fact occurred.

Evolution is a change in gene frequencies, and the first prediction of the theory that language is an evolutionary adaptation is that there should be genes that have as one of their distinctive effects the development of normal human language abilities. Such a gene would be identifiable as an allelic alternative to a gene that leads to an impairment in language. Since pleiotropy is ubiquitous, one need not expect that such a gene would affect *only* language; but its effects on language should not be consequences of some more general deficit such as a hearing disorder, dysarthria, or retardation.

Clinical psycholinguists have long known of the collection of syndromes called Specific Language Impairment (SLI), in which a child fails to develop language on schedule and struggles with it throughout life (Bishop et al. 1995; Leonard 1998; van der Lely et al. 1998). By definition SLI is not a consequence of autism, deafness, retardation, or other non-linguistic problems, though it may co-occur with them. In one form of the syndrome, sometimes called 'Grammatical SLI', the children are normal in intelligence, auditory perception, and the use of language in a social context, but their speech is filled with grammatical errors and they are selectively deficient in detecting ungrammaticality and in discriminating meaning based on a sentence's grammar (van der Lely et al. 1998; van der Lely and Stollwerck 1996). Though it was once thought that SLI comes from a deficit in processing rapidly changing sounds, that theory has been disproven (Bishop et al. 1999; Bishop et al. 2001; van der Lely et al. 1998).

SLI runs in families and is more concordant in monozygotic than in dizygotic twins, suggesting it has a heritable component (Bishop et al. 1995; Stromswold 2001; van der Lely and Stollwerck 1996). But the inheritance patterns are usually complex, and until recently little could be said about its genetic basis. In 1990 investigators described a large multi-generational family, the KEs, in which half the members suffered from a disorder of speech and language, distributed within the family in the manner of an autosomal dominant gene (Hurst et al. 1990). Extensive testing by psycholinguists showed a complex phenotype (Bishop 2002). The affected family members on average have lower intelligence test scores (perhaps because verbal coding helps performance in a variety of tasks), but their language impairment cannot be a simple consequence of low intelligence, because some of the affected members score in the normal range, and some score higher than their unaffected relatives (Bishop 2002; Lai et al. 2001). And though the affected members have problems in speech articulation (especially as children) and in fine movements of the mouth and tongue (such as sticking out their tongue or blowing on command), their language disorder cannot be reduced to a motor problem, because they also have trouble with identifying phonemes, understanding sentences, judging grammaticality, and other language skills (Bishop 2002).

In 2001, geneticists identified a gene on Chromosome 7, FOXP2, that is perfectly associated with the syndrome within the KE family and in an unrelated individual (Lai et al. 2001). They also argued on a number of grounds that the normal allele plays a causal role in the development of the brain circuitry underlying language and speech, rather than merely disrupting that circuitry when mutated.

A second crucial prediction of the language-as-adaptation theory is that there should be *many* genes for language. If human language can be installed by a single gene, there would be no need to invoke natural selection, because it is not staggeringly improbable that a single gene could have reached fixation by genetic drift or hitchhiking. But if a large set of co-evolved genes is necessary, probability considerations would militate against such explanations. The more genes are required for normal language, the lower the odds that our species could have accumulated them all by chance.

It seems increasingly likely that in fact many genes are required. In no known case of SLI is language wiped out completely, as would happen if language was controlled by a single gene which occasionally is found in mutated form. Moreover, SLI is an umbrella term for many distinct syndromes (Leonard 1998; Stromswold 2001; SLI Consortium 2002). Grammatical SLI, for example, is distinct from the syndrome affecting the KE family, which in turn is distinct from other cases of SLI known to clinicians (van der Lely and Christian 1998). In yet another syndrome, language delay, children are late in developing language but soon catch up, and can grow up without problems (Sowell 1997). Language delay is highly heritable (Stromswold 2001), and its statistical distribution in the population suggests that it is a distinct genetic syndrome rather than one end of a continuum of developmental timetables (Dale et al. 1998). There are yet other heritable disorders involving language (Stromswold 2001), such as stuttering and dyslexia (a problem in learning to read which may often be a consequence of more general problems with language). Both have been associated with specific sets of chromosomal regions (Stromswold 2001).

With recent advances in genomics, the polygenic nature of language is likely to become more firmly established. In 2002, an 'SLI Consortium' discovered two novel loci (distinct from FOXP2) that are highly associated with SLI but not associated with low non-linguistic intelligence (SLI Consortium 2002). Moreover, the two loci were associated with different aspects of language impairment, one with the ability to repeat non-words, the other with expressive language, further underscoring the genetic complexity of language.

The most important prediction of the adaptation theory is that language should show evidence of a history of selection. The general complaint that evolutionary hypotheses are untestable has been decisively refuted by the recent explosion of quantitative techniques that can detect a history of selection in patterns of statistical variation among genes (Kreitman 2000; Przeworski et al. 2000). The tests depend on the existence of neutral evolution: random substitutions of nucleotides in non-coding regions of the genome, or substitutions in coding regions that lead to synonymous codons. These changes have no effect on the organism's phenotype, and hence are invisible to natural selection. The genetic noise caused by neutral evolution can thus serve as a baseline or null hypothesis against which the effects of selection (which by definition reduces variability in the phenotype) can be measured.

For example, if a gene has undergone more nucleotide replacements that alter its protein product than replacements that do not, the gene must have been subject to selection based on the function of the protein, rather than having accumulated mutations at random, which should have left equal numbers of synonymous and amino-acid-replacing changes. Alternatively, one can compare the variability of a gene among the members of a given species with the variability of that gene across species; a gene that has been subjected to selection should vary more between species than within species. Still other techniques compare the variability of a given gene to estimates of the variability expected by chance, or check whether a marker for an allele is found in a region of the chromosome that shows reduced variation in the population because of a selective sweep. About a dozen such techniques have been devised so far. The calculations are complicated by the fact that recombination rate differences, migrations, population expansions, and population subdivisions can also cause deviations from the expectations of neutral evolution, and therefore can be confused with signs of selection. But techniques to deal with these problems have been developed as well.

It is now obvious how one can test the language-as-adaptation hypothesis (or indeed, any hypothesis about a psychological adaptation). If a gene associated with a trait has been identified, one can measure its variation in the population and apply the tests for selection. The day that I wrote this paragraph, the first of such tests has been reported in *Nature* (Enard et al. 2002). A team of geneticists examined the FOXP2 protein (the cause of the KE family's speech and language disorder) in the mouse, several primate species, and several human populations. They found that the protein is highly conserved among mammals: the chimpanzee, gorilla, and monkey versions of the protein are identical to each other and differ in only one amino acid from the mouse version and two from the human version. But two of the three differences between humans and mice occurred in the human lineage after its separation from the common ancestor with the chimpanzee. And though the variations in the gene sequence among all the non-human animals produce few if any functional differences, at least one of the changes in the human lineage significantly altered the function of the protein. Moreover, the changes that occurred in the human lineage have become fixed in the species: the team found essentially no variation among forty-four chromosomes originating in all the major continents, or in an additional 182 chromosomes of European descent. The statistical tests showed that these distributions are extremely unlikely to have occurred under a scenario of neutral evolution, and therefore that the FOXP2 gene has been a target of selection in human evolution. The authors further showed that the selection probably occurred during the last 200,000 years, the period in which anatomically modern humans evolved, and that the gene was selected for directly, rather than hitchhiking on an adjacent selected gene. Alternative explanations that rely on demographic factors were tested and at least tentatively rejected.

This stunning discovery does not *prove* that language is an adaptation, because it is possible that FOXP2 was selected only for its effects on orofacial movements, and that its effects on speech and language came along for the ride. But this is implausible given the obvious social and communicative advantages that language brings, and the fact that the deficient language in SLI is known to saddle the sufferers with educational and social problems (Beitchman et al. 1994; Snowling et al. 2001).

The studies I reviewed in this section are, I believe, just a beginning. I predict that evolutionary game theory will assess the selective rationale for an increasing number of universal properties of human language, and that new genes for language disorders and individual variation in language will be discovered and submitted to tests for a history of selection in the human lineage. In this way, the theory that language is an adaptation, motivated originally by the design features and natural history of language, will become increasingly rigorous and testable.

## FURTHER READING

For general introductions to the structure and function of language, see Baker (2001); Bickerton (1990); Jackendoff (1994; 2002); Lightfoot and Anderson (2002); Miller (1991); Pinker (1994).

Good overviews of natural selection and adaptation include Dawkins (1986); Dawkins (1996); Maynard Smith (1986; 1989); Ridley (1986); Weiner (1994); Williams (1966). The debate over whether language is a product of natural selection may be found in the target article, commentaries, and reply in Pinker and Bloom (1990).

Specific Language Impairment is explained in Leonard (1998); van der Lely et al. (1998). An overview of the genetics of language can be found in Stromswold (2001). Evolutionary game theory is explained by its founder in Maynard Smith (1982). Methods for detecting natural selection in molecular genetic data are reviewed in Aquadro (1999); Kreitman (2000); Przeworski et al. (2000).

# 3

# The Language Mosaic and its Evolution

*James R. Hurford*

## Introduction

It is natural to ask fact-demanding questions about the evolution of language, such as 'Did *Homo erectus* use syntactic language?', 'When did relative clauses appear?', and 'What language was spoken by the first *Homo sapiens sapiens* who migrated out of Africa?' One function of science is to satisfy a thirst for such answers to questions comprehensible in everyday terms, summarized as 'What happened, and when?' Such questions are clearly genuinely empirical; there is (or was) a fact of the matter. A time-travelling investigator could do fieldwork among the *Homo erectus* and research the first question, and then make forward jumps in time and research the other questions. I believe, however, that study in the evolution of language will not yield answers to such questions in the near future. Therefore, finding answers to such empirical-in-principle questions cannot be the purpose of language evolution research. The goal is, rather, to explain the present.

Evolutionary linguistics does not appeal to an apparatus of postulated abstract principles specific to the subject to explain language phenomena. Language is embedded in human psychology and society, and is ultimately governed by the same physical principles as galaxies and mesons. Not being physicists, or even chemists, we can take for granted what those scientists give us. In the hierarchy of sciences 'up' from physics, somewhere around biochemistry, and, on a parallel track, in mathematics and computational theory, facts begin to appear which can be brought to bear on the goal of explaining language. These facts are not in themselves linguistic facts, but linguistic facts are distantly rooted in them. The basic linguistic facts needing explanation are these: there are thousands of different languages spoken in the world; these languages have extremely complex structure; and humans uniquely (barring a tiny minority of pathological cases) can learn any of these languages. These broad facts subsume an army of more detailed phenomena pertaining to individual languages. Such facts are, of course, the standard goals of linguistics. But modern mainstream linguistics has ignored the single most promising dimension of explanation, the evolutionary dimension.

Linguistic facts reflect acquired states of the brains of speakers. Those brains were bombarded in childhood with megabytes of information absorbed from the environment through various sensory channels, and influencing (but not wholly determining) neurogenesis. The grown neurons work through complex chemistry, sending information at various speeds and with varying fidelity buzzing around the brain and out into the muscles, including those of the vocal tract. This is a synchronic description, reduced almost to caricature, of what happens in an extremely complex organism, a human, giving rise to linguistic facts, our basic *explananda*. Facts of particular languages are themselves partly the result of specific historical contingencies which we cannot hope to account for in detail. Collecting such facts is work at the indispensable descriptive coalface of linguistics. The theoretical goals of linguistics must be set at a more general level, accounting for the range, boundaries, and statistical distribution of language-specific facts.

Much of biology, like most of linguistics, is devoted to wholly descriptive synchronic accounts of how living organisms work. But at one time in the world's history there were no living organisms. The evolutionary branch of biology aims to explain how the observed range of complex organisms arose. With the unravelling of the structure of DNA, evolutionary theory began the reductive breakthrough, still incomplete, from postulating its own characteristic abstract principles to a sound basis in another science, chemistry. Any evolutionary story of how complex organisms arose must now be consistent with what we know about the behaviour of molecules. The evolutionary story must also be consistent with knowledge from another new and independent body of theory, represented in the early work of D'Arcy Thompson (1961) and recently by such work as Kauffman (1993; 1995) and West et al. (1997). This work emphasizes that the environment in which natural selection operates is characterized by mathematical principles, which constrain the range of attractor states into which evolution can gravitate.

The evolutionary biologists Maynard Smith and Szathmáry (1995) have identified eight 'major transitions in evolution'. Their last transition is the emergence of human societies with language. Chomsky has stressed that language is a biological phenomenon. But prevalent contemporary brands of linguistics neglect the evolutionary dimension. The present facts of language can be understood more completely by adopting an evolutionary linguistics, whose subject matter sits at the end of a long series of evolutionary transitions, most of which have traditionally been the domain of biology. With each major transition in evolution comes an increase in complexity, so that a hierarchy of levels of analysis emerges, and research methods necessarily become increasingly convoluted, and extend beyond the familiarly biological methods. Evolution before the appearance of parasitism and symbiosis was simpler. Ontogenetic plasticity, resulting in phenotypes which are not simply predictable from their genotypes, and which may in their turn affect their own environments, further complicates the picture. The advent of social behaviour necessitates even more complex methods of analysis, many not susceptible to mathematical modelling, due to the highly non-linear nature of advanced biosocial systems. With plasticity (especially learning) and advanced social behaviour comes the possibility of culture, and a new channel of information transfer across generations. Cultural evolution, mediated by learning, has a different dynamic from biological evolution; and, to make matters even more complex, biological and cultural evolution can intertwine in a co-evolutionary spiral.

The key to explaining the present complex phenomena of human language lies in understanding how they could have evolved from less complex phenomena. The fact that human language sits at the end (so far!) of a long evolutionary progression certainly poses a methodological challenge. Nevertheless, it is possible to separate out components of the massively complex whole, and to begin to relate these in a systematic way to the present psychological and social correlates of language and to what we can infer of their evolutionary past. Modern languages are learned by, stored in, and processed online by evolved brains, given voice by evolved vocal tracts, in evolved social groups. We can gain an understanding of how languages, and the human capacity for language, came into existence by studying the material (anatomical, neural, biochemical) bases of language in humans, related phenomena in less evolved creatures, and the dynamics of populations and cultural transmission.

A basic dichotomy in language evolution is between the biological evolution of the language capacity and the historical evolution of individual languages, mediated by cultural transmission (learning). In the next section I will give a view of relevant steps in the biological evolution of humans towards their current fully-fledged linguistic capacity.

Biological Steps to Language-Readiness:
Pre-Adaptations

In this section, I review some of the cognitive pre-adaptations which paved the way for the enormously impressive language capacity in humans. While these pre-adaptations do not in themselves fully explain how the full, uniquely human ability finally emerged, they do give us a basis for beginning to understand what must have happened.

A pre-adaptation is a change in a species which is not itself adaptive (i.e. is selectively neutral) but which paves the way for subsequent adaptive changes. For example, bipedalism set in train anatomical changes which culminated in the human vocal tract. Though speech is clearly adaptive, bipedalism is not itself an adaptation for speech; it is a pre-adaptation. This example involves the hardware of language, the vocal tract. Many changes in our species' software, our mental capacities, were necessary before we became language-ready; these are cognitive pre-adaptations for language. Pre-adaptations for language involved the following capacities or dispositions:

1. A *pre-phonetic* capacity to perform speech sounds or manual gestures.
2. A *pre-syntactic* capacity to organize longer sequences of sounds or gestures.
3. *Pre-semantic* capacities:
   a. to form basic concepts;
   b. to construct more complex concepts (e.g. propositions);
   c. to carry out mental calculations over complex concepts.
4. *Pre-pragmatic* capacities:
   a. to infer what mental calculations others can carry out;
   b. to act cooperatively;
   c. to attend to the same external situations as others;
   d. to accept symbolic action as a surrogate for real action.
5. An *elementary symbolic* capacity to link sounds or gestures arbitrarily with basic concepts, such that perception of the action activates the concept, and attention to the concept may initiate the sound or gesture.

If some capacity is found in species distantly related to humans, this can indicate that it is an ancient, primitive capacity. Conversely, if only our nearest relatives, the apes, possess some capacity, we can conclude that it is a more recent evolutionary development. Twin recurring themes in the discussion of many of these abilities are *learned*, as opposed to innate, behaviour and *voluntary control* of behaviour.

Voluntary control is a matter of degree, ranging from involuntary reflexes to actions whose internal causes are obscure to us. Probably all vertebrates can be credited with some degree of voluntary control over their actions. In some sense, and in some circumstances, they 'decide' what to do. In English, 'voluntary' is reserved for animate creatures. Only jokingly do we say of a machine that it 'has a mind of its own', but this is precisely when we do not know what complex internal states lead to some unpredicted behaviour. 'Voluntary' is used to describe whole actions. If actions are simple, they may, like reflex blinking, be wholly automatic, and involuntary. If an action is complex, although the whole action may be labelled 'voluntary', it is likely to have an automatic component and a non-automatic component. Both the automatic and the non-automatic component may be determined by complex processes obscure to us. What singles humans out from other species is the capacity to acquire automatic control, in the space of a few years, of the extremely complex syntactic and phonological processes underlying speaking and understanding language. Such automatization must involve the laying down of special neural structures. It seems reasonable to identify some subset of these neural structures with what linguists call a grammar. The sheer size of the information thus encoded (languages are massive) testifies to the enormous plasticity, specialized to linguistic facts, of the human brain.

Human languages are largely learned systems. The more ways a species is plastic in its behaviour, the more complex are the cultural traditions, including languages, that can emerge. Our nearest relatives, the chimpanzees, are plastic in a significantly wider range of behaviours than any other nonhuman animals; their cultural traditions are correspondingly more multifaceted, while falling far short of human cultural diversity and complexity. Combined with plasticity, voluntary control adds more complexity, and unpredictability, to patterns of behaviour. Much of the difference between humans and other species can be attributed to greatly increased plasticity and voluntary control of these pre-adaptive capacities.

### Pre-Phonetic Capacity

Chimpanzees cannot speak. They typically have little voluntary breath control. To wild chimpanzees, voluntary breath control does not come naturally. On the other hand, chimpanzees have good voluntary control over their manual gestures, although they are not as capable as humans of delicate manual work. A pre-adaptation that was necessary for the emergence of modern spoken language was the extension of voluntary control from the hands to the vocal tract.

Learning controlled actions by observation entails an ability to imitate. Imitation involves an impressive 'translation' of sensory impressions into motor commands. Think of a smile. Without mirrors or language, one has no guarantee that muscle contractions produce the effect one perceives in another's face. Given the required voluntary control, and the anatomical hardware, imitation of speech sounds should be easier than imitation of facial gestures, because one can hear one's own voice. A capacity for imitation is found in a perplexing range of species. Some birds can imitate human speech, and many other sounds as well. Dolphins can be trained to imitate human movements. A capacity for imitation can evolve separately in different species, with or without the other necessary pre-adaptive requirements for human language. A neural basis of imitation has been found in monkeys in the form of 'mirror neurons', which fire both when an animal is carrying out a certain action, such as grasping, and when it observes that same action carried out by another animal. A recurrent theory in phonetics is the 'motor theory of speech perception', which claims that speech sounds are represented in the brain in terms of the motor commands required to make them.

Although they cannot speak, our ape cousins have no trouble in recognizing different spoken human words. The capacity to discriminate the kinds of sounds that constitute speech evidently preceded the arrival of speech itself.

### Pre-Syntactic Capacity

Syntax involves the stringing together of independent subunits into a longer signal. We are concerned in this section with what Marler (1977) calls 'phonological syntax', as opposed to 'lexical syntax'. In phonological syntax the units, like the letters in a written word, have no independent meaning. In lexical syntax the units, such as the words in an English sentence, have meanings which contribute to the overall meaning of the whole signal. Many bird species can learn songs with phonological syntax. Oscine birds, which learn complex songs, are very distant relatives of humans. Many other birds, and more closely related species, including most mammals, do not produce calls composed of independent sub-units. Our closest relatives, the apes, do produce long calls composed of sub-units. The long calls of gibbons are markers of individual identity, for advertising or defending territory. The subunit notes, used in isolation, out of the context of long calls, are used in connection with territorial aggression, and it is not clear whether the meanings of these notes can be composed by any plausible operation to yield the identity-denoting meaning of the whole signal.

Male gibbon singing performances are notable for their extreme versatility. Precise copies of songs are rarely repeated consecutively, and the song repertoires of individual males are very large. Despite this variability, rules govern the internal structure of songs. Male gibbons employ a discrete number of notes to construct songs. Songs are not formed through a random assortment of notes. The use of note types varies as a function of position, and transitions between note types are nonrandom.   (Mitani and Marler 1989: 35)

Although it is fair to call such abilities in apes 'pre-syntactic', they are still far removed from the human ability to organize sequences of words into complex hierarchically organized sentences. Little is known about the ability of apes to learn hierarchically structured behaviours, although all researchers seem to expect apes to be less proficient at it than humans; see Byrne and Russon (1998) and Whiten (2000) for some discussion.

### Pre-semantic capacities

### Basic Concept Formation

Many species lead simple lives, compared to humans, and even to apes, and so may not possess very many concepts, but they do nevertheless possess them. 'Perceptual categorization and the retention of inner descriptions of objects are intrinsic characteristics of brain function in many other animals apart from the anthropoid apes'. (Walker 1983: 378). The difference between humans and other animals in terms of their inventories of concepts is quantitative. Animals have the concepts that they need, adapted to their own physiology and ecological niche. What is so surprising about humans is how many concepts they have, or are capable of acquiring, and that these concepts can go well beyond the range of what is immediately useful. Basic concrete concepts, constituting an elementary pre-semantic capacity, were possessed by our remote ancestors. (A good survey appears in Jolly 1985: ch. 18; see also Allen and Hauser 1991.)

Something related to voluntary control is also relevant to pre-semantic abilities. We need not be stimulated by the presence of an object for a concept of it to be evoked. Some animals may have this to a limited degree. When an animal sets off to its usual foraging ground, it knows where it is going, because it can get there from many different places, and even take new routes. So the animal entertains a concept of a place other than where it currently is. But for full human language to have taken off, a way had to evolve of mentally reviewing one's thoughts in a much more free-ranging way than animals seem to use.

## Complex Concept Formation

The ability to form complex conceptual structures, composed systematically of parts, is crucial to human language. Logical predicate–argument structure underlies the messages transmitted by language. The words constituting human sentences typically correspond to elements of a conceptual/logical representation. While apes may perhaps not be capable of storing such complex structures as humans, it seems certain that they have mental representations in predicate–argument form. Simply attending to an object is analogous to assigning a mental variable to it, which functions as the argument of any predicate expressing a judgement made by the animal. The two processes of attending to an object and forming some judgement about it are neurologically separate, involving different pathways (dorsal and ventral) in the brain. This is true not only for humans but also for apes and closely related monkeys (see argument and references in Hurford 2003.) It seems certain that all species closely related to humans, and many species more distantly related, have at least this representational capacity, which is a pre-semantic pre-adaptation for language.

## Mental Calculation

Humans are not the only species capable of reasoning from experienced facts to predictions about non-experienced states of affairs. There is a large literature on problem-solving by animals, leading to ranking of various species according to how well they perform in some task involving simple inference from recent experience (see Krushinsky 1965 for a well-known example). Apes and monkeys perform closest to humans in problem-solving, but their inferential ability falls short of human attainment.

## Pre-Pragmatic Capacities

### Mind-reading and manipulation

When a human hears an utterance, he has to figure out what the speaker intended; this is mind-reading. When a human speaks, she does so with some estimation of how her hearer will react; this is social manipulation. Humans have especially well-developed capacities for social manipulation and mind-reading, and these evolved from similar abilities in our ancestors, still visible in apes. Social intelligence, a well-developed ability to understand and predict the actions of fellow members of the group, was a necessary prerequisite for the emergence of language. Recent studies amply demonstrate these manipulation and mind-reading abilities in chimpanzees (Byrne and Whiten 1988; de Waal 1982; 1989; Hare et al. 2001).

### Cooperation

People can understand the intended import of statements whose literal meanings are somehow inappropriate, such as *It's cold in here*, intended as a request to close the window. To explain how we cope with such indirectness, traditional logic has to be supplemented by the Cooperative Principle (Grice 1975), which stipulates that language users try to be helpful in specified ways. The use of language requires this basis of cooperativeness. No such complex communication system could have evolved without reliable cooperativeness between users.

Humans are near the top of the range of cooperativeness. The basis of cooperation in social insects is entirely innate, and the range of cooperative behaviours is small. In humans, building onto a general natural disposition to be cooperative, cooperation on a wide range of specific group enterprises is culturally transmitted. Children are taught to be 'team players'. No concerted instruction in cooperation exists outside humans, but there are reports of cases where an animal appears to be punished for some transgression of cooperativeness (Hauser 1996: 107–9). So the basis for cooperative behaviour, and for the instilling of such behaviour in others, exists in species closely related to humans. Chimpanzees and bonobos, in particular, frequently engage in reconciliation and peacemaking behaviour (de Waal 1988; 1989). Dispositions to cooperation and maintenance of group cohesion are pragmatic cognitive pre-adaptations for language.

## Joint Attention

Cats are inept at following a pointing finger; dogs are better. Language is also used to 'point at' things, both directly and indirectly. Linguists and philosophers call this 'reference'. When a speaker refers to some other person, say by using a personal pronoun, the intention is to get the hearer to attend to this other person. Successful use of language demands an ability to know what the speaker is talking about. A mechanism for establishing joint attention is necessary. Human babies and children are adept at gaze- and finger-following (Franco and Butterworth 1996; Morales et al. 2000; Charman et al. 2000). The fact that humans, uniquely, have whites to their eyes probably helps us to work out what other people are looking at.

Primates more closely related to humans are better at following the human gaze than those less closely related (Itakura 1996). Chimpanzees follow human gaze cues, while non-ape species such as macaques fail to follow human gaze cues. But experiments on rhesus macaques interacting with other rhesus macaques show that these animals do follow the gaze of conspecifics (Emery et al. 1997). Spontaneous pointing has also been observed in captive common chimpanzees (who had not received language training) (Leavens et al. 1996) and in young free-ranging orangutans (Bard 1992). It thus appears that animals close to humans possess much of the cognitive apparatus for establishing joint attention, which is the basis of reference in language.

## Ritualized Action

Short greetings such as *Hello!* and *Hi!* are just act-performing words; they do not describe anything, and they cannot be said to be true or false. We can find exactly such act-performing signals in certain ritualized actions of animals. The classic example of a ritualized action is the snarling baring of the teeth by dogs, which need not precede an imminent attack, and is a sign of hostility. Human ritualized expressions such as *Hello!* are relics of ancient animal behaviour, mostly now clothed in the phonemes of the relevant language. But some human ritualized expressions, such as the alveolar click, 'tsk', indicating disapproval, are not assimilated into the phonology of their language (in this case English). The classic discussion of ritualization in animal behaviour is Tinbergen (1952), who noted the signal's 'emancipation' from its original context. This process of dissociation between the form of the signal and its meaning can be seen as the basis of the capacity to form arbitrary associations between signals and their meanings, discussed in the next section. (See Haiman 1994 for a more extended argument that ritualization is a central process in language evolution.)

### Elementary Symbolic Capacity

The sound of the word *tree*, for instance, has no iconic similarity with any property of a tree. This kind of arbitrary association is central to language. Linguistic symbols are entirely learned. This excludes from language proper any possible universally instinctive cries, such as screams of pain or whimpers of fear. In the wild, there are many animals with limited repertoires of calls indicating the affective state of the animal. In some cases, such calls also relate systematically to constant aspects of the environment. The best-known example is the vervet monkey alarm system, with distinctive calls for different classes of predator. There is no evidence that such calls are learned to any significant degree. Thus no animal calls, as made in the wild, can as yet be taken as showing an ability to learn an arbitrary mapping from signal to message.

Trained animals, on the other hand, especially apes, have been shown to be capable of acquiring arbitrary mappings between concepts and signals. The acquired vocabularies of trained apes are comparable to those of 4-year-old children, with hundreds of learned items. An ape can make a mental link between an abstract symbol and some object or action, but the circumstances of wild life never nurture this ability, and it remains undeveloped.

The earliest use of arbitrary symbols in our species was perhaps to indicate personal identity (Knight 1998; 2000). They replaced non-symbolic indicators of status such as physical size, and involuntary indexes such as plumage displays. In gibbons, territorial calls also have features which can indicate sex, rank and (un)mated condition (Cowlishaw 1992; Raemaekers et al. 1984).

The duetting long call behaviour of chimpanzees and bonobos, where one animal matches its call to that of another, indicates some transferrability of the calls between individuals, and an element of learning. But such duetting is probably 'parrot-like', in that the imitating animal is not attempting to convey the 'meaning' (e.g. rank, identity) of the imitated call. The duetting behaviour is not evidence of transfer of *symbolic* behaviour from one individual to another. Probably the duetting behaviour itself has some social/pragmatic significance, perhaps similar to grooming.

In humans the ability to trade conversationally in symbols comes naturally. Even humans have some difficulty when the symbol clashes with its meaning, for example if the word *red* is printed in green. Humans can overcome such difficulties and make a response to the *symbol* take precedence over the response to the *thing*. But chimpanzees apparently cannot suppress an instinctive response to concrete stimuli in favour of response to symbols. With few exceptions, even trained apes only indulge in symbolic behaviour to satisfy immediate desires. The circumstances of wild chimpanzee life have not led to the evolution of a species of animal with a high readiness or willingness (as with humans) to use symbols, even though the rudiments of symbolic ability are present.

All these pre-adaptations illustrate cases where some ability crucial to developed human language was present, if to a lesser degree, in our prelinguistic ancestors. Note that the levels of linguistic structure where language interfaces with the outside world, namely phonetics, semantics and pragmatics, were (apart from motor control of speech) in all likelihood relatively closer to modern human abilities than the 'core' levels of linguistic structure, namely phonology and morphosyntax. The elaborated phonology and syntax so characteristic of full human language came late to the scene. In modern humans, syntactic and phonological organization of utterances, though learned, is largely automatic, not under conscious control. In a sense, then, language evolved 'from the outside in'; the story is of a widening gap, bridged by learnable automatic processes, between a signaller's intentions (meanings) and the signal itself. Near the beginning, there were only simple calls analogous to English *Hello*, in which an atomic signal is directly mapped onto an atomic social act. Every human utterance is still a speech act of some sort. We now have the possibility of highly sophisticated speech acts, whose interpretation involves decoding of a complex signal into a complex conceptual representation, accompanied by complex calculations to derive the likely intended social force of the utterance. The crucial last biological step towards modern human language capacity was the development of a brain capable of acquiring a much more complex mapping between signals and conceptual representations, giving rise to the possibility of the signals and the conceptual representations themselves growing in complexity. In the first generations after the development of a brain capable of acquiring such a complex mapping, communication was not necessarily much more complex. The actual complex structures that we now find in the communication systems (i.e. languages) of populations endowed with such brains may have taken some time to emerge. The mechanisms by which languages grew in biologically language-ready populations will be discussed in the next Section.

## Cultural Evolution of Languages

### The Two-Phase Nature of Language Transmission

I have referred earlier to the 'phenomena of human language'. Modern linguistics focuses equally, if not more, on the noumena of language. A noumenon/phenomenon distinction pervades linguistics from Saussure's *langue* and *parole*, through Chomsky's competence and performance to his later I(nternal)-language and E(xternal)-language. Chomsky's postulation of competence attributes psychological reality to the language system, held in individual minds. This contrasts with Saussure's characterization of *langue* as an entity somehow belonging to the language community. The move to individual psychological reality paved the way for an explanatory link between the evolution of language and biological evolution. Modern linguistics, preoccupied with synchronic competence, has yet to realize the potential for explaining both linguistic phenomena and linguistic noumena in terms of a cyclic relationship between the two, spiralling through time.

Spoken utterances and particular speech acts located in space and time are produced by speakers guided by knowledge of grammatical well-formedness, paraphrase relations, and ambiguity. This knowledge was in turn formed in response to earlier spoken utterances and speech acts, as users acquired their language. Modern linguistics has tended to characterize the overt phenomena of language, the spatio-temporal events of primary linguistic data (PLD), as 'degenerate', and of little theoretical interest. The burden of maintaining the system of a language, as it is transmitted across generations, has been thrust almost wholly onto the postulated innate cognitive apparatus, which makes sense of the allegedly chaotic data in similar ways in all corners of the globe, resulting in linguistic universals.

Clearly humans are innately equipped with unique mental capacities for acquiring language. Language emerges from an interaction between minds and external events. The proportions of the innate cognitive contribution and the contribution due to empirically available patterns in the stimuli remain to be discovered. Methodologically, it is much harder to study performance data systematically, as this requires copious corpus-collecting, and it is not a priori obvious what to collect and how to represent it. In transcribing the linguistic data input to a child, it is highly probable that the transcriber imposes decisions informed by his own knowledge, and thus the true raw material which a child processes is not represented. This difficulty contrasts with the study and systematization of adult linguistic intuitions, accomplished from the armchair. But the intractability of the data giving rise to adult linguistic intuitions does not imply that the only proper object of study is linguistic competence. Because language emerges from the interaction of minds and data, linguistics *must* concern itself with both phases in this life-cycle.

This view of language as a cyclic interaction across generations between I-language and E-language, has been taken up by historical linguists. Rather than postulating abstract laws of linguistic change, they (e.g. Andersen 1973; Lightfoot 1999) appeal to principles relating the spoken output of one generation to the acquired knowledge of the next. This is a healthy development. Historical linguistics, however, is concerned with explaining language change as it can be observed (or deduced) from extant data, either ongoing changes or reconstructed changes hypothesized from comparing related languages and dialects. Historical linguistics is not, in general, concerned with accounting for the emergence of modern complex forms of language from earlier simpler forms. As such, historical linguistics typically makes 'uniformitarian' assumptions (see Newmeyer 2002; Deutscher 1999 for discussion of uniformitarianism). By contrast, one task of evolutionary linguistics is to work out how modern complex linguistic systems could have arisen from simpler origins, using the cyclic interaction between spatio-temporal data and acquired grammars as its central explanatory device. This task has been undertaken from two quite different directions, by theorists of grammaticalization and computer modellers working with the 'iterated learning model' (ILM). I discuss them briefly below.

### Grammaticalization

At the heart of the grammaticalization theory is the idea that syntactic organization, and the overt markers associated with it, emerges from nonsyntactic, principally lexical and discourse, organization. The mechanism of this emergence is the spiralling interaction of the two phases of a language's existence, I-language and E-language. Through frequent use of a particular word, that word acquires a specialized grammatical role that it did not have before. And in some cases this new function of the word is the first instance of this function being fulfilled at all, in the language concerned. Clear examples are seen in the emergence of Tok Pisin, the Papua New Guinea creole. In Tok Pisin, *-fela* (or *-pela*) is a suffix indicating adjectival function, as in *niupela* 'new', *retpela* 'red', *gutpela* 'good'. This form is clearly derived from the English noun *fellow*, a noun not originally identified with any particular grammatical function, other than those associated with all nouns. Grammaticalization occurs in the histories of all languages, not just in the creolization process.

Grammaticalization theory has largely been pursued by scholars concerned with relatively recent changes in languages (Traugott and Heine 1991; Hopper and Traugott 1993; Traugott 1994; Pagliuca 1994). In keeping with a general reluctance to speculate about the remote past, most grammaticalization theorists have not theorized about the very earliest languages and the paths from them to modern languages. Nevertheless, a recurrent central theme in grammaticalization studies is *unidirectionality*. The general trend of grammaticalization processes is all in one direction. Occasionally there may be changes in the opposite direction, but these are infrequent, and amply outnumbered by changes in the typical direction. It follows that the general nature of languages must have also changed over time, as languages accumulated more and more grammaticalized forms. Heine is one of the few grammaticalization theorists who has speculated about what this implies for the likely shape of the earliest languages.

... on the basis of findings in grammaticalization studies, we have argued that languages in the historically non-reconstructible past may have been different—in a systematic way—from present-day languages. We have proposed particular sequences of the evolution of grammatical structures which enable us to reconstruct earlier stages of human language(s). ... such evolutions lead in a principled way from concrete lexical items to abstract morphosyntactic forms. [This] suggests, on the one hand, that grammatical forms such as case inflections or agreement and voice markers did not fall from heaven; rather they can be shown to be the result of gradual evolutions. Much more importantly, [this] also suggests that at the earliest conceivable stage, human language(s) might have lacked grammatical forms such as case inflections, agreement, voice markers, etc. so that there might have existed only two types of linguistic entities: one denoting thing-like time stable entities (i.e. nouns), and another one for non-time stable concepts such as events (i.e. verbs).    (Heine and Kuteva 2002: 394)

To stimulate discussion, I will be at least as bold as Heine, and offer the following suggestions about what earlier stages of human languages were like, based on the unidirectionality of grammaticalization processes. The origin of all grammatical morphemes (function words, inflections) is in lexical stems. This leads one to hypothesize that the earliest languages had: no articles (modern articles typically originate in demonstratives, or the number *one*); no auxiliaries (these derive from verbs); no complementizers (which may originate from verbs); no subordinating conjunctions (also likely to derive from verbs); no prepositions (deriving from nouns); no agreement markers (deriving from pronouns); no gender markers (deriving from noun classifiers, which in their turn derived from nouns); no numerals (from adjectives and nouns); no adjectives (from verbs and nouns).

In addition, I speculate that the earliest languages had: no proper names (but merely definite descriptions); no illocution markers (such as *please*); no subordinate clauses, or hypotaxis; no derivational morphology; less differentiation of syntactic classes (perhaps not even noun and verb); and less differentiation of subject and topic. All this is characteristic of (unstable) pidgins and reminiscent of Bickerton's construct 'protolanguage'; a crude pidgin-like form of communication with no function words or grammatical morphemes. Still in the syntactic domain, Newmeyer (2000) has theorized that all the earliest languages were SOV (once they had the noun/verb distinction).

In keeping with ideas from grammaticalization theory about meaning, the earliest languages would have had, in their semantics: no metaphor; no polysemy; no abstract nouns; fewer subjective meanings (e.g. epistemic modals); less lexical differentiation (e.g. *hand/arm, saunter/stroll/amble*); fewer hyponyms and superordinate terms.

One can apply similar ideas in phonology. Probably the earliest languages had simple vowel systems and only CV syllable structure. See the next subsection for mention of computer modelling of the emergence of phonological structure, via the cyclic two-phase mechanism of language transmission.

### Computer Modelling of Language Evolution

Grammaticalization theorists work backward from modern languages, via known processes of linguistic change, toward earlier, simpler stages of language. By contrast, computer modellers of emerging language start from simulated populations with *no language at all*, and their simulations can lead to interesting results in which the populations have converged on coordinated communicative codes which, though still extremely simple, share noteworthy characteristics with human language. Some examples of such work are Batali (1998; 2002), Kirby (2000; 2002), Hurford (2000), Teal and Taylor (1999), and Tonkes and Wiles (2002). A survey of some of these works, analysing their principal dimensions, and the issues they raise, appears in Hurford (2002). Hurford refers to this class of computer models as 'expression/induction' (E/I) models; Kirby has rechristened this general class 'iterated learning models' (ILMs), a term which seems likely to gain currency. There is a noticeable trend in recent computer simulations of language evolution away from modelling of the biological evolution of features of the language acquisition device (e.g. Hurford 1989; 1991; Batali 1994). More recent simulations (such as those cited earlier in this paragraph) typically model the evolution of languages, via iterated learning. Such studies, moreover, do not typically attempt to 'put everything together' and reach a full language-like outcome; rather they explore the interactions between pairs of strictly isolated factors relevant to the iterated learning model (e.g. Brighton and Kirby 2001).

Language has not always existed. Hence there is a puzzle concerning what behaviour the first speakers of a language used as a model in their learning. Computer modelling studies have addressed this problem, using simulations in which individuals have a limited capacity for random invention of linguistic forms corresponding to given (pre-existing) meanings. Massive advances in computing power make it possible to simulate the complex interactive dynamics of language learning by children and their subsequent language behaviour as adults, which in turn becomes the model for learning by the next generation of children. It is now possible not only to simulate the learning of a somewhat complex communication system by a single individual, on the basis of a corpus of presented examples of meaning–form pairs, but to embed such individual learning processes in a population of several hundred individuals (each of whose learning is also simulated) and to simulate the repetition of this population-wide process over many historical generations.

The cited research has implemented such simulations with some success in evolving syntactic systems which resemble natural language grammars in basic respects. This research can be seen as a step up from the preceding paradigm of generative grammar. In early generative grammar, the researcher's task was to postulate systems of rules generating all and only the grammatical sentences of the language under investigation. Early generative grammars were somewhat rigorously specified, and it was possible in some cases to check the accuracy of the predictions of the grammar. But, whether rigorously specified or not, the grammars were always postulated. How the grammars themselves came to exist was not explained, except by the quite vague claim that they were consistent with the current theory of the innate Language Acquisition Device. The recent simulation studies, while still in their infancy, can legitimately claim to embody rigorous claims about the precise psychological and social conditions in which grammars themselves evolve.

This strand of computational simulation research has the potential to clarify the essentials of the interaction between (a) the psychological capacities of language learners and (b) the historical dynamics of populations of learners giving rise to complex grammars resembling the grammars of real natural languages. In such simulations, a population of agents begins with no shared system of communication. The agents are 'innately' endowed with certain competencies, typically including control of a space of possible meanings, an inventory of possible signals, and a capacity for acquiring grammars of certain specified sorts on exposure to examples of meaning–signal pairs. The simulations typically proceed with each generation learning from its predecessor, on the basis of observation of its communicative behaviour. At first, there is no coherent communicative behaviour in the simulated population. Over time, a coherent shared syntactic system emerges. The syntactic systems which have been achieved in this research paradigm are all, of course, simpler than real attested languages, but nevertheless possess many of the central traits of natural language syntactic organization, including recursivity, compositionality of meaning, asymmetric distribution of regular and irregular forms according to frequency, grammatical functional elements with no denotational meaning, grammatical markers of passive voice and of reflexivity, and elementary partitioning into phrases.

There has been less computer simulation of the evolution of phonological systems, but what exists is impressive. De Boer (2001) manages to approximate to the distribution of vowels systems in the languages of the world through a model in which individual agents exchange utterances and learn from each other. An early computational study (Lindblom et al. 1984) can be interpreted as modelling the processes by which syllables become organized into structured CV sequences of segments, where the emergent selected consonants and vowels are drawn from economical symmetrical sets, as is typical of actual languages.

Computer simulations, within the iterated learning framework, starkly reveal what Keller (1994) has called 'phenomena of the third kind', and Adam Smith (1786) attributed to an 'Invisible Hand'. Languages are neither natural kinds, like plants and animals, nor artefacts, deliberate creations of humans, like houses and cars. Phenomena of the third kind result from the summed independent actions of individuals, but are not intentionally constructed by any individual. Ant trails and bird flocks are phenomena of the third kind, and so, Keller persuasively argues, are languages. Simulations within the ILM framework strip the interaction between individuals down to a bare minimum from which language-like systems can be shown to emerge. The key property of these models is that each new generation learns its language from a restricted set of exemplars produced by the preceding generation.

One of the most striking results of this work is this: in a population capable of both rote-learning and acquisition of rules generalizing over recurrent patterns in form–meaning mapping, a pressure exists toward an eventual emergent language that expresses meanings compositionally. No calculation of an individual agent's fitness is involved, nor does any consideration of the communicative efficacy of the language play a part. The convergence on 'efficient' languages is essentially a mathematical outcome of the framework, analogous to the hexagonal cells of honeycombs. At least some of the regular compositional patterning we see in languages is the result, not of humans having an inbuilt bias towards learning languages of a certain type, but of the simple fact that languages are passed on from one generation to the next via a limited channel, a 'bottleneck'. As Daniel Dennett has remarked (personal communication), this turns the familar 'poverty of the stimulus' argument in relation to language acquisition on its head. The poverty of stimulus argument appealed to an intuition that human languages are learned from surprisingly scanty data. Work in the iterated learning framework shows that in order for regular compositional language to emerge, a bottleneck between the adult providers of exemplary data and the child learner is *necessary*. Interesting experiments show that in these models, overprovision of data (i.e. practically no bottleneck) results in no convergence on a regular compositional language.

These two strands of research, grammaticalization studies and computer modelling within the ILM, are at present quite distinct, and followed by non-overlapping research communities. Computer modellers typically come from backgrounds in artificial intelligence, and know little Latin and less

Greek (to put it kindly); grammaticalization theorists come predominantly from humanities backgrounds, and have difficulty conceptualizing computer models. These two research strands will ultimately converge. When they do converge, they should also converge on the attested facts of historical change and creolization.

## Last Words

In the last two decades, new techniques, such as gene sequencing, massive computer simulation, and the various brain imaging methods, have flashed light on intriguing features scarcely contemplated before. But these flashlights are highly selective in their illumination, each gathering reflections from only a few dimensions of the hugely multidimensional space of language structure and use. Language evolution research must continue to feed, voraciously and eclectically, on the results from a very wide range of disciplines. The study of language origins and evolution *is* harder than molecular biology, physical anthropology, or language acquisition research, for example, because at various levels it draws on all of these, and more. We now understand far more about questions of language origins and evolution than has ever been understood before. But precisely because we can now begin to grasp the nature of the questions better, we also know that good answers are even more elusive than we thought.

### FURTHER READING

As background readings relevant to many of the issues raised under the heading of 'pre-adaptation' in this chapter, I suggest the chapters in the three edited collections, Hurford et al. (1998), Knight (2000), and Wray (2002). For work on grammaticalization and related theoretical positions, I suggest de Boer (2001), Hopper and Traugott (1993), Pagliuca (1994), and Traugott and Heine (1991). For work on computational simulations of language evolution, I suggest Briscoe (2002) and Parisi and Cangelosi (2001).

# 4

## What Can the Field of Linguistics Tell Us About the Origins of Language?

*Frederick J. Newmeyer*

### Introduction

To a non-linguist, the question raised in the title of this chapter must sound nothing less than bizarre. One's first reaction would undoubtedly be to wonder *what other field*, if not linguistics, would be in a position to theorize about language origins and evolution. After all, one would hardly expect to find survey articles entitled 'What Can Botany Tell Us About the Origins of Plants?' or 'What Can Geology Tell Us About the Origins of Rocks?' Nevertheless, until quite recently at least, linguists have not seen it as within their purview to address the origins of the faculty that forms their object of study. The task has been taken up by individuals from a potpourri of fields, ranging from anthropology to neuropsychology to zoology. For example, a typical collection of articles on the topic, *Studies in Language Origins*, Vol. 1 (Wind et al. 1989), contains sixteen contributions, only three by individuals who describe themselves in the 'Notes on Contributors' as 'linguists'.

The purpose of this chapter is twofold. First, it will explain why over the years linguists have taken a sceptical view of the possibility that their expertise might shed light on language origins. Second, it will discuss the reasons why many, though certainly not most, linguists have had a change of mind on the suitability of the topic for scholarly investigation.

The title of the chapter carries an implicit assumption that the 'field of linguistics' is well defined. However, like all academic subdivisions, linguistics has a prototype structure, with clear exemplars of practitioners at the centre and individuals of more uncertain status at the periphery. A fairly

I would like to thank Morten Christiansen, James Hurford, and Simon Kirby for their comments on an earlier draft of this chapter.

encompassing stance will be taken here; one will be considered a member of the field of linguistics if he or she has one or more linguistics degrees or, by virtue of relevant sociological markers (self-identification, academic programme membership, publication in linguistics journals, etc.), can be regarded as a contributor to the field.

## Linguists' Reluctance to Theorize About the Origins of Language

It has by now passed into the traditional lore of linguistics, repeated at every conference devoted to language origins and evolution, that in 1866 the Société Linguistique de Paris issued an outright injunction against speculation on the topic at its conferences and in its publications. Rumour has it that the Linguistic Society of America considered the same ban upon its founding in 1924, but settled instead for a 'gentlemen's agreement' (as such things were then known) prohibiting papers devoted to this area. Whether this story is true or not, no full-length article appeared in the society's journal, *Language*, until the publication of Briscoe (2000). The remainder of this section will attempt to explain why linguists have traditionally shied away from formulating hypotheses about language evolution. The first subsection will point to the resistance of theoretical linguists to discuss the topic, and the second to the reasons for their belief that the biological sciences have little to contribute to it. The third subsection will explain why the field of historical linguistics seemingly has little relevance to the understanding of language origins, and the fourth will stress that the uniformitarian hypothesis has discouraged speculation on the topic.

### *Theoretical Linguistics*

Theoretical linguists, that is, linguists whose interest is the nature of the human language faculty, have been very reluctant to enter into discussions of language origins. Such might seem surprising, given that current linguistic theory hypothesizes the existence of an innate universal grammar (UG) common to all humans. But it goes without saying that if one is interested in exploring the question of the origins of a faculty, then one has to have a pretty good idea of precisely what that faculty *is*. The properties of UG are, however, anything but clear. There are more than a dozen different and competing theories, each of which necessarily presents a different explanandum for the language evolution researcher. The views of Noam Chomsky on the nature of UG have long been dominant within the field of linguistics, but they themselves have undergone marked changes from decade to decade (or more frequently). Hence the logically prior task of elucidating precisely *what* evolved has taken research priority over elucidating *how* it evolved.

To the extent that Chomsky has been willing to speculate on language origins at all, his remarks have only served to discourage interest in the topic among theoretical linguists. He has adamantly opposed, for example, the idea that the principles of UG arose by virtue of their utility in fostering the survival and reproductive possibilities of those individuals possessing them. For example, he has this to say about the adaptive utility of the structure-dependence of grammatical rules, a central property of UG:

> Following what I take to be Searle's suggestion, let us try to account for [structure-dependence] in terms of communication. ... But a language could function for communication (or otherwise) just as well with structure-independent rules, so it would seem.   (Chomsky 1975: 57–8)

Along the same lines, Lightfoot (2000) and Martin and Uriagereka (2000) argue that the UG principles of trace binding and feature attraction respectively are at least partly dysfunctional.

Chomsky has been most sympathetic to an exaptationist scenario, in which UG arose as a by-product of other changes, such as an increase in brain size. For example, he raised the possibility that UG might have arisen to solve what is, essentially, a 'packing problem' (on this point see also Jenkins 2000):

> We know very little about what happens when $10^{10}$ neurons are crammed into something the size of a basketball, with further conditions imposed by the specific manner in which this system developed over time. It would be a serious error to suppose that all properties, or the interesting properties of the structures that evolved, can be 'explained' in terms of natural selection.   (Chomsky 1975: 59)

Needless to say, if the properties of UG are what they are as a result of physical principles (whatever those principles might be), then it falls to the physicist and molecular biologist to unravel language origins, not to the theoretical linguist.

## The Biological Sciences

Linguists have traditionally regarded the results of fields such as evolutionary biology, animal communication, and others branches of the biological sciences as being of little help in understanding language origins and evolution. The long-time consensus among linguists is that the two principal tools of evolutionary biology, the comparative method and the interpretation of the fossil record, are inapplicable for this purpose. A biologist interested in exploring the evolution of some structural property of some species will first of all avail himself or herself of the comparative method, which involves the identification of homologues to the relevant property in some related species. From examination of the differences that the property manifests in each species, it is often possible to build plausible conjectures about its evolution. Conjectures about the origins of 'the panda's thumb' (Gould 1980) are possible because there exist homologous structures in other species closely related to the panda. Furthermore, for most structures the fossil record provides a guide to the evolutionary biologist. Even if bears and raccoons had never evolved, there might well be million-year-old skeletons of 'proto-pandas' to provide a key to the origins of their thumbs.

However, at first blush the comparative method seems to be utterly inapplicable to language origins. Organisms that do (or might) have sophisticated systems of communication (honey bees, dolphins) are orders of magnitude removed from the branch of the tree of life on which our species evolved. Our closest relatives—the great apes — have no obvious 'language' outside of a few calls and grunts. In other words, the central aspects of language—syntax and phonology—have no evident homologues. In that sense, language is an emergent trait (or 'key innovation') and poses, along with all such traits, particularly difficult problems for evolutionary biology. Likewise, there are no archeological digs turning up specimens of the language of 100,000 years ago. While the fossil record has given us a reasonably clear picture of the evolution of the vocal tract, grammatical structure, needless to say, is not preserved in geological strata.

For these reasons, a course of study in linguistics rarely involves a class in animal communication. The consensus has been that the study of species other than our own, no matter how close they might be genetically, can contribute little to an understanding of how language evolved.

## Historical Linguistics

Historical linguists have been as sceptical as theoretical linguists that their subfield might shed light on language origins. The primary method by which they reconstruct the ancestor of genetically related languages is (also) known as the 'comparative method'. By comparing sound–meaning correspondences in related languages, it is possible to make reasonable hypotheses about the sound structure of the 'proto-language' from which the related languages descended. One might imagine, then, that it would be possible to apply the comparative method to all languages of the world and thereby reconstruct 'Proto-World', the original human language. This is not an option, however, because we have no evidence that all languages of the world share a common ancestor. Estimates of the number of distinct (unrelatable) language families run as high as 250, more than half of them in the Americas (Campbell 1998: 163). Many historical linguists posit considerably fewer; one extreme (and almost universally rejected) position holds that that there are identifiable sound–meaning correspondences linking all the world's languages (Ruhlen 1994). But even Ruhlen admits that there are insufficient numbers holding across the totality of language families to reconstruct the sound system of Proto-World.

There is no way that the comparative method can establish that two languages are unrelated, and therefore no way that it can disconfirm the hypothesis of a single Proto-World. The problem is that languages change so fast that (by most estimates) the method allows a time depth for reconstruction of about 6,000–8,000 years. Past this point, cognates are unrecognizable. So if Proto-World coincided with the appearance of *Homo sapiens*, let us say arbitrarily 200,000 years ago, then the intervening 192,000–194,000 years would have been more than enough time to obliterate any signs of what it might have looked like (see Kiparsky 1976).

While the comparative method has limitations for phonological change it is virtually worthless for syntactic change (and therefore for the possibility of a full reconstruction of Proto-World). For example, most modern Indo-European languages have SVO (subject–verb–object) word order; most ancient Indo-European languages were SOV. The comparative method would therefore not lead to the correct reconstruction. The comparative method fails for syntax because we do not have an analogue to the across-the-board replacement of a sound in a particular environment by another sound in that same environment. Rather, we have gradual, wholesale pattern replace ment. For example, Old English had a predominantly SOV word order pattern with SVO as a minor alternant. Over the centuries, SVO became more and more utilized and SOV eventually died out. So it makes no sense to say that some particular (SVO) sentence in Modern English corresponds to some particular (SOV) sentence in Old English. And because of that, the possibility of comparing some sentence of Modern English to its translations in Latin, Greek, Sanskrit, and Irish, in order to reconstruct its Proto-Indo-European 'ancestor', is out of the question.

Another technique of historical linguistics is known as 'internal reconstruction'. One compares alternating forms in a single language and, taking a number of factors into account, such as the phonological structure of the language, the plausibility that forms like A are more likely to change into forms like B, rather than vice versa, and so on, one reconstructs an ancestral form. In general, internal reconstruction could not be expected to bear on language origins at all, given its limited scope of operation. However, if reconstructions led, in language after language, to a similar *sort* of ancestral form, then one might have some justification in positing that 'sort of form' as predominating at an earlier stage of language development. The changes that go under the heading of 'grammaticalization' seem to have such a property. Most researchers of this topic (see e.g. Heine et al. 1991; Hopper and Traugott 1993), basing their conclusions on internally reconstructed evidence, posit a unidirectional change over time whereby affixes and function words (such as auxiliaries) develop from full lexical categories such as nouns and verbs. Given the unidirectionality of grammaticalization changes, might one posit, then, that Proto-World had lexical categories, but no affixes or function words? Unfortunately, no such conclusion seems possible. The entire progression from full lexical category to affix can take fewer than 2,000 years to run its course. Clearly, if there were no processes creating new lexical categories, we would be in the position of saying that languages remained constant from the birth of *Homo sapiens* until a couple of millennia ago, at which point the unidirectional grammaticalization processes began. There is, however, no reason to believe that such might have been the case.

### The Uniformitarian Hypothesis

Another reason that linguists have felt that they have little to contribute to an understanding of the origins and evolution of language derives from the practically universal adoption of the principle of uniformitarianism, namely the idea that all languages are in some important sense equal. First and fore most, linguists have rejected the idea, prevalent throughout most of the nineteenth century, that one language can be characterized as 'more primi tive' or 'more advanced', grammatically speaking, than another. By the 1920s if not earlier, it had become apparent that grammars of all languages are composed of the same sorts of units—phonemes, morphemes, and so on— and therefore all grammars can be analysed by means of the same theoret ical tools. As far as the correct theory of grammar is concerned, a vowel alter nation in English is no more or no less relevant than one in Sierra Popoluca Furthermore, since the oldest languages that we have on record—the earli est samples of Babylonian, Chinese, Greek, and so on—manifest the same grammatical devices as modern languages, it has been typically concluded that there is no overall *directionality* to language change—that is, it is gener ally assumed that human languages have always been pretty much the same in terms of the typological distribution of the elements that compose them.

Most linguists hold an ever stronger version of the hypothesis than that characterized above: they maintain that for any given language, there is no correlation between aspects of that language's grammar and properties of the *users* of that grammar. That is, there is no correlation between the gram mar of a particular language and the culture, personality, world-view, and so on of the speakers of that language. Uniformitarian assumptions have, in general, been carried over into research on language evolution. Specifically most scenarios of language evolution take uniformitarianism for granted in that they equate the origins of human language with whatever genetic event created human beings. The implication, of course, is that, at least in its grammatical aspects, language has not changed a great deal since that even While such scenarios leave open the question of whether a more 'primitive form of language existed prior to the genetic event that gave birth to the species, they provide no place for the evolution of language after its birth But since, virtually by definition, linguists study human language, uniform tarianism would seem to leave no place for their expertise in discussions of its origins.

## The Linguistic Turn to Language Evolution Research

The reasons for shying away from speculation about language origins and evolution that were outlined in the previous section still hold sway in th field of linguistics as a whole. The number of linguists who have published on the topic is tiny, and one rarely hears of a conference on general linguistic topics whose programme includes a paper on evolution. Nevertheless, the situation has begun to change. Linguists of various persuasions have written books or scholarly essays on the topic, and a biennial conference on language evolution organized primarily by linguists has now met four times. This section will both explain why the topic is less taboo than it once was and outline briefly some of the more important contributions by linguists. In brief, I will first point to more refined theories of Universal Grammar which, combined with the availability of sophisticated techniques of formal modelling, have led some theoreticians to propose hypotheses about language origins. Second, increased understanding of the evolution of the vocal tract and of the cognitive capacities of the higher apes have served to inform these hypotheses. The last two subsections discuss how more profound understanding of language typology, combined with a more measured view of uniformitarianism have fed hypotheses about how human language may have looked tens and even hundreds of thousands of years ago.

### Theoretical Linguistics Reconsidered

We start to find publications by theoretical linguists devoted to language evolution in the late 1980s. Most of the initial research did not take on the origins of UG itself, but focused rather on more general—and uncontroversial—properties of human language. For example, nobody would question the reproductive advantage of being able to convey an unbounded number of stimulus-free messages. Brandon and Hornstein (1986) suggest that evolutionary pressure for phenotypic transmission of information, which demands a system with such properties, was especially acute in the capricious and rapidly changing environment in which our ancestors lived. This paper was soon followed by two important proposals by James Hurford. The first (Hurford 1989) attempted to account for a construct assumed by all theories of language, namely, the 'Saussurean sign', i.e. the existence of an arbitrary bidirectional mapping between a phonological form and some representation of a concept. Hurford mathematically modelled an evolutionary scenario that is likely to have engendered this construct. The second (Hurford 1991), again using mathematical modelling, demonstrated that the critical period for language acquisition (i.e. the period before puberty) is not in and of itself an evolutionary adaptation. Rather, it is a consequence of selective pressure to have one's full language acquired as early in life as possible, rather than being spread out over the course of one's lifetime.

In a landmark paper published at around the same time, Pinker and Bloom (1990) present a forceful defence (*contra* Chomsky) that UG must have originated as an adaptation, and in particular that 'there is every reason to believe that a specialization for grammar evolved by a conventional neo-Darwinian process' (p. 707). Their principal argument is based on the adaptive complexity of grammar, and is modelled on what Darwin himself took to be the only scientific explanation for adaptive complexity, natural selection. Before Darwin, the intricate aggregate of interlocking parts that constitute the human eye was taken as irrefutable proof of divine creation. Darwin stood this argument on its head, and showed that natural selection could achieve the same result. Whatever one might feel about some particular principle internal to UG, nobody could deny that the entire package of properties is adaptive. UG, therefore, according to Pinker and Bloom, demands an explanation in terms of Darwinian natural selection as well.

Naturally, one would like to know why a *particular* model of UG arose in the course of evolution, say, one with structure-dependence, the Empty Category Principle, and Subjacency, as opposed to some other set of logically possible components. Pinker and Bloom's position is that even if the particular subcomponents themselves are 'arbitrary' (i.e. not in and of themselves adaptive), that fact is irrelevant to whether UG as a whole can be attributed to natural selection. As they note, arbitrary conventions can be adaptive as long as they are shared. Thus the (possibly tongue-in-cheek) remark attributed to Jerry Fodor that language is 'the worst possible system for communication' because of its 'syntactic baggage' which is 'enormously complex, mathematically absurd, and uneconomical' (Smith and Miller 1966: 270) does not threaten an adaptive account of UG, even if it is true. The seemingly dysfunctional complexity of the 1966-style phrase structure and transformational rules is consistent with UG being an adaptation, as long as all humans are born 'expecting' the same sorts of rules.

The following year saw the first publication devoted to the explanation of the origins of the autonomy of syntactic processes from other aspects of grammar. Newmeyer (1991) began by asking how sound and meaning might be linked in a grammar. Since humans can conceptualize many thousands of distinct meanings and can produce and recognize a great number of distinct sounds, one's first thought might be that this relation could be

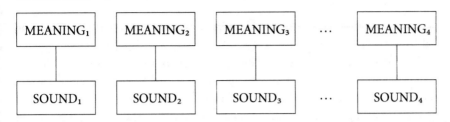

Fig. 4.1 Pairings of sound and meaning

expressed in large part by a simple pairing of individual sounds with individual meanings, as in Fig. 4.1.

At the domain of lexical meaning, no such one-to-one pairing exists; a vastly greater number of words can be stored, retrieved, and used efficiently if sequences of a small number of distinctive sounds are paired with meanings than by a direct mapping between individual meanings and individual sounds. But what about propositional meaning, where the question of a one-to-one pairing is rarely, if ever, raised? The infinitude of possible messages that can be conveyed cannot in and of itself be the explanation; while humans can formulate an indefinite number of propositions, they can also produce and perceive an indefinite number of sound sequences. Thus a one-to-one pairing between proposition and sound sequence is at least within the realm of logical possibility.

Newmeyer speculates that the most plausible answer is that the evolutionary origins of our capacity to vocalize and our capacity to form cognitive representations of the world are too distinct from each other for this to have ever been a workable possibility. A major evolutionary step toward vocal communication was the development of an intermediate level between sound and meaning, a 'switchboard', if you will, which had the effect of coordinating the two. Only at that point could propositional meanings be conveyed vocally with any degree of efficiency. Newmeyer went on to argue that many of the properties of human language syntax, including its autonomy, followed as a result.

Newmeyer also attempted to argue for the selective advantage of *individual* UG constraints. Such a view has found little support in the community of linguists interested in language evolution. As commentators on the paper dryly observed (see e.g. Lightfoot 1991), it seems unlikely that a protohominid endowed with the principle of Subjacency would be more likely to attract a desirable mate, and thereby leave plentiful offspring, than one without this principle.

Nevertheless, recent years have seen publications by linguists that attempt to provide an account of the origins of principles internal to UG. Two rather different considerations have facilitated such a move. First, in the mainstream 'Chomskyan' direction in syntactic theory, the long list of individual UG principles of the Government-Binding approach has been replaced, in the Minimalist Program (MP), by one formal operation, 'Merge', and metalinguistic economy principles governing its operation. In the view of Berwick (1998), the parsimonious inventory of basic elements in the MP makes it realistic to pose the question of why syntax has the architecture that it has.

Second, a number of linguists have suggested, following a trend in evolutionary biology, that extra-adaptive mechanisms might account for the incorporation into the genome of specific principles of UG. For example, Kirby (1998) points to the Baldwin Effect—the evolution of neural mechanisms that encode beneficial behaviour in a population which has a fair amount of plasticity at birth in brain wiring. He argues at length that the Baldwin Effect provides a mechanism by which UG principles might have become biologized without increasing the survival and reproductive possibilities for any particular individual who, by chance, happened to have acquired one or more of them. Indeed, it is predicted that even arbitrary (i.e. non-functional) principles can be biologized, as long as they are shared by a population.

Finally, it should be mentioned that phonologists and phoneticians have been, in general, less reluctant than syntacticians to speculate on the evolutionary origins of their topic of inquiry. The consensus seems to be that phonological differentiation was a response to lexical pressure. As Studdert-Kennedy (1998a: 171) notes, 'We do not have to postulate phonemes and features as the innate axioms of the structuralist tradition, [but rather] as emerging from prior constraints on perception, articulation, and learning, according to general biological principles of self-organization'. In different ways, for example, this is the position advocated by Studdert-Kennedy (1998b), MacNeilage (1998), and Lindblom (1998). And de Boer (2001), by means of computational modelling, demonstrates how at least some phonological universals might emerge from the anatomy and physiology of the vocal tract and the capacity for imitation, rather than from Darwinian selection.

*The Biological Sciences Reconsidered*

The most comprehensive accounts of language origins and evolution published by linguists in the past decade have attempted to integrate their hypotheses with results in evolutionary biology, neurology, primatology, and other life sciences. Three individuals stand out in this regard—Derek Bickerton, Andrew Carstairs-McCarthy, and Philip Lieberman. Their work will be discussed in the following paragraphs.

Derek Bickerton's reputation in linguistics has its roots in his groundbreaking work, both empirical and theoretical, on pidgin and creole languages (see especially Bickerton 1975; 1981). He was struck by the remarkably similar structural features that creoles around the world manifest, and by their at times breathtakingly swift transition from pidgins under conditions of highly degenerate input. His conclusion was that such languages might furnish a 'key' to UG in a way that languages that have evolved gradually over historical time cannot. Bickerton's next step was to speculate that that pidgin-to-creole transition might well be a 'replay' of the evolutionary transition from a form of communication not embodying UG to full language. His two following books (Bickerton 1990; 1995), were devoted to furnishing evidence for this hypothesis. What emerged was a bold, detailed, and internally consistent—albeit highly controversial—scenario for the evolutionary origins of human language.

Bickerton sees human language as the product of a two-stage process. The first took place with the rise of *Homo erectus*, approximately 1.6 million years ago. *Erectus* developed 'protolanguage', a mode of linguistic expression that attaches vocal labels to pre-existing concepts. Hence, the ultimate origins of language lie in cognitive representation. Protolanguage was within the envelope of biological possibility long before *erectus*. Indeed any animal with a representational system developed enough to analyse the world into a broad range of categories shows a readiness for it. But with *erectus* came sufficient evolutionary motivation for its development, and a rudimentary vocal channel facilitating its expression (see especially Bickerton 1990: ch. 6).

Protolanguage is different from language in that it lacks a fully developed syntax. Interestingly, Bickerton posits that protolanguage was not, in a sense, wholly replaced by language; rather, language builds on its predecessor, which remains latent in modern humans. So the language of children under the age of 2, the language of children deprived of linguistic input during the critical period, and pidgin languages all exemplify protolanguage. Also, it is the language of signing apes, whose inability to progress further linguistically is a function of their restricted system of cognitive representation.

A single mutation, coincident with the transition from *H. erectus* to *H. sapiens*, created true language from protolanguage. The major linguistic consequence of this mutation was the imposition of recursive hierarchical structure on pre-existing thematic structure, in one swoop transforming protolanguage to true modern human language. In his 1990 book, Bickerton implied that this mutation was responsible for introducing syntax, improving the vocal tract to facilitate vocal communication, and modifying the skull—an idea that some commentators reacted to with derision (see e.g. Pinker 1992). However, in the following book (Bickerton 1995), the effects of the mutation are far less evolutionarily catastrophic, in that he explicitly separates out the mutation involved in the linking of pre-existing cognitive subsystems involved in syntax with the other anatomical changes relevant to our full repertoire of linguistic abilities. And Bickerton (1998) posits the single-step creation of a new neural pathway linking thematic structure (now regarded as itself a product of primate reciprocal altruism) and phonetic representation, rather than an actual mutation. Bickerton's views are developed still further, and given a firmer neurophysiological footing, in Calvin and Bickerton (2000).

Carstairs-McCarthy (1999) takes three essential yet seemingly unrelated properties of human language, and argues that they are inextricably linked evolutionarily. The first is the enormous number of words with distinct meanings possible in any given language. The second is the principle of language organization known as 'duality of patterning', in which languages are describable in terms of two combinatory systems—one involving meaningless sounds and syllables and one involving meaningful words and phrases. The third is the universal distinction between the syntactic categories S and NP. He ties the three properties of language ultimately to the descent of the larynx. During the *Homo erectus* period the larynx began to lower from its standard mammalian position—high up in the mouth—to its anatomically modern one. Our ancestors made use of the consequent increased vocalization potential by developing a larger call vocabulary. Duality of patterning now follows as a matter of course. Given an expanded call system, it would not be long before individual calls would be strung together, preserving their component meanings, thereby creating complex calls. At this point, calls are now analysable into recurring constituents at two levels. Now, what about the S/NP distinction? Again, human anatomy shaped human destiny. In a nutshell, this syntactic distinction was modelled on the structure of the phonological syllable, whose structure, in turn, is largely based on an acoustically characterizable hierarchy of sonority. The S/NP distinction, then, is no more than a syntactic reflection of the distinction between syllables as a whole and their marginal elements.

Philip Lieberman's general approach could not be more different from that of Bickerton and Carstairs-McCarthy. The former attempt to develop a theory of language evolution compatible with the theoretical constructs developed within generative grammar over the past forty or so years. Lieberman, a phonetician by training, devoted an early book (Lieberman 1975) to the evolution of the vocal tract, and tacitly assumed the basic ideas of generative syntax and phonology. But in the series of books that followed (Lieberman 1984; 1991; 2000), he has attempted to show that the innate UG posited within generative grammar is *incompatible* with what we know about evolution. For example, in his 1991 book he first challenged the idea that humans were innately equipped for syntax *per se*. Rather, syntactic abilities emerged as a by-product of the adaptation for rapid vocal communication, which itself was grounded in the collateral development of the supralaryngeal vocal tract and the neural circuitry needed to automatize speech motor activity and to decode encoded speech. The attempt to derive syntax from motor control is a feature of his later books as well. As he puts it in Lieberman (1998), the 'neural bases of human speech, motor control, and syntax appear to be linked together in a functional language system' characterized by 'neural circuits' that are learned, rather than transmitted as genetic blueprints.

It remains to be seen to what extent the (mostly incompatible) proposals discussed in this and the previous section will come to be accepted by the broad community of linguists. As Rudolf Botha has stressed in a series of articles in the journal *Language and Communication* (see e.g. Botha 1997; 2002), both neo-Darwinian and extra-adaptive accounts of the origins of language face conceptual and empirical challenges that do not plague such accounts of purely physical features.

### Historical Linguistics Reconsidered

For the reasons alluded to earlier, no respectable historical linguist believes that the standard tools of the field—the comparative method and internal reconstruction—can shed any light on language origins. However, there have been a few speculative attempts to extrapolate from observed changes in languages over time to provide generalizations about what language might have looked like at the dawn of our species. For example, Comrie (1992) suggests that certain complexities of modern language such as morphophonemic alternations, phonemic tone and vowel nasalization, and fusional morphology were not present in the earliest human languages, but arose over historical time. Hombert and Marsico (1996) come to the same conclusion. In their view, complex vowel systems are fairly recent historical developments. In particular, they present evidence that seems to suggest that front rounded vowels and nasalized vowels have shown a tendency to increase over the centuries; few reconstructed proto-systems show any evidence of having had them.

Heine and Kuteva (2002*a*) do not dispute the fact that that the course of the changes observed in grammaticalization takes place too rapidly to allow an early human language reconstruction based on the distribution of data from languages currently spoken. However, they feel that the nature of grammaticalization does allow the conclusion that the earliest human language lacked grammatical items such as functional categories and inflectional morphology. As they note, even if we accept the existence of a cycling of changes, there must have been a starting point from which linguistic material was fed into the cycle before it started evolving. The simplest assumption, given the observed changes in grammaticalization, was that at this starting point there were only lexical categories such as nouns and verbs.

Finally, Newmeyer (2000) argues that if one adopts a set of assumptions that are held by many language typologists, it might follow that the original human language had OV word order. The assumptions are that SOV order predominates among the languages spoken today, that the historical change OV > VO is both more common than the change VO > OV and more natural, and that SOV languages are more likely to have alternative orderings of S, V, and O than do SVO languages. From the conjunction of these assumptions, it would seem to follow that SOV order was once much more typologically predominant than it is now. A further assumption of Newmeyer's is that protolanguage (in Bickerton's sense) had thematic structure, but lacked quantificational structure. Now, given the commonplace observation that VO languages are 'good at' representing quantification directly, but 'bad at' representing thematic structure directly, while OV languages are 'good at' representing thematic structure directly, but 'bad at' representing quantification directly, it might follow that the earliest human language had OV order.

## Uniformitarianism Reconsidered

There has always been some linguistically informed opposition to uniformitarianism. For example, the shibboleth that grammar and culture are *completely* independent might well be too strong. Perkins (1992) argues that the more complex the culture, the less complex the grammatical system for expressing deixis. He maintains that while European languages may have words for expressing the concepts 'this', 'that', and (possibly) 'that over there', languages spoken by hunter-gatherers tend to have vastly more ways of expressing deictic concepts. In a less ambitious study, Webb (1977) argues that the transitive use of *have* (as in *I have a car*) is correlated with ownership of property. Other studies have tied properties of language structure to the degree of isolation of the speakers of the language or the amount of contact that speakers have with speakers of other languages. Trudgill (1992), for example, has argued that the languages of small, isolated communities tend to be overrepresented by grammatical constructs that pose difficulty for non-native speakers, such as complex inflectional systems that are characterized by multiple paradigms for each grammatical class. They also tend to manifest the sorts of feature that one might expect to be present in a tight social network, such as weird rules with enough exceptions to mystify non-native learners. Nettle (1999), by means of computer simulations, has come to very much the same conclusion.

Along the same lines, some grammatical features are more characteristic of literate than of pre-literate societies. Givón (1979), for example, suggests that use of referential indefinite subjects is such a case. More importantly, Givón and many others have suggested that the use of subordinate clauses increases dramatically with literacy. The major study along these lines is Kalmár (1985), which maintains that Samoyed, Bushman, Seneca, and various Australian languages rarely employ subordination. Mithun (1984) has made the same point. She carried out text counts on a number of languages with respect to the amount of subordination that one finds in discourses carried out in those language. All languages manifest *some* subordination (in fact Tlingit manifests a lot of it) but there is a strong correlation between its rare use and the pre-literate status of their speakers. Hence it might seem reasonable to consider subordination to be a later evolutionary development.

The basic problem for language evolution research, as far as the p
ciple of uniformitarianism is concerned, is that if grammar is tailorec
the needs and properties of language users (to whatever degree), and
needs of language users now are not what they used to be, then it follc
that grammar is probably not what it used to be. So we should examine
functional factors that might affect syntactic structure and see if the
ance among them might have changed over time. Functional linguists ;
functionally minded generative grammarians have generally pointed to
following three factors as the most important determinants of gramr
ical form: pressure to process language rapidly (parsing pressure), press
to keep form and meaning in alignment (structure–concept iconicity), ;
pressure for syntactic structure to mimic the structure of discourse.

Now, what is the 'balance' between these forces? Newmeyer (1998)
gues that parsing and iconicity predominate, but with the former pres
ing a somewhat stronger influence on grammars than the latter. Hawk
(1994) has shown that pressure from discourse plays a relatively mi
role in shaping syntax. Many computer simulations of language evolut
have explicitly or implicitly made the same assumptions. But such assui
tions are based on an examination of language spoken today. What rea:
do we have for thinking that the balance of functional forces was the sa
10,000 or 50,000 years ago? The answer to that question is that we have
way of knowing. Indeed, the balance might have been very different. For
ample, one popular scenario for language evolution holds that the root
language structure are in conceptual representation, and that language ｜
only later coopted for vocal communication (see Pinker and Bloom 1ᶜ
Newmeyer 1990; Wilkins and Wakefield 1995). If so, then the effects of
nicity might well have been more evident in early human language thar
day. Parsing pressure would have begun to shape language only gradu
over time.

Parsing pressure would certainly have been weaker if—as seems pl.
ible—subordination was a rarely used grammatical device in the earl
period of human language. So if Subjacency is a product of genetic
similation (as in Kirby and Hurford 1997—see above), then it must h
appeared rather late in the course of language evolution. Such is a r
uniformitarian conclusion, to be sure. Another scenario—and one that
have already pointed to—holds that proto- or early language was subjec
what Givón calls the 'pragmatic mode', that is, one in which discourse p
sure was the primary determinant of the ordering of grammatical eleme

If so, effects of parsing and iconicity would have been less manifest than those deriving from pressure to convey information in some orderly and unambiguous fashion.

In other words, we have no reason to believe, and every reason to doubt, that the functionally motivated aspects of grammar have remained constant over time (for more discussion on the general question of uniformitarianism in linguistics, see Christy 1983; Deutscher 2000; and Newmeyer 2002).

## Conclusion

Until quite recently, linguists were extremely reluctant to speculate about the origins and evolution of language. However, as a result of more refined theories of Universal Grammar, an increased understanding of the evolution of the vocal tract and the cognitive capacities of the higher apes, and a less rigid view of uniformitarianism in language than once held sway, linguists have begun to contribute to the language evolution literature. The answer to the question raised in the title of this chapter, 'What can the field of linguistics tell us about the origins of language?', is 'More with each passing year'.

### FURTHER READING

This chapter has presupposed a basic knowledge of approaches to linguistics as a prerequisite to the discussion of how they have been extended to speculations about the origin and evolution of language. Below are some selected references for the reader who wishes to explore in more depth the basic conceptions that underlie these approaches.

The most recent comprehensive technical exposition of the Minimalist Program, the approach to formal linguistics being developed by Chomsky and his associates, is Chomsky (1995). However, considerably more readable overviews are provided by Radford (1997) and Culicover (1997). Jackendoff (2002) ties together work in linguistics with that in cognitive and biological science in a *tour de force* that presents an approach to the architecture of grammar that differs somewhat from Chomsky's. A chapter of the book is devoted to the evolution of language. Other important frameworks for grammatical analysis (which themselves have spawned a small literature on language origins) are Head-Driven Phrase Structure Grammar (Pollard and Sag 1994) and Lexical-Functional Grammar (Bresnan 2001). Laver (1994) and Roca and Johnson (1999) are good recent introductions to phonetics and phonology respectively.

As far as historical linguistics is concerned, perhaps the best current overvie᷎ Campbell (1998). The articles in Durie and Ross (1996) debate the success of comparative method. Two important works explore the possibility of establ ing links of language relatedness going back tens of thousands of years: Greenb (1987) and Nichols (1992). However, it needs to be stressed that the former ha᷎ ceived a far more favourable reading outside the field of linguistics than within addition to the references cited in the chapter itself, the following works prov a look at current approaches to grammaticalization: Traugott and Heine (19⁹ 1991*b*), Bybee et al. (1994), and Heine and Kuteva (2002*b*).

Finally, there is a rich literature developing diverse functionalist approache᷎ grammar. Haiman (1985), Givón (1995), Van Valin and Lapolla (1997), and ] (1997) are a small sample. Newmeyer (1998) is devoted to the reconciliatior᷎ formal and functional approaches. Functional explanations for the typolog᷎ distribution of languages are presented in Comrie (1989) and Croft (1990). Th᷎ works attempt to link typological generalizations to principles governing langu᷎ change, and are therefore relevant to an understanding of language evolution.

# 5

## Symbol and Structure: A Comprehensive Framework for Language Evolution

*Derek Bickerton*

### Speaking as a Linguist[1]

I approach the evolution of language as a linguist. This immediately puts me in a minority, and before proceeding further I think it's worth pausing a moment to consider the sheer oddity of that fact. If a physicist found himself in a minority among those studying the evolution of matter, if a biologist found himself in a minority among those studying the evolution of sex, the world would be amazed, if not shocked and stunned. But a parallel situation in the evolution of language causes not a hair to turn.

Why is this? Several causes have contributed. Linguists long ago passed a self-denying ordinance that kept almost all of them out of the field, until quite recently. Since nature abhors a vacuum, and since the coming into existence of our most salient talent is a scientific question that should concern anyone seriously interested in why humans are as they are, other disciplines rushed to fill that vacuum. Then again, language doesn't look as if it should be all that complex, not like genetics or quantum mechanics. We all speak at least one, that one we acquired without a lick of conscious effort, and most non-linguists, in the unlikely event that they opened a copy of *Linguistic Inquiry* or *Natural Language and Linguistic Theory* only to find stuff every bit as hard going as genetics or quantum mechanics, would in many cases react by saying 'What's all this nonsense about? Why are they making such a fuss about something that's perfectly simple and straightforward?' And they

---

[1] Scout's honour, I hadn't read Fritz Newmeyer's contribution to the present volume when I wrote this chapter, nor had he read mine when he wrote his. The impressive similarity of our introductions was quite spontaneous, a highly natural reaction to the circumstances, and should serve as a wake-up call to linguists and non-linguists alike.

would probably go on to say, 'What do *I* need this stuff for? I'm a systematic biologist/palaeo-anthropologist/ evolutionary psychologist/computational mathematician [strike out whichever do not apply]—I do not need this'.

Well, the reason they do need this is simple and straightforward. Language is a means of communication (among many other functions) that differs radically from the means of communication of any other species. At the same time, we are a species that differs radically from other species in our creativity and variability of behaviour (anyone who confuses this statement of plain fact with the claim that we are the pinnacle of evolution or divinely created is herself seriously confused). There is a good chance that these two uniquenesses are not coincidental; in other words, we most likely are the way we are because we have language and no one else does. If that is so, then it must be because there are specific properties of language which, if other species had had them first, would have produced similar results. What we have to do is determine what these properties are, which of them are essential and which accidental. We have to determine how we came to have not just 'language' in the abstract—whatever that might mean—but the precise set of linguistic properties that happens to correlate with, and most probably causes, our unique nature. We have to do this if we are ever going to explain how humans evolved. But we can't explain why language has the set of properties that it does have, and no other set, if we do not know what those properties are.

## The Interdisciplinary Stance

This is not, yet, a particularly widespread view. A lot of writers believe one can treat language as a given, a black box, in effect, and account for its evolution simply by selecting the selective pressure that gave rise to it. Was it grooming substitute (Dunbar 1993; 1996)? Or maybe setting up a menstruation ritual for female bonding (Power 1998; 2000)? Or letting men know if their women had cheated on them (Ridley 1993: 229)? Or initiating marriage, so men would know who they weren't supposed to cheat with (Deacon 1997)? The fact that these and similar explanations flourish side by side tells one immediately that not enough constraints are being used to limit possible explanations.

Simply taking into account what we know about language should form an adequate constraint, since all these proposals run up against some of guage feature quite incompatible with them. Take the grooming proposal. It is far from clear why, if language simply substituted for grooming when group size became too large, language should invariably convey factual information, indeed be incapable of *not* giving factual information, even in flattering someone (*That outfit really suits you—matches your eyes*). Surely a similar result could have been achieved simply by using pleasant but meaningless noises. Lovers often do just that, even now, with all of language at their disposal. Or take the proposal that the driving force behind the emergence of language was gossip and/or some sort of Machiavellian manipulation. Since there undoubtedly was a time when the vocabulary was zero, there must also have been a time when the vocabulary was vanishingly small, no more than three or four units/signs/words, whatever those may have been. The question is simply whether the gossip one could transmit with such a vocabulary would be of any interest whatsoever. It seems unlikely. Similarly, it is highly implausible that, with a small initial stock of symbols, one could do much in the way of social manipulation.

Both gossip and manipulation require a vocabulary of some size, but such a size could hardly have been achieved unless earlier and smaller vocabularies had already served some useful purpose. However, the issue of the minimal vocabulary size required for implementing functions such as these is simply not addressed by those who claim a social-intelligence source for language (see Bickerton 2002a for a fuller treatment of these and related questions).

Ignorance of both language and linguistic theory seemed to me for a long time to be the most serious deficiency among writers on the evolution of language. Then I reviewed two books by linguists (Loritz 1999; Jenkins 2000; see Bickerton 2001) and I'm no longer so sure. Such ignorance now appears as simply a special case (perhaps the most serious, though by no means the only serious one) of a much more widespread tendency. There are at least six fields–linguistics, palaeo-anthropology, evolutionary biology, neurology, psychology and primatology–that cannot possibly be ignored in any study of language evolution, and a number of others, such as genetics or palaeo-climatology, that bear on it perhaps somewhat less directly. All too often, a writer whose home is in one or other of these disciplines will make a proposal that is unacceptable in terms of one or more of the other relevant disciplines. This is not inevitable. It certainly does not result from the impossibility of acquiring the necessary knowledge, since anyone of average intelligence should, given goodwill and a little effort, be able to master enough of the literature in all relevant disciplines to avoid making gross errors.[2]

Having mastered linguistics and all the other relevant disciplines, are we now ready to make sense of language evolution? I don't think so. There's something that is implicit in much of my own writing which I fully realized only quite recently. It is that the biggest obstacle to understanding the evolution of language is thinking about it as 'the evolution of language'.

## Divide and Rule

Language as we know it today involves the coming together of three things: modality, symbolism, and structure. I can see no reason for supposing that all three evolved as a package deal, and good reasons for supposing that they evolved separately. Let's look at each of these three things in turn.

First modality: that includes speech and sign. For many, 'speech' and 'language' are interchangeable; how depressingly often one turns to the index of a book on human evolution to find the damning entry: 'Language: see Speech'. Most recently Mithen (2000: 216) has claimed that Neanderthals had probably acquired 'a degree of vocalization that is most appropriately described as language'. And even some linguists take a similar approach: Lieberman (1984; 1991; 1996), for example, assumes that once speech was there, the rest followed.

However, I take the arguments in Burling (2000) and Sperber and Origgi (2001) to be quite unanswerable. Before any of the three components of language could exist, let alone come together, there had to be comprehension of some kind, however primitive; pre-humans at some stage had to start trying to figure out one another's intentions. This largely solves the problem of what I once called the 'magic moment' (Bickerton 1990): how did the first hearer of a meaningful signal know that it was a meaningful signal (as opposed to a cough, a grunt of pain, or whatever)? Answer: if our ancestors were already trying to interpret the behaviour of their conspecifics, even

---

[2] I speak here with all the zeal of the converted, having myself violated biological probabilities with the 'macro-mutation' scenario of Bickerton (1990) and neurological probabilities with the 'different bits of the brain getting linked' scenario of Bickerton (1995; 1998). Although I know there are neurologists who do not buy them, I know of no evidence, neurological or other, that rules out the proposals of Calvin and Bickerton (2000), which have now replaced those earlier ones, and to which I return later.

perhaps to the extent of reading meanings where none was intended, they surely wouldn't take long to recognize intentional meanings. It also neatly solves the problem of whether language began as sign or speech. The answer is that it probably began as both—a mixture of anything that might serve to convey meaning. The original mixture of isolated grunts and gestures may have eventually settled on the vocal mode merely through the exigencies of communicating at night, over distance. or in dense vegetation. For a small initial vocabulary, no vocal improvement would have been needed.

Afterwards, as more (and more complex) information gradually came to be exchanged, attempts to convey it would have strongly selected for improved vocal capacities. I think there can be no doubt that the capacity to transmit information was what selected for improved speech capacity, rather than vice versa. Being able to speak more clearly does not, in and of itself, give you more to say. There is thus good reason to believe that the speech modality, far from being the driving force behind language, was entirely contingent on the two other components, the symbolic and the structural, and was developed and refined in response to their development.

Those components, the symbolic and the structural, are also distinct and can also be dissociated from one another. They are actually dissociated in several forms of development than can still be observed in the world around us: in early-stage pidgins, in early stage second-language learning, and in the productions of trained apes and other animals.[3] Of course you can't get a double dissociation: syntactic structure without symbolic content, the kind of thing you saw in old-fashioned phrase-structure rules, cannot be used by animate beings to communicate with one another. But symbolic representation does not require any kind of structure—telegrams and headlines are immediately comprehensible with relatively little grammatical structure, and we can perfectly well understand utterances that have no syntactic structure at all, like Pinker's *skid–crash–hospital* (Pinker 1994). Indeed a variety of factors I have discussed at length in earlier work suggest that, in the evolution of our species, symbolism may have preceded syntax by as much as two million years.

---

[3] I have previously claimed early-stage first-language learning as an example (Bickerton 1990). This needs caveatting, if you'll pardon the Haigism. Children learning inflected languages can and do acquire morphological affixes (they would have a hard time not doing this, in languages where bare stems are virtually or completely non-appearing) and at least sometimes use them correctly. They also acquire some basic facts about word order in the target language. But these are not syntax (see 'Fear of syntax' for what is).

Perhaps the clearest evidence for phylogenetic dissociation lies in the fact that while symbolic representation (at some level; and under instruction, at a near-human level, see Savage-Rumbaugh 1986) is within the reach of a number of non-human animals, syntax, regardless of the quantity or quality of instruction, remains beyond the reach of any other species ('putting symbols in a regular order' does not, of course, come anywhere close to being syntax). This inevitably suggests that the genetic and neural substrates for the two are quite distinct, and that they therefore must have distinct evolutionary histories. Consequently, explaining how language evolved requires us to answer two separate and quite distinct scientific questions. The first is: how and why one particular primate species, or one primate line of descent, developed a system of communication involving symbolic representation that allowed the transfer of (potentially) unlimited factual information, and the basic principles of which differed from those of all previous systems of communication. The second is how such a system acquired the very specific structural characteristics that the syntax of modern human languages exhibits. If one abbreviates these questions to 'How did meaningful units (words or signs) evolve?' and 'How did syntax evolve?', little is lost. But if only one question is answered, or if the two issues are mixed in together and confused with one another, we will continue to get the conflicting and unsatisfying accounts of language evolution that have predominated to this date. Let us therefore keep them clearly separate and deal with each in turn.


## Symbols

The most crucial thing to grasp about the emergence of symbolic representation is that it must have been primarily a cultural rather than a biological event. This idea, again implicit in some of my earlier work, has not previously been stated in quite these terms. However, it follows inevitably from the fact that a neural substrate adequate for some level of symbolic representation exists not simply in other great apes but among creatures as phylogenetically distant from us as African Grey parrots (Pepperburg 2000). This widespread nature of potential for symbolic representation suggests an analogous rather than a homologous development, the kind of development that produced fins in sharks, ichthyosaurs, and cetaceans. Probably the potential for symbolism exists in any animal with a brain of sufficient complexity, and this would hardly be surprising, given the still wider spread of iconic and indexical precursors of symbolism (for Pavlov's dogs, the ringing of a bell was an indexical representation of food, for example). This view may be disturbing to some who, with Deacon, see the Rubicon between us and other animals as being symbolism rather than syntax. Part of the reason may be that when people think of symbolism, they think of the sophisticated version we enjoy today—a vast branching network of symbols each of which is interpretable in terms of other symbols—and not of the primitive version, compounded mainly of indexical and iconic associations, that may have come into existence two million years ago. Another part may be disbelief that, given the benefits of communication—so visible to us, with the twenty-twenty vision of hindsight—any animal capable of communication would fail to use it under natural conditions. Such a view ignores the essential unreliability of language. Words require little energy to produce; they are 'cheap tokens' and can be used with little or no risk or cost to deceive, just as easily as to inform. Body language is much more reliable for most animal purposes.[4]

I suspect that the only things missing for any relatively large-brained species were the two Ms–modality and motivation. Modality was perhaps the lesser problem, given the low requirement of an initial minimal vocabulary. Any modality capable of differentiating a half-dozen or so symbols would do for a start. Motivation was another matter. To us, able to appreciate the myriad potential uses of language, its possession seems too obvious a boon. But since no species has the gift of foresight, we should ask what benefits this minimal vocabulary would have bestowed on any other species. The answer is simple: none. Solitary species do not need to communicate. Other social species get along fine with non-linguistic methods. Since only one social species has even begun to develop the language mode, the logical place to look for motivation is in circumstances unique to that species. Elsewhere (Bickerton 2002a) I have argued in detail that the initial protolanguage arose through the exigencies of extractive foraging in mainly dry savannah-type environments. I will not repeat those arguments here; suffice it to say that hunger and a high risk from predation would have engendered social

---

[4] An actual illustration may be relevant here. Karl Muller, a part-Hawaiian who runs a shelter for the homeless in Honolulu, had a confrontation with the shelter's cook. Karl removed his false teeth. The cook ran away. He knew that the removal of the teeth (an expensive set) meant that Karl, an impressive street fighter, was ready to accept serious damage in going the limit with him. No mere verbal threat would have deterred the cook, a skilled amateur boxer ('He could have been a contender', according to Karl).

systems in which individuals were more interdependent than they are in most primate societies, and where, accordingly, a degree of trust sufficient to overcome the 'cheap tokens' problem would necessarily be engendered.

Symbolism arose culturally, then, because the minimal necessary bio logical equipment was already in place and the exploitation of symbolism directly benefited both individuals and groups (groups by optimizing forag ing under the fission–fusion constraints that a wide day-range plus vulner ability to predation imposed, individuals by enhancing the status of those who located and led the group to the best food sources). The only question that remains is what the earliest symbols were like.

## Holistic Versus Synthetic

Until quite recently, it was generally assumed that ontogeny and phylogeny, though far from indissolubly wedded, were at least alike to this extent: the earliest units of pre-human utterances were pretty similar to the earliest units of contemporary infants. That is to say, they were basically single units with ostensively definable referents, perhaps somewhat broader in mean ing than the units of an adult vocabulary. Recently, however, this notion has been challenged from a variety of perspectives, all of which converge on the assumption that holistic utterances, semantically equivalent to one-clause sentences in modern languages, formed the earliest linguistic utterances.

Wray (2000) sees this proposal as solving the 'continuity paradox' (Bick erton 1990: 7). Calls can be interpreted holistically, so there could be a seamless transition from a non-linguistic communication system to some form of protolanguage. Carstairs-McCarthy (1999; 2000), assuming syntax to be modelled on the syllable (but see the next section), has to have a holis tic protolanguage for the syntactic equivalents of onset, nucleus, and coda to be factored out of. Computational linguists (Batali 1998; Kirby 2000; Briscoe 2002)[5] likewise assume initial holistic utterances, so that, by a pro cess described as 'self-organization', vast quantities of variable and random

---

[5] Hurford (2000*b*: 225–6) claims that his approach to computational evolution is syn thetic, rather than analytic. This is somewhat disingenuous, in light of his own statement that 'speakers were prompted to express atomic meanings (e.g. BERTIE, SAY, or GIVE) 50% of the time, and simple or complex whole propositions (e.g. HAPPY (CHESTER), LIKE (JO, PRU DENCE) or SAY (BERTIE (HAPPY (JO))) 50% of the time' (Hurford 2000: 334). The other three authors cited here appear to make their 'speakers' produce holistic propositions 100 per cent of the time.

utterances could gradually converge on fixed forms with fixed meanings. This holophrastic approach to initial symbols can even claim some kind of history, since holistic beginnings are implicit in the 'singing ape' conjectures of Darwin, Jespersen, and others.

The approach is, however, beset by a variety of problems. Initially, there is the problem of comprehension: while one can deduce many things about others' intentions from their behaviour, anyone who has visited a country with a strange language knows that such understanding does not extend to linguistic behaviour. Since Quine's discussion (1960; see also Premack 1986) the problem of the speaker who says *Gavagai!* to the researcher when a rabbit runs past has been well known. We may believe, unlike Quine, that the word is rather unlikely to mean 'undissociated rabbit parts'. But that may be largely because we benefit from a history, both ontogenetic and phylogenetic, of learning what words are likely to mean. It is a task of considerable difficulty, although well worth attempting, to try to imagine oneself as not knowing any words at all, not even knowing what words were or could do. Certainly the passage of a rabbit at the moment of utterance is no guarantee that *Gavagai!* has anything to do with the rabbit: events do not occur in a vacuum; something else that might be being referred to is always going on. Even if we could somehow know that *Gavagai!* and the rabbit were connnected, they could merely be connected in the way that *God bless you!* is connected with a sneeze.

Now these are problems that would affect understanding even if words or other symbols referred to isolated objects or events and were initially used in the presence of those objects and events (an unlikely proviso—there is little point in talking about what people can see for themselves, and indeed the whole point of language as opposed to animal communication systems in general is that the former, but not the latter, can be used to inform about things that are not physically present). Those problems of understanding are compounded infinitely if the initial utterances of a language do not correspond to anything tangible or easily identifiable, but refer to some set of circumstances that may or may not be apparent from the surrounding context. If the intended meaning is apparent from that context—if, say, initial utterances were things like *Give that to me!* or *Stay away from her!*—you wouldn't need language to express them. Such things are much more unambiguously expressed by behaviour already in an animal's communicative repertoire, such as begging gestures or threat gestures. If the intended meaning is not apparent from that context, the receiver would never be able to select, from a potentially infinite range of possible meanings, the one that the sender meant to express.

It is no accident that in most, if not all, computer simulations of language evolution, the self-organizing 'agents' already *know what their interlocutor means to say.* If the problem space were not limited in this way, the simulations simply wouldn't work—the agents would never converge on a workable system. But such unrealistic initial conditions are unlikely to have applied to our remote ancestors.

Let us suppose, counter to probability, that our ancestors somehow developed a holistic language. They would then have confronted the problem of how to go from such a language to languages that are built up from discrete units with single meanings (as all languages are today). Even today, modern children, equipped with all the bells and whistles of full human language, have a hard time segmenting adult utterances which, though they may sound holistic to the child, consist already of ready-made word units. How much more difficult for creatures with no experience of language to segment strings that were genuinely holistic!

There are two logical possibilities, one of which must be fulfilled by any such holistic utterances. One is that the units that would eventually dissolve into discrete words already contained regularities within the holistic utterance—a phonetic sequence like *-meg-*, for instance, might occur in any holophrase that made reference to 'meat'. This would remove the problem at the same time as it removed any possible justification for supposing that language began in this way. For if such utterances could be straightforwardly decomposed into the equivalent of words, then words as we know them already existed and there would have been no point in starting out with holophrastic units.

The other possibility is that the sequences were truly holistic, in other words that their sound structures bore no relation to one another: 'Give the meat!' might then be *megalup* and 'Take the meat!' might be *kokubar*. From these, or any similar examples, it would simply be impossible to factor out a single symbol for 'meat'.

A holophrastic account has yet more difficulties to face. It may seem easy to translate a holophrase (given that we understand it correctly) into an equivalent sentence of discrete words. But suppose some large male hominid, in the presence of a female, aggressively utters *\*\*&\*&x@\*\*!* We may reasonably take this to mean, 'Stay away from her!' But it could just as easily mean 'Do not go near her!' Or, 'Stay right where you are!' Or even 'You——,

you, get out of my sight!' How smaller units can be factored out from holo-phrases when even their global meanings are so potentially ambiguous remains unclear.

But perhaps the biggest problem with the holistic approach is that it doesn't explain anything worth explaining. All the substantive problems in language evolution—how symbolism got started and fixed, how, when, and why structure emerged, where and how and to what extent any of this got instantiated in neural tissue—remain to be solved, whether one accepts a holistic account or not.

Accordingly, it is more parsimonious to assume that language began as it was to go on—that discrete symbols, whether oral or manual, were there from the beginning. I do not know of a single coherent argument why they shouldn't have been. If they were, it is most likely that, once these symbols exceeded the merest handful, they began to be strung together in some ad hoc fashion. One hears frequently of 'proto-syntax', which seems to mean one-clause sentences with fixed word order, and there is a widespread but wholly erroneous belief that this does not merely constitute a step in the direction of real syntax, but that once one has achieved such a level of structure, real syntax follows automatically. In other words, they account for *The cat sat on the mat* and then cross their fingers,[6] confident that 'self-organization' will take care of the rest.


Fear of Syntax

Perhaps the most depressing aspect of language evolution studies is fear of syntax, which, the present collection suggests, is as widespread as ever. I know of no other field of study in which the work of a large body of highly intelligent specialists is so systematically misinterpreted, ignored, or even trashed. As a matter of plain fact, we have learned more about syntax in the last forty years than in the preceding 4,000, but you'd never guess that from reading most books on language evolution, including, alas, this one. Syntax forms a crucial part, arguably the most crucial part—since no other species is capable of it—of human language. If we are going to explain how language evolved, we have to explain how syntax evolved. If we are going

[6] I am not the originator of this sentence, but I have sought in vain for years to find its true begetter. I believed it to be Lila Gleitman, but she (p.c.) has denied authorship; if its real author contacts me, I will be happy to make full acknowledgment.

to explain how syntax evolved, we have to explain how it came to have the peculiar properties it has, and no others. It just will not do to dismiss it as due to self-organization, or the grammaticization of discourse, or analogies with motor systems or syllabic structure, or any of the other one-paragraph explanations that writers on language evolution typically hope to get away with.

The trouble is that most non-syntacticians think syntax is just a matter of regular word order (I wonder what they think syntacticians do all day!) plus perhaps a few prefixes and suffixes. As a corrective to this view, I offer the following brief test on Real Syntax (an asterisk before a word or sentence means that it is ungrammatical):

(1)  (a)   Bill wants someone to work for.
     (b)   Bill wants someone to work for him.

Why does a pronoun at the end of the sentence change the understood subject of *work*?

(2)  (a)   Who was it you said you didn't wanna/want to see?
     (b)   Who was it you said you didn't *wanna/ want to see you?

Why does a pronoun at the end of the sentence stop you from contracting *want to* to *wanna*?.

(3)  (a)   Which letter did you throw away without opening?
     (b)   That letter you threw away without opening contained anthrax.
     (c)   *You were wise to throw away that letter without opening.

Why is it okay to leave out an *it* after *opening* in the first and second sentence but not the third?

(4)  (a)   We wanted the chance to vote for each other.
     (b)   *We wanted the champ to vote for each other.

Why, given that the second sentence is perfectly logical and comprehensible—I wanted the champ to vote for you and you wanted him to vote for me—is it ungrammatical?

Two things need to be emphasized here. First, these sentences do not exhibit weird quirks peculiar to English or other western European languages (it is worth noting that the first full-length generative grammar of any language dealt with Hidatsa (a native-American language of North America)—Matthews (1961)—and that there is a vast generative literature on Australian, Austronesian, Native American and countless other non-western European languages that unfortunately doesn't seem to have had a wide readership). To the contrary, the phenomena these sentences illustrate arise from broad general principles familiar to anyone who is up to speed on generative syntax (for those who aren't, I can only refer them to said literature). Second, it should be apparent that phenomena of the type illustrated in these sentences are vanishingly unlikely to have come about through social factors, or self-organization, or the streamlining of discourse, or any of the many alternative explanations currently on offer.

This does not, however, mean that they must remain mysterious. It is a good bet that they are as they are because that is the way the brain works—that when syntax is finally and fully understood, it will become apparent that the algorithms the brain actually uses to produce sentences will necessarily produce (as epiphenomena, one may suppose) the features of (1)–(4). It should therefore be the task of anyone seriously interested in the evolution of language to work at either one end or both ends of the mystery: finding out the most parsimonious description of syntax that will satisfy the syntactic facts, or trying to determine (through neuro-imaging or any other available means) how the brain actually puts sentences together. Once we know exactly what evolved, we may begin to approach a final answer as to how it evolved.

As a linguist, I can only attempt the first course. One promising avenue of inquiry, briefly sketched in the appendix to Calvin and Bickerton (2000), further developed in yet to be published work (Bickerton 2003; in preparation) and now described as 'surface minimalism', would reduce syntax to only three components:

(5) (a) Conditions on the attachment of words to one another.
    (b) Cycles of attachment yielding domains that consist of heads and their modifiers (phrases and clauses).
    (c) Principles derived from the order in which constituents are attached to one another.

If language can run on these resources and these only, nothing like the massive amount of task-specific innate equipment many researchers have very reasonably feared would be required. (5a), or a great deal of it, can be derived directly from semantics. (5b) can be derived via a shift in function of a kind of social score-keeping device such as may have developed in several primate lines: the mapping of every event in episodic memory into a simple schema incorporating who did what to whom (see further Bickerton 2000; Calvin and Bickerton 2000). (5c) can be derived from the way the brain processes any kind of material. The brain is adept at merging series of discrete inputs into coherent wholes (it does this every time you look at anything), and it can keep track of the sequence of its own operations through the gradualness with which neuronal activity decays (Pulvermuller 2002). All that is needed to run such a system is a far higher number of neurons and more of both cortico-cortical and cortico-cerebellar connections than we find in the brains of other primates.

Hurford (2000*a*: 223) has, very reasonably, expressed doubts as to whether such a stripped-down system could handle 'many of the examples given by Lightfoot [2000]'. Consider Lightfoot's *pièce de résistance*, to which he devotes almost half his paper: the asymmetry between subjects and objects that allows much freer extraction of the former than the latter (Lightfoot's (11)–(13)):

(6)  (a)   Who do you think that Ray saw?
     (b)   *Who do you think that saw Ray?

(7)  (a)   Which problem do you wonder how John solved?
     (b)   *Who do you wonder how solved which problem?

(8)  (a)   This is the sweater which I wonder who bought.
     (b)   *This is the student who I wonder what bought.

According to Lightfoot, this asymmetry represents a serious dysfunction in language (it 'conflicts with the desire/need to ask questions about the subjects of tensed clauses': Lightfoot 2000: 240), so cannot in itself be adaptive, but must result from some more general condition 'that presumably facilitates parsing' (p. 244). However, he has no explanation for such a condition beyond the suggestion that 'complex, dynamical systems can sometimes go spontaneously from randomness to order' (p. 245).

In surface minimalism, order of attachment (5c) yields two crucial principles, priority and finality (see Bickerton 2002*b*), which are involved in many different syntactic relations. Only the second of these need concern us here. A constituent X is final in a domain Y if there is no constituent Z such that X could be attached within Y before Z is attached. In other words, final attachments (in Lightfoot's examples, final referential attachments, or final arguments) mark the boundaries of domains (phrases or clauses). Non-final arguments (as in the (a) sentences) can be moved freely, since no information about domain boundaries is lost if they are moved. But if final arguments are moved, as they are in the (b) sentences above, those boundary markers are removed and the sentences consequently become harder to parse, because it is less clear where one clause ends and another begins, and therefore more difficult to assign (unambiguously *and automatically*, as syntax must do) an argument to the domain to which it belongs. Accordingly, restrictions are placed on the mobility of final arguments: the *that* which introduces the Theme argument of a verb can't be attached unless a final argument is in place (6b), and question words (*who, which,* etc.) that are final arguments cannot be moved at all if any other question word has been moved (7b, 8b).

## Timing Syntactic Emergence

One question that remains is when syntax emerged. If it emerged gradually, as many (see especially Pinker and Bloom 1990) think it did, there is no problem. A gradually enlarging brain, by providing, not greater intelligence *per se*, but more available neurons and more specialized connections between neurons, could have gradually provided more and more syntax—sentences just got longer and longer, as I once naively supposed (Bickerton 1981: ch. 5).

At least two things are seriously wrong with this. First, the principles involved are across-the-board principles: they apply everywhere, to all structures. At any given time, either they were in place or they weren't. Once they were in place, what was to stop syntax becoming immediately like it is today?

The other involves cognitive development. If we can measure cognitive development by the artefacts our ancestors produced (and what other way do we have?), there was something close to cognitive stagnation over the two million or so years that preceded the appearance of our species (if you doubt this, check out Iain Davidson's deconstruction of the palaeo-anthropological progress myth, chapter 8 below). Then, suddenly, creativity blossomed.

Somehow, there is a threshold effect. Somehow, it has to be explained. The advent of fully syntacticized language is the best candidate explanation so far. If anyone can think of a better alternative, or can explain (instead of merely explaining away) the suddenness of the transition, I'll be delighted

to hear it. Until then, with all its problems (e.g. why Neanderthals, with l
ger brains than ours, didn't win out against us), the best explanation is s
that syntax as we know it developed in our species but in no other.

A second timing problem associated with the origins of syntax invol
the connection between fully syntacticized language and what has be
called the 'Great Leap Forward' (the explosion of human culture that al
edly took place some 30,000–40,000 years ago). If syntax emerged with
human species (say,120,000 years or more ago), what accounts for the lc
delay before any tangible consequences appeared?

The answer is, of course, that syntacticized language enables but does i
compel. Even today, in the Amazon and Congo jungles, there exist (bare
we are killing them off as quickly as we can) human societies whose tool!
in an age of spacecraft and supercomputers, show relatively little adva
over those of Cro-Magnons. What human language confers is not a tech
logical straitjacket, but freedom—freedom to develop any way you car
think you want to (into societies where you work ever-lengthening hour
some servile and soul-destroyingly repetitive job so that you can afford
buy labour-saving devices, or into societies where you gather all you ne
for subsistence in fifteen to twenty hours a week and hang out the rest of
time drinking and gossiping). Instead of wondering why culture didn't
plode the moment language emerged, maybe we should be wondering v
having acquired language, we chose the path that led to mass poverty,
ploitation, perennial warfare, and perennial injustice.

## Conclusion

I have tried in this chapter to present a framework adequate to include
the processes that uniquely produced language in the human species. T
framework may be summarized as follows. Driven by climate changes i
habitats where predators were fierce and common but food was scarce
least one primate species began to exchange basic information about
environment in order to survive. But proto-language is not bee langu
Once invented symbols begin to be used, they can be used to describe a
thing, consequently language can be adapted for social or any other p
poses. So symbols multiplied but structure probably did not, preven
from developing by inadequate numbers of neurons and the right k
of connectivity. Once both of these had developed, things the brain co

already do enabled protolanguage to develop quite rapidly into language as we know it today. The first group to cross some threshold that allowed un-limited combinations of words *and ideas* happened to be ours. And the rest, as they say, is history.

### FURTHER READING

The attitude of many non-linguists to linguists who concern themselves with language evolution is expressed by Ingold (1993). The collection that contains his article (Gibson and Ingold 1993) is itself fairly representative of a variety of non-linguistic approaches to the subject. Nowak et al. (2002) illustrates the problems that may arise when a high level of non-linguistic sophistication (in this case, in computer science) mixes with a lower level of linguistic sophistication; linguists (and some biologists too) may boggle at the assertion that 'during primate evolu-tion there was a succession of U(niversal) G(rammar)s that finally led to the UGs of human beings' (p. 615). However, some non-linguists do show a more sophisticated level of understanding, among them Maynard Smith and Sazthmary (1995) and Szathmary (2001).

The latter source develops the idea that although the language faculty must be biologically determined, it does not rely on hard-wired modules as sensory and motor faculties do. This position is related closely to the position on brain size held by Calvin (1996*a*; 1996*b*), who takes as critical the development of 'excess' neurons, which are not committed to any specific function but can be recruited for a number of tasks (including linguistic tasks), depending on what the brain is concerned with at any given moment. This view is consistent with results derived from both brain imaging and lesion studies (Damasio et al. 1996; Crosson 1993; Indefrey et al. 2001), It is also consistent with a view of brain activity held by Dennett (1991; 1997), in which there is no central homunculus or 'executive suite' in the brain. Rather than an individual thinking a thought and expressing it in words (the conventional view), sentences are constantly forming and reforming in the mind, but only the ones that can recruit enough neurons get to be consciously thought or spoken.

Certain developments within the minimalist programme suggest possibilities of reconciling generative syntax with Darwinian evolution. Berwick (1998) is among the few writers who tackle this explicitly, but although they make no reference to evolution, protagonists of the derivational approach to minimalism (Epstein et al. 1998) are producing analyses that are easier to reconcile with biological and neuro-logical constraints than alternative theories.

# 6

## On the Different Origins
## of Symbols and Grammar

*Michael Tomasello*

Human communication is most clearly distinguished from the communication of other primate species by its use of (1) symbols and (2) grammar. This means that progress on questions of language origins and evolution depends crucially on a proper understanding of these two phenomena. To state my own bias up front, I believe that symbols and grammar need to be investigated from a more explicitly psychological perspective than has been the case in the past. Further, I believe that we will make the most rapid progress if we investigate first and in detail extant systems of communication that are simpler than full-blown human language—specifically, those of human children and those of our nearest primate relative, the chimpanzee. What I would like to do in this chapter, therefore, is to look first at symbols and then at grammar from the point of view of the most basic social-cognitive and communicative processes involved—to see if this can help us in our thinking about how language might have evolved in the human species.

### Symbols

This is not the place to review all of the many approaches to linguistic symbols that exist on the current intellectual scene. Suffice it to say that criteria such as arbitrariness and spatial–temporal displacement are decidedly unhelpful when looking at actual communicative processes. After all, Pavlov's dog associated the arbitrary sound of a bell with food that was not at the time perceptually present. And the expressions 'duality of patterning' and 'stands for' (as in the locution 'symbols stand for their referents') simply put a new name on the phenomenon without providing any further insights.

My proposal is that symbolic communication is the process by which one individual attempts to manipulate the attention of, or to share attention with, another individual. In specifically linguistic communication, as one form of symbolic communication, this attempt quite often involves both (a) reference, or inviting the other to share attention to some outside entity (broadly construed), and (b) predication, or directing the other's attention to some currently *unshared* features or aspects of that entity (in the hopes of sharing attention to the new aspect as well). Comprehension of an act of symbolic communication thus consists in understanding that 'She is attempting to direct my attention to X' or 'She is attempting to direct my attention to Y with respect to X'.

We may make this account a bit more concrete by looking at some very young human infants. Six-month-old infants interact dyadically with objects, grasping and manipulating them, and they interact dyadically with other people, expressing emotions back-and-forth in a turn-taking sequence. If people are around when they are manipulating objects, the infants mostly ignore the objects. If objects are around when they are interacting with people, they mostly ignore them. But at around 9–12 months of age a new set of behaviours begins to emerge that are not dyadic, like these early behaviours, but triadic in the sense that they involve infants coordinating their interactions with objects and people, resulting in a referential triangle of child, adult, and the object or event to which they share attention. Most often the term 'joint attention' has been used to characterize this whole complex of social skills and interactions (see Moore and Dunham 1995). Most prototypically, it is at this age that infants for the first time begin flexibly and reliably to look where adults are looking (gaze following), to engage with them in relatively extended bouts of social interaction mediated by an object (joint engagement), to use adults as social reference points (social referencing), and to act on objects in the way adults are acting on them (imitative learning). In short, it is at this age that infants for the first time begin to 'tune in' to the attention and behaviour of adults on outside entities.

Not unrelatedly, at around this same age infants also begin actively to direct adult attention and behaviour to outside entities using deictic gestures such as pointing or holding up an object to show it to someone. These communicative behaviours represent infants' attempts to get adults to tune in to *their* attention and interest to some outside entity. Also important is the fact that among these early deictic gestures are both imperatives, attempts to get the adult to do something with respect to an object or event, and declaratives, attempts to get adults simply to share attention to some object or event. Declaratives are of special importance because they indicate especially clearly that the child does not just want some result to happen, but that she really desires to share attention with an adult as an end in itself. It is thus the contention of some theorists, including myself, that the simple act of pointing to an object or event for the sole purpose of sharing attention with someone else is a uniquely human communicative behaviour, the lack of which is also a major diagnostic for the syndrome of childhood autism (e.g. Tomasello 1995; Gómez et al. 1993; Baron-Cohen 1993).

Carpenter et al. (1998) followed a group of infants longitudinally from 9 to 15 months of age as they engaged in nine different joint attentional activities, everything from following gaze to imitating actions on objects to declarative pointing. They found that for any individual child all nine skills emerged together as a group within just a few months, with some predictable orderings among individual skills and with correlations among the ages of emergence as well. The almost simultaneous ontogenetic emergence of these different joint attentional behaviours, and the fact that they emerge in a correlated fashion, suggests that they are not just isolated cognitive modules or independently learned behavioural sequences. They are all reflections of a single cognitive change: infants' dawning understanding of other persons as intentional agents. Intentional agents are animate beings who have goals and who make active choices among behavioural means for attaining those goals, including active choices about what to pay attention to in pursuing those goals. This new understanding of other persons represents a veritable revolution in the way infants relate to their social and cultural worlds.

This social-cognitive revolution at 1 year of age sets the stage for the infants' second year of life, in which they begin imitatively to learn the use of all kinds of tools, artefacts, and symbols—that is, cultural entities with an intentional dimension, things that point beyond themselves to other outside entities. Thus, tools point to the problems they are designed to solve and linguistic symbols point to the phenomena they are designed to indicate. Therefore, socially to learn the conventional use of a tool or a symbol, children must come to understand why, toward what outside end, the other person is using the tool or symbol, that is to say, the intentional significance of the tool use or symbolic practice—what it is 'for', what 'we', the users of this tool or symbol, do with it (Tomasello 1998a). For example, in a study by Meltzoff (1988) 14-month-old children observed an adult bend at the waist and touch his/her head to a panel, thus turning on a light. They followed suit, presumably thinking, 'This is how this artefact works'. Infants thus engaged in this somewhat unusual and awkward behaviour, even though it would have been easier and more natural for them simply to push the panel with their hand. One interpretation of this behaviour is that infants understood that (a) the adult had the goal of illuminating the light and then chose one means for doing so, from among other possible means, and (b) if they had the same goal they could choose the same means.

Cultural learning of this type thus relies fundamentally on infants' tendency to identify with adults, and on their ability to distinguish in the actions of others the underlying goal and the different means that might be used to achieve it. This interpretation is supported by the more recent finding of Meltzoff (1995) that 18-month-old children also imitatively learn actions that adults intend to perform, even if they are unsuccessful in doing so. Similarly, Carpenter et al. (1998) found that 16-month-old infants will imitatively learn from a complex behavioural sequence only those behaviours that appear intentional, ignoring those that appear accidental. Young children do not just mimic the limb movements of other persons, they attempt to reproduce other persons' intended actions in the world.

The main point in the current context is this. Although it is not obvious at first glance, something like this same imitative learning process must happen if children are to learn the symbolic conventions of their native language. Although it is often assumed that young children acquire language as adults stop what they are doing, hold up objects, and name these objects for them, this is empirically not the case. Linguistics lessons such as these are characteristic of only some parents in some cultures and for only some kinds of words (e.g. no one names for children acts of 'giving' or prepositional relationships such as 'on' or 'for'). In general, for the vast majority of their language children must find a way to learn a new word in the ongoing flow of social interaction, sometimes even from speech not addressed to them (Brown 2001). In some recent experiments something of this process has been captured. For example, in the context of a finding game, an adult announced her intentions to 'find the toma' and then searched in a row of buckets all containing novel objects. Sometimes she found it in the first bucket searched, smiling and handing the child the object. Sometimes, however, she had to search longer, rejecting unwanted objects by scowling at them and replacing them in their buckets until she found the one she wanted (again indicated by a smile and the termination of search). Children learned the new word for the object the adult intended to find regardless of whether or how many objects were rejected during the search process (Tomasello and Barton 1994; see Tomasello 2000, for a review of other similar studies). Indeed, a strong argument can be made that children can only understand a symbolic convention in the first place if they understand their communicative partner as an intentional agent with whom one may share attention—since a linguistic symbol is nothing other than a marker for an intersubjectively shared understanding of a situation (Tomasello 1998*b*; 2000).

Let us now turn to our cousins, the non-human primates, and to their attempts (if that is what they are) to share attention with others. Many scientists outside the field take as the paradigm case of non-human primate communication the alarm calls of vervet monkeys. The basic facts are set out in Cheney and Seyfarth (1990). In their natural habitats in East Africa, vervet monkeys use three different types of alarm call to indicate the presence of three different types of predator: leopards, eagles, and snakes. A loud, barking call is given to leopards and other cat species, a short cough-like call is given to two species of eagle, and a 'chutter' call is given to a variety of dangerous snake species. Each call elicits a different escape response on the part of vervets who hear the call: to a leopard alarm they run for the trees; to an eagle alarm they look up in the air and sometimes run into the bushes; and to a snake alarm they look down at the ground, sometimes from a bipedal stance. These responses are just as distinct and frequent when researchers play back previously recorded alarm calls over a loudspeaker, indicating that the responses of the vervets are not dependent on their actually seeing the predator. On the surface, it seems as if the caller is directing the attention of others to something they do not perceive or something they do not know is present. These alarm calls would thus seem to be referential and therefore good candidates for precursors to human language.

But several additional facts argue against this interpretation. First, no ape species has such specific alarm calls (Cheney and Wrangham 1987). Since human beings are most closely related to apes, it is not possible that vervet monkey alarm calls could be the direct precursor of human language unless apes at some point used them also. Indeed, the fact is that predator-specific alarm calls are used by a number of non-primate species—from ground squirrels to domestic chickens—who must deal with multiple predators requiring different types of escape responses (Marler 1976), although no one considers these as direct precursors to human language. Second, vervet monkeys do not seem to use any of their other vocalizations referentially.

They use 'grunts' in various social situations (and some ape species have similar 'close' calls as well: Cheney and Seyfarth 1990), but these mainly serve to regulate dyadic social interactions such as grooming, play, fighting, and sex, not to draw attention to outside entities. Alarm calls thus are not representative of other monkey calls and so they do not embody a generalized form of communication. Third, primate vocalizations are almost certainly not learned, as monkeys and apes raised outside of the normal social environments still call in much the same way as those who grow up in normal social environments (although some aspects of call comprehension and use may be learned: Tomasello and Zuberbühler 2002). And one would think that language could only have evolved from socially learned and flexibly used communicative signals.

Our nearest primate relatives, the chimpanzees, actually communicate in more flexible and interesting ways with gestures rather than with vocalizations. Although they have a number of more or less involuntary postural and facial displays that express their mood, (e.g. piloerection indicating an aggressive mood and 'play-face' indicating a playful mood), they also use a number of gestures intentionally, that is, in flexible ways tailored for particular communicative circumstances. What marks these gestures as different from involuntary displays and most vocalizations is: (1) they are clearly learned, as different individuals use different sets of them; (2) they are used flexibly, both in the sense that a single gesture may be used in different contexts and in the sense that different gestures may be used in the same context; and (3) they are clearly sensitive to audience, as the signaller typically waits expectantly for a response from the recipient after the gesture has been produced (Tomasello et al. 1985).

In their natural communication with conspecifics, chimpanzees employ basically two types of intentional gesture. First, 'attractors' are imperative gestures aimed at getting others to look at the self. For example, when youngsters want to initiate play they often attract the attention of a partner to themselves by slapping the ground in front of, poking at, or throwing things at them (Tomasello et al. 1989). Because their function is limited to attracting attention, attractors most often attain their specific communicative goal from their combination with involuntary displays; for example, the specific desire to play is communicated by the involuntary 'play-face', with the attractor serving only to gain attention to it. The second type of intentional gestures are 'incipient actions' that have become ritualized into gestures (see Tinbergen 1951 on 'intention-movements'). These gestures are

100 *Michael Tomasello*

also imperative in function, but they communicate more directly what specifically is desired. For example, play hitting is an important part of the rough-and-tumble play of chimpanzees, and so many individuals come to use a stylized 'arm-raise' to indicate that they are about to hit the other and thus initiate play. Many youngsters also ritualize signals for asking their mother to lower her back so they can climb on, for example, a brief touch on the top of the rear end, ritualized from occasions on which they attempt to push her rear end down. Interestingly, in using their gestures chimpanzees demonstrate an understanding that the bodily and perceptual orientation of the recipient is an important precondition for the gesture to achieve its desired goal; for example, they use their visually based gestures only when the recipient is already looking at them (Tomasello et al. 1994).

Nevertheless, chimpanzees still do not use their gestures referentially. This is clear because (1) they almost invariably use them in dyadic contexts—either to attract the attention of others to the self or to request some behaviour of another toward the self (e.g. play, grooming, sex)—not triadically, to attract the attention of others to some outside entity; and (2) they use them exclusively for imperativepurposes to request actions from others, not for declarative purposes to direct the attention of others to something simply for the sake of sharing interest in it or commenting on it. Thus, perhaps surprisingly, chimpanzees do not point to outside objects or events for others, they do not hold up objects to show them to others, and they do not even hold out objects to offer them to others (Tomasello and Call 1997).

A number of scholars have recently cautioned against using human language as an interpretive framework for non-human primate communication (Owings and Morton 1998; Owren and Rendall 2001). According to these theorists, non-human primate communicative signals are not used to convey meaning or to convey information or to refer to things or to direct the attention of others, but rather to affect the behaviour of others directly. If this interpretation is correct—and it is certainly consistent with the facts outlined above—then the evolutionary foundations of human language lie in the attempts of individuals to influence the behaviour of conspecifics, not their mental states. Attempting to influence the attention and mental states of others is a uniquely human activity, and so must have arisen only after humans and chimpanzees split from one another some six million years ago.

To summarize, what we are seeing at 9–12 months of age in human infants is the ontogenetic emergence of the species-unique social-cognitive adaptation that made possible human culture and, as a special case of that, human symbolic communication. Other non-human primates do not seem to have this same adaptation. Not only do non-human primates not seem to use linguistic or other types of symbols, they do not even use non-linguistic means of directing and sharing attention with others. Although it is unknown when the social-cognitive adaptation that enabled symbolic communication emerged in human evolution, one plausible hypothesis is that it emerged only very recently with modern humans, and that it was indeed this very adaptation that enabled them to out-compete other hominids (Tomasello 1999).

## Grammar

It is currently popular to believe that grammar also (or perhaps only grammar) derives from a species-unique biological adaptation. Discussion of this difficult issue would take us far afield into theoretical linguistics, but some key points can nevertheless be made. The proposal that there is a biologically determined Universal Grammar—which contains specific linguistic content—rests crucially on the hypothesis that there are indeed contentful similarities, or identities, among the grammatical structures of all of the languages of the world. As we investigate more and more of the world's 6000+ languages, this hypothesis is proving more and more difficult to maintain. Of course we can take the grammar of Standard Average European and impose it on these other languages. But when we look at, for example, Austronesian languages, on their own terms, we find that they work in some quite unexpected ways: they simply do not have some categories and constructions that appear in European languages, and they of course have some of their own categories and constructions as well (see Dryer 1997; Croft 2002). I do not mean to imply that there are no linguistic universals; of course there are. But these do not consist of specific linguistic categories or constructions; they consist of general communicative functions such as reference and predication, or cognitive abilities such as the tendency to conceptualize objects and events categorically, or information-processing skills such as those involved in dealing with rapid vocal sequences (Tomasello 1995).

The question thus arises: if grammatical structures do not come from the genes, where do they come from? The answer is quite well known to typologists and historical linguists: from processes of grammaticalization operating over historical time. These processes take loose discourse sequences, comprising linguistic symbols for concrete items of experience such as objects and actions, and turn them into coherent grammatical constructions with various specialized symbols that perform grammatical functions with respect to these concrete symbols, such as marking case, tense, or constituency. And this is a point that cannot be stressed too much (although I will not have time to elaborate it here): grammatical constructions are themselves symbolic. To take a very mundane example, if I say *X floosed Y the Z*, native speakers of English will immediately understand some sort of event or transfer—even without the aid of any recognizable content words. As Adele Goldberg has demonstrated in her 1995 book, and as Ron Langacker (1987; 1991) has been arguing for years, grammatical constructions are kind of linguistic gestalts that themselves function symbolically. Children hear and learn them in the same basic way (with some twists of course) that they hear and learn lexical items. Children's contribution is that they then discern patterns across the various grammaticalized utterances they hear, and thus form the kinds of linguistic abstraction that become the hallmark of mature linguistic competence. Let us look briefly at each of these two time-frames—historical and ontogenetic—in just a bit more detail.

Each of the thousands of languages of the world has its own inventory of linguistic symbols and constructions that allow its users to share experience with one another symbolically. This inventory of symbols and constructions is grounded in universal structures of human cognition, human communication, and the mechanics of the vocal-auditory apparatus—and so all languages share some features—but the particularities of specific languages each have their own cultural histories. These particularities come from differences among the various peoples of the world in the kinds of things they think it important to talk about and the ways they think it useful to talk about them—along with various historical 'accidents', of course. The crucial point for current purposes is that all the symbols and constructions of a given language are not invented at once, and once invented they often do not stay the same for very long. Rather, linguistic symbols and constructions evolve and change and accumulate modifications over historical time, as human beings use them with one another and adapt them to changing circumstances.

The most important dimension of the historical process in the current context is grammaticalization or syntacticization, which involves loose and redundantly organized discourse structures congealing into tight and less redundantly organized syntactic constructions (see Traugott and Heine 1991*a*; 1991*b*; Hopper and Traugott 1993, for some recent research). For example, (1) loose discourse sequences such as *He pulled the door and it opened* may become syntacticized into *He pulled the door open* (a resultative construction); (2) loose discourse sequences such as *My boyfriend ... He plays piano ... He plays in a band.* may become *My boyfriend plays piano in a band*; (3) a sequence such as *My boyfriend ... He rides horses ... He bets on them.* may become *My boyfriend, who rides horses, bets on them*; and (4) complex sentences may also derive from discourse sequences of initially separate utterances, as in *I want it ... I buy it.* evolving into *I want to buy it.* In the process, free-standing, contentful words often turn into grammatical morphemes (e.g. auxiliaries, prepositions, tense markers, case markers), as a kind of 'glue' that holds the new construction together. Children now learn these constructions as symbolic wholes, which also have functionally significant constituent parts.

Systematic investigation into processes of grammaticalization and syntacticization is still in its infancy; indeed, the suggestion that languages may have evolved from structurally simpler to structurally more complex forms by means of processes of grammaticalization and syntacticization is somewhat speculative—these processes are most often thought of by linguists as sources of change only. But grammaticalization and syntacticization are able to effect serious changes of linguistic structure in relative short periods of time—for example, the main diversification of the Romance languages took place during some hundreds of years—and thus there is no reason why they could not also work to make a simpler language more complex syntactically in some thousands of years. Exactly how grammaticalization and syntacticization happen in the concrete interactions of individual human beings and groups of human beings, and how these processes might relate to the other processes of sociogenesis by means of which human social interaction ratchets up the complexity of cultural artefacts, requires more psychologically based linguistic research into processes of linguistic communication and language change.

Turning now to ontogeny, the essential point—made by Chomsky many years ago—is that children do not hear linguistic abstractions, but only concrete utterances; they must supply the abstractions themselves. But how they do this is a point of dispute. The best-known answer—first proposed by Chomsky and more recently popularized by Pinker (1994) and others—is that children do not have to learn or construct abstract syntactic structures, but rather already possess them as a part of their innate language faculty (Universal Grammar). Recent research suggests, however, that most of young children's early language is not based on adult-like abstractions, innate or otherwise. For example, in a detailed diary study Tomasello (1992) found that most of his English-speaking daughter's early multi-word speech revolved around specific verbs and other predicative terms. That is to say, at any given developmental period each verb was used in its own unique set of utterance-level schemas, and across developmental time each verb began to be used in new utterance-level schemas (and with different Tense–Aspect–Modality morphology) on its own developmental timetable, irrespective of what other verbs were doing during that same time period. There was thus no evidence that once the child had mastered the use of, for example, a locative construction with one verb she could then automatically use that same locative construction with other semantically appropriate verbs. Generalizing this pattern, Tomasello (1992) hypothesized that children's early grammars could be characterized as an inventory of verb-island constructions (utterance schemas revolving around verbs), which then defined the first syntactic categories as lexically based things such as 'hitter', 'thing hit', and 'thing hit with' (as opposed to subject/agent, object/patient, and instrument; see also Tomasello and Brooks 1999).

Lieven et al. (1997; see also Pine et al. 1998) found some very similar results in a sample of twelve English-speaking children, and a number of systematic studies of children learning languages other than English have also found basically item-based organization. For example, in a study of young Italian-speaking children Pizzuto and Caselli (1992; 1994) found that of the six possible person–number forms for each verb in the present tense, about half of all verbs were used in one form only, and an additional 40 per cent were used with two or three forms. Of the 10 per cent of verbs that appeared in four or more forms, approximately half were highly frequent, highly irregular forms that could only have been learned by rote, not by application of an abstract rule. In a similar study of one child learning to speak Brazilian Portugese, Rubino and Pine (1998) found adult-like subject–verb agreement patterns only for the parts of the verb paradigm that appeared with high frequency in adult language (e.g. first person singular), not for low-frequency parts of the paradigm (e.g. third person plural). The clear implication of these findings is that Romance-speaking children do not master the whole verb paradigm for all their verbs at once, but only master some endings with some verbs—and often different ones with different verbs. It should also be noted that syntactic overgeneralization errors such as *Do not fall me down*—which might be seen as evidence of more general and categorical syntactic knowledge—are almost never produced before about 3 years of age (see Pinker 1989).

Finally, experiments using novel verbs have also found that young children's early productivity with syntactic constructions is highly limited. For example, Tomasello and Brooks (1998) exposed 2–3-year-old children to a novel verb used to refer to a highly transitive and novel action in which an agent was doing something to a patient. In the key condition the novel verb was used in an intransitive sentence frame such as *The sock is tamming* (to refer to a situation in which, for example, a bear was doing something that caused a sock to 'tam'—similar to the verb *roll* or *spin*). Then, with novel characters performing the target action, the adult asked children the question: *What is the doggie doing?* (when the dog was causing some new character to tam). Agent questions of this type encourage a transitive reply such as *He's tamming the car*—which would be creative since the child has heard this verb only in an intransitive sentence frame. The outcome was that very few children produced a transitive utterance with the novel verb, and in another study they were quite poor at two tests of comprehension as well (Akhtar and Tomasello 1997). It is important that 4–5-year-old children are quite good at using novel verbs in transitive utterances creatively, demonstrating that once they have acquired more abstract linguistic skills children are perfectly competent in these tasks (Pinker et al. 1987; Maratsos et al. 1987; see Tomasello 2000 for a review). Finally, Akhtar (1999) found that if 2.5–3.5-year-old children heard such things as *The bird the bus meeked*, when given new toys they quite often repeated the pattern and said such things as *The bear the cow meeked*—only consistently correcting to canonical English word order at 4.5 years of age. This behaviour is consistent with the view that when 2–3-year olds are learning about *meeking* they are just learning about *meeking*; they do not assimilate this newly learned verb to some more abstract, verb-general linguistic category or construction that would license a canonical English transitive utterance.

This general approach may be extended to more complex structures. For example, Diessel and Tomasello (2001) looked at seven children's earliest utterances with sentential complements and found that virtually all of them were composed of a simple sentence schema that the child had already mastered combined with one of a delimited set of matrix verbs (see also Bloom 1992). These matrix verbs were of two types. First were epistemic verbs such as *think* and *know*. In almost all cases children used *I think* to indicate their own uncertainty about something, and they basically never used the verb *think* in anything but this first-person, present tense form; that is, there were virtually no examples of *He thinks . . ., She thinks . . .*, etc., virtually no examples of *I do not think . . ., I can't think . . .*, etc., and virtually no examples of *I thought. . ., I didn't think . . .*, etc. And there were almost no uses with a complementizer (virtually no examples of *I think that . . .*). It thus appears that for many young children *I think* is a relatively fixed phrase meaning something like *Maybe*. The child then pieces together this fixed phrase with a full sentence as a sort of evidential marker, but not as a 'sentence embedding' as it is typically portrayed in more formal analyses. The second kind of matrix verbs are attention-getting verbs like *Look* and *See* in conjunction with full finite clauses. In this case, children use these 'matrix' verbs almost exclusively in imperative form (again almost no negations, no non-present tenses, no complementizers), suggesting again an item-based approach not involving syntactic embedding.

A similar story may be told with respect to children's utterances with relative clauses (Diessel and Tomasello 2000). Virtually all of English-speaking children's earliest relative clauses, in the period before 3–3.5 years of age, occur in presentational constructions of the type: *Here's the chair that broke, There's the drink I want,* and *It's the toy that spins.* Each consists of (1) a main clause with the verb *to be* that is a highly practised sequence for children of this age (i.e. children have said by this time many thousands of times *Here's the X, There's the X,* and *It's the X:* Lieven et al. 1997), and (2) a relative clause, usually with an intransitive verb, that conveys new information about the topic introduced with the presentational main clause. Relative constructions of this type thus express a single proposition, with topic introduced in a common topic-introducing construction and comment expressed in the newly learned relative clause construction. After 3 or 3.5, the children began to use more complex relative constructions in which a relative clause, including an intransitive or transitive verb, was attached to a noun in a fully-fledged main clause. It is only at this point that we may accurately speak of children producing utterances with subordinate relative clauses.

The point is that young children are learning the specific linguistic items and constructions of the language they hear around them. Initially, they do not operate on the basis of any linguistic abstractions, innate or otherwise. Fairly quickly, however, they find some patterns in the way concrete nouns are used and form something like a category of noun, but schematization across larger constructions goes more slowly. The process of how children find patterns in the ambient language and then construct categories and schemas from them is not well understood at this point. But some progress has recently been made.

Children begin the abstraction process first by creating 'slots' in otherwise item-based schemas (Tomasello et al. 1997). It is not known precisely how they create these slots, but one possibility is that they observe adult speech variation in that utterance position and so induce the slot on the basis of 'type frequency' (Langacker 1988; Krug 1998; Bybee and Schiebman 1999). The nature and extent of type variation needed for different kinds of productivity is not known at this time, and indeed after a certain point in development it may be that type variation in the slots of constructions becomes less important as these slots come to be more precisely defined functionally. Another possibility—not mutually exclusive but rather complementary to the above—is that abstract constructions are created by a relational mapping across different verb island constructions (Gentner and Markman 1997). For example, in English the several verb island constructions that children have with the verbs *give, tell, show, send,* and so forth, all share a 'transfer' meaning and they all appear in a structure: NP+V+NP+NP (identified by the appropriate morphology on NPs and VPs). Children may thus make constructional analogies based on similarities of both form and function: two utterances or constructions are analogous if a 'good' structure mapping is found both on the level of linguistic form and on the level of communicative function. Precisely how this might be done is not known at this time, but there are some proposals that a key element in the process might be some kind of 'critical mass' of exemplars, to give children sufficient raw material from which to construct their abstractions (Marchman and Bates 1994).

One relevant study of the early stages is that of Childers and Tomasello (2001). They trained 2–2.5-year-old English-speaking children with English transitive verbs of varying degrees of familiarity to children of this age. More importantly, they also varied the type variation in the nominal slots around these verbs in the utterances the children heard. Thus, some children heard these verbs with full nouns only in the slots; for instance, as new characters were used to act out the event they heard *The bunny's striking the tree* (repeat), *The bear's striking the cat* (repeat), and so forth—and they heard a similar pattern for sixteen different verbs. Other children heard these same verbs with both pronouns and full nouns in the slots. That is, for each pair of characters they heard something like *The bear's striking the tree. He's striking it*, with the pronouns used across all models (and across all sixteen verbs) being *He's VERBing it*. These latter models thus provided children both with a familiar utterance frame (*He's VERBing it*) and with type variation, as different nouns were used as well. And indeed it was models of this latter type that best facilitated children's later ability to use a nonce verb (e.g. *dacking*) in a transitive frame (which they mostly did with pronouns).

Overall, the main point is that young children begin by imitatively learning specific pieces of language in order to express their communicative intentions. As they attempt to comprehend and reproduce the utterances produced by mature speakers—along with the internal constituents of those utterances—they come to discern certain patterns of language use, and these patterns lead them to construct a number of different kinds of (at first very local) linguistic categories and schemas. As with all kinds of categories and schemas in cognitive development, the conceptual 'glue' that holds them together is function: children categorize together things that do the same thing communicatively (see Tomasello 1992 on functionally based distributional analysis). For people who are generally sceptical that structures as complex and abstract as those embodied in the syntax of natural languages could evolve historically, we need only to point to other cultural products, such as algebra. Although, of course, the analogy with language is not perfect, algebra is a cultural product that is clearly not in the genes (the majority of cultures and people in the world do not know or use algebra and indeed we know its history). Of course human beings are quantitatively inclined biologically, but the specific structures of algebra are a product of a particular historical evolution. Languages are the same way.

## Conclusion

I am afraid that I have no real evolutionary fairy tale with which to conclude. At some point, hominids were communicating with one another in the typical great ape fashion—something like modern-day chimpanzee gestures and vocalizations, perhaps. For some reason—I know not what—the individuals of one population began to understand one another as intentional agents whose attention and other psychological states to the outside world could be actively followed into, manipulated, and shared. This enabled the conventionalization—symbolization, if you prefer—of a set of communicative behaviours premised on this intersubjective understanding. Of course a good candidate for this special population is modern humans—but I really have no idea.

One obvious implication of this scenario is that there was no specific adaptation 'for' symbolic or linguistic communication. The adaptation was 'for' a particular kind of social cognition—understanding others on analogy with the self—and symbols then developed as a kind of natural consequence. When you know that someone else has psychological experiences like your own, it is just natural for an intelligent primate to want to manipulate those states for various cooperative and competitive purposes.

Grammatical constructions then emerged from discourse patterns over historical time with no further biological adaptations—except perhaps further adaptations of the vocal-auditory apparatus to enable the ever-faster processing of real-time speech, which may then have had grammatical consequences. When children grow up in the midst of utterances embodying these constructions, they find their own patterns and make their own abstractions—which underlie the awesome creativity of mature linguistic competence and also some of the changes in languages that occur across generations historically.

And so we may see the origins and emergence of human language as one part of the much larger process of human culture, that is, as one more instance of what I have in other contexts called 'the ratchet effect'. Like many cultural products language is a complex outcome of human cognitive and social processes taking place in evolutionary, historical, and ontogenetic time. And different aspects of language—for example, symbols and grammar—may have involved different processes at different evolutionary times.

## FURTHER READING

There is a large and complex philosophical literature on the nature of symbols. In the modern context this begins with the work of the philosopher Charles Peirce, whose work is not so very accessible. He is the first to distinguish, for example, between icons, indices, and symbols. One reasonable collection of some of his papers is Buchler (1955). An interesting extension of his ideas may be found in Suzanne Langer's (1957) *Philosophy in a New Key*. The work of Ernst Cassirer (1944) on the *Philosophy of Symbolic Forms* is also useful, and Nelson Goodman's (1978) work has also been influential. In an evolutionary context, the main theoretical work is that of Sebeok (1990). In a developmental context, the work of Piaget (1962) and Werner and Kaplan (1963) are classic. Sinha (1988) also is very useful.

The main sources for learning more about processes of grammaticalization are Traugott and Heine (1991*a*; 1991*b*), Hopper and Traugott (1993), Heine (1991), Keller (1994), Trask (1996), and Croft (2000). Grammaticalization theory is well established for explaining language change, but its application to language origins has not been systematically pursued—for the obvious reason of a lack of data about exactly what kinds of thing people said early in the process.

# 7

# Universal Grammar and Semiotic Constraints

*Terrence W. Deacon*

## Neither Nature Nor Nurture

It has become an unquestioned dictum in modern linguistics that all human languages share a core set of common grammatical principles: a Universal Grammar (UG). What is to be included among these universals is not universally agreed upon, nor are the elements all of the same type (e.g. some candidate universals are rule-like, some are constraint-like, and some are structural), nor is there agreement on the source of these universals (e.g. nature/nurture). Over time, there has also been some erosion of the features once considered categorically universal and the expansion of features that are considered the variable expression of universal biases and constraints.

For the most part, theories of language structure, language processing, and language origins all take many of the most common regularities of language as givens. But the universality of words, the major constituent class distinctions, the complementarity of noun-like and predicate-like constituents in the formation of grammatical sentences, and the ubiquity of recursive relationships, among many other basic universals, cannot be taken as self-evident axioms. The existence of even vaguely analogous counterparts in non-human communication and cognition is questionable, and even theories that assume them to be innately prespecified must ultimately face the question of why such a system evolved this way and not in some other way. Finding the proper joints at which to cut language into its universal and variable components and understanding why there are language universals in the first place, are key mysteries of human cognition. But in one sense they are not linguistic questions. To probe them we must ultimately consider infralinguistic factors: the semiotic, functional, neurological, and evolutionary constraints and biases that may have played a role.

By far the most influential proposal concerning the origins of language universals derives from arguments originally put forth by Noam Chomsky. In a series of progressively refined theoretical positions, Chomsky and his followers have argued that the core language universals derive from an innate language-specific 'mental faculty' (e.g. Chomsky 1968; 1980; 1994; Pinker 1994). Human language competence is, in this view, a set of biologically inherited language principles that specify possible grammars and their variable components. This complex claim has been the focus of one of the most heated debates in the history of the study of language. Although most linguists agree that there are universals, there is considerable disagreement concerning their origin: whether they are innate and biologically evolved or else culturally constructed conventions that must be learned. These two sources have been treated as though they are exhaustive alternatives. Whatever the ultimate answer turns out to be, it is assumed by proponents on both sides of the debate that it will need to be stated in terms of these two options, or some combination.

I disagree with this last assumption, and will argue that the nature/nurture debate has diverted attention from the more general problem of the ultimate origins of design principles in language. I believe that that there are major aspects of UG that are of neither biological nor cultural origin. In this chapter I will suggest that many of these core language universals reflect *semiotic* constraints, inherent in the requirements for producing symbolic reference itself. As a result, though both evolved innate predispositions and culturally evolved regularities are relevant, neither can be considered the ultimate source of these constraints. With respect to both so-called 'symbolic' approaches (which merely invoke features of UG by fiat and assume their pre-formation in the human mind) and various inductive, functional, and 'neural network' approaches (that attempt to demonstrate how they can spontaneously emerge as optimizing solutions in response to communicative or learning constraints), this represents a third paradigm of explanation, with affinities with both sides of the debate. It posits universality of a kind that is prior to language experience and yet also argues that this is only expressed functionally, as these constraints shape the self-organization and evolution of communication in a social context.

The main goal of this essay will be to explain what I mean by 'semiotic constraints', give an account of the sense in which they must be prior to language and yet not strictly speaking innate, and finally hint at how they may influence the emergence of universals of language structure. As a non-linguist sticking his nose into a complex field where I lack sophistication in identifying specific constructional regularities and in citing examples of their consequences in different languages, I can do little more than offer predictions and plausibility arguments that linguists may or may not choose to flesh out. I only hope this exercise of thinking outside the usual boxes can provide a useful expansion of vision.

## Other Kinds of Universals

In domains other than language we encounter universals that are neither biological nor cultural in origin. Prototypical examples are found abundantly in mathematics and have for centuries prompted philosophical debates concerning the origins of abstract form. Consider the operations of addition, subtraction, multiplication, and division in elementary arithmetic. These are, in one sense, cultural creations. They are conventional operations using culturally created tokens that could be embodied in a seeming infinite variety of different patterns of tokens and manipulations. But it is not quite right to say that these operations were 'invented'. Specific individuals over the history of mathematics did indeed invent the various notation systems we now use to represent mathematical relationships, but often these inventions came as a result of *discoveries* they made about the representation of quantitative relationships. Mathematical 'facts' have a curious kind of existence that has fascinated philosophers since the ancient Greeks. Being represented seems to be an essential constitutive feature of a mathematical entity, but this does not mean that anything goes. Mathematical representations are precisely limited in form. Generalizations about the representation and manipulation of quantity, once expressed in a precise symbolic formalization, limit and determine how other mathematical generalizations can be represented in the same formal system. For my purpose it is irrelevant whether these kinds mathematical entity are 'latent in the world' in some Platonic sense or whether they emerge anew with human efforts to formalize the way we represent quantitative relationships. What matters is that symbolic representations of numerical relationships are at the same time arbitrary conventions and yet subject to non-arbitrary combinatorial consequences. Because of this deep non-arbitrariness, we feel confident that mathematics done anywhere in the universe will have mostly the same form, even if the medium of notation were to differ radically.

Consider the concept of 'prime number'. It is a generalization about division (or more specifically about limitations on division). Division is a manipulation of symbols that abstractly represents the process of breaking up wholes into regular smaller units. It is a symbol-manipulation operation defined with respect to numerical symbols, and in this way is analogous to a class of syntactic operations defined with respect to the phrasal relationships it can parse. The class of quantities that are 'divisibile without remainder' is determined by applying this syntactic operation again and again on different numbers, essentially applying this particular syntactic rule across different numerical contexts. The complementary class of primes might be compared to a class of phrases whose reference causes the application of a given syntactic rule to produce equivocal results.

Primeness is a culturally constructed symbolic abstraction about this class of mathematical outcomes. Construction of this category was contingent on a number of cultural historical factors. Most notably, it was brought into existence only after mathematicians had devised a suitably flexible symbolic notation system for numbers and for representing arbitrarily repeated counts of identical subtractions (i.e. division). Comprehending the concept of prime number also required a certain level of neural sophistication that non-human species appear to lack, as far as we know. But primeness itself did not exactly get invented, nor did it evolve. Despite the fact that the means to represent it was invented, the concept itself is defined with respect to culturally constructed conventions, and its notational form may be arbitrary, primeness is universal. And yet it is not an innate idea. It is universal in the most literal sense because it is implicit in any sufficiently complex mathematical system.

Because the sequence of prime numbers represents an abstraction about the representation of quantities—the exceptions to a conventional rule about division—no natural process we know of generates a series of prime numbers. This property has been employed by SETI (Search for Extra-Terrestrial Intelligence) scientists to send unmistakable symbolic signals into deep space. Signals consisting of pulses counting out prime numbers would be immediately recognized as only able to be emitted by intelligent symbol users, and not by any natural astronomic object. This scenario is presented in the movie *Contact* based on the book of the same name by one of SETI's originators, Carl Sagan. Ultimately, this amounts to a universally interpretable cultural convention (an oxymoron according to current assumptions).

Irrespective of which side one takes in the long-running philosophical debate about whether mathematical forms actually 'exist' independent of being represented, recognizing the possibility of this kind of universal is the first step in escaping the pointless debate over the nature or nurture of language universals. It makes no sense to attribute the source of the universality of a symbolic abstraction like primeness to either nature or nurture. Its universality derives instead from the fact that the source of its emergence in mathematical theories (however we conceive of it) is due to factors 'prior' to any representation of it or any condition of its discovery.

One way to conceive of this 'priority' is that it comes about because of the constraints involved in representing quantity. For example, mathematical operations must avoid undermining constancy of quantitative reference across every allowable representational manipulation, or else they fail us as tools of representation. In other words, however convoluted the transformations from one expression to another, we want 5 to still equal 5 at the end of the process. More generally, any symbolic compositions that are equivalent transforms of one another must have identical quantitative reference. But symbolic reference (as compared to iconic and indexical reference, discussed below) is particularly susceptible to producing equivocation of reference. This is because the dissociation between token features and features in the referent domain make it possible for multi-symbol constructions and relations to confound reference while obscuring how this has occurred. Over the course of intellectual history, discoveries concerning these constraints have often been key to the development of whole new branches of mathematical theory, and have even provided insights into previously unrecognized aspects of the physical phenomena they represent.

This has its counterpart in language. Similar to mathematical constraints on quantitative equivocation, avoiding equivocation about who did what to whom, when, with what, in which way, is a critical requirement in language communication (along with the need to ground these symbols in pragmatic concerns). But the symbolic referential domain of language includes a vastly larger number of dimensions of reference than mere quantity. With this comes an enormous increase in the ways symbols can be combined, and a correspondingly large number of ways reference can be made ambiguous or even self-undermining by combinatorial manipulations. As a result, constraints on symbolic referential form in language should be more abstract, more complex, and more subtle than in the narrow domain of quantitative reference. There should, nevertheless, be analogous a priori universal constraints on the forms of linguistic combinatorial relationships, if they are to avoid confusion of reference *by construction.*

Because they apply to such a narrow range of reference, mathematical representations can be made arbitrarily precise and categorical. Because consistency is paramount, whereas ease of expression and interpretability is secondary, it does not matter that mathematical formalisms are in many cases clumsy to produce and may demand hours or even years of reflection and consideration to understand. Linguistic reference, on the other hand, is open in referential scope and yet must be produced and interpreted in real time with limited capacity for reflection. Luckily, linguistic communication does not often require the absolute precision of reference that mathematics requires. It need only be sufficiently precise to be of current use, which allows speakers and hearers to take advantage of numerous ancillary sources of information implicit in the conditions of communication and cognition.

The constraints on referential consistency should thus be expressed primarily at highly abstract semiotic levels of analysis in which even the most basic linguistic constituent categories are derived cases. But if linguistic theories take these universals as primitives, their causes will remain unrecognized, and UG will appear ad hoc and inexplicable except by appeal to evolutionary fiat and innate foreknowledge.

## The Fallacy of Simple Correspondence

I believe that the most serious source of confusion about the nature of language universals derives from assumptions implicit in what might be called the standard computational conception of language, sometimes referred to as the 'symbolic' or formal approach as opposed to those that are more functional or process oriented. Characterizing it as 'symbolic' reflects the fact that the constituent elements in these analyses are symbol tokens standing for abstract linguistic categories and operations, such as noun phrase and determiner and rules about their necessary and optional compositional relationships in a language; e.g. NP → (det) + (adj) + N + (PP). The rules determining allowable operations of a language are in this way defined with respect to these abstract linguistic 'roles'. The success of this computational analogy when applied to natural language suggests that syntactic rules can be specified largely (if not wholly) irrespective of the specific semantic or referential mapping of the specific words or phrases that take these roles, as though syntax and semantics were logically orthogonal dimensions of language.

Computation can generally be defined as the mechanized implementation of a symbolic manipulation. In the remapping, say, of an addition process into the workings of some determinate machine, the specific physical tokens that stand for the numerals and the physical changes that stand for the operation can be arbitrarily chosen so long as there is an isomorphism maintained between the corresponding state changes of the mechanism and the way the symbols themselves would be manipulated if a person were to perform a 'real' addition. Determining an effective isomorphism to any given class of symbol manipulations is not a trivial enterprise, though it is aided by the arbitrariness and conventionality of the symbolic reference which ensures that physical features of the tokens need have nothing to do with which physical properties determine the mechanistic isomorphism. The remarkable extent to which language syntax is describable in algorithmic formalism, clearly motivates the intuition that this mapping is reversed for language, imposing the structure of a neural mechanism on language syntax. But even if different assignments allow different mechanisms to perform the same computation (the basis of what is called functionalism in philosophical discussions), we cannot infer from this that symbolic reference contributes nothing to the particular mapping choices. That workable mappings are drawn from a very limited set of options seems almost self-evident for mathematical computation because the referential domain is simple. The assignment of any one token can be entirely unconstrained by physical details, *but* only if a highly constrained mapping between the domains as a whole can be identified with respect to it. I believe that the same is true for language and its reference, but in this case the referential domain is to all intents and purposes unbounded.

There is probably no term in cognitive science more troublesome than the word 'symbol'. Curiously, philosophers still puzzle over the problem of symbolic reference while linguists of all stripes often take for granted that it is simple. The standard caricature of the symbolic reference relationship is a code in which a random correspondence is set up between elements of two kinds, symbol tokens and objects. But this caricature, accepted as a starting point for both symbolic and connectionist models of language, assumes sets of discrete referents of the same kind. The world of referents is not, however, merely a set of discrete sign tokens, and the mapping is neither simple nor direct. The very different realms of words and physical world phenomena must be linked via a complex and multilayered intervening web of semiotic relationships. This web is often described as a set of 'conventions', but much is hidden in this concept.

So although symbolic reference is characterized in negative terms, i.e. in terms of the independence of token properties of signs and physical properties of their referents, it is a mistake to treat it like a token-to-token correspondence or to treat the *convention* that creates this reference as mere social fiat. The conventionality behind symbolic reference is a *system* of relationships, not a mere collection of mappings, and consequently there are *systemic constraints* affecting what can and cannot constitute higher-order forms of symbolic reference. Though symbolic reference allows tremendous flexibility of association between sign vehicle and the object or property or event it symbolizes, this flexibility is only possible because of this highly structured system of *sub*-symbolic relationships. So even though words are relatively unconstrained in the ways they can be mapped onto referents, combinations of words inherit constraints from the lower order mediating relationships that give words their freedom of mapping. These constraints limit the classes of referentially consistent higher-order symbol constructions.

Much of the confusion over the meaning of the term 'symbol' appears to derive from the fact that the use of the word has diverged along at least two distinctive disciplinary lines during the last century. These distinguish the humanities and social sciences on the one hand from recent philosophy, mathematics, and the cognitive sciences on the other. Over time these divergent uses have inherited many theory-internal assumptions, to the point where it would appear on the surface that we are dealing with radically different, mutually exclusive concepts, merely referred to using the same term. But both the technical etymological history and the differences in use suggest a different interpretation: that these two uses of 'symbol' are complementary, referring to two different aspects of the same phenomenon.

These two definitions of symbol can be summarized as follows:

*Humanities*: A symbol is one of a conventional set of tokens that mark a node in a complex web of interdependent referential relationships and whose specific reference is not obviously discernible from its token features. Its reference is often obscure, abstract, multifaceted, and cryptic and tends to require considerable experience or training to interpret.
*Computation*: A symbol is one of a conventional set of tokens manipulated with respect to certain of its physical characteristics by a set of substitution, elimination, and combination rules, and which is arbitrarily correlated with some referent.

The humanities definition suggests the most sophisticated form of referential relationship, whereas the computational definition suggests the most elementary, concrete, stripped-down, and minimalistic form.

These discipline-correlated definitions of 'symbol' sound as though they refer to entirely different sorts of entity, and yet their historical derivation from a common origin hints at a deeper kinship. In general, the computational definition is about the production and manipulation of the physical symbol tokens, whereas the humanities and social sciences definition is about the interpretation and the significant effects of symbolic reference. This apparent opposition reflects the way symbolic reference unequally distributes the semiotic 'work' required for sign interpretation versus sign production.

This difference can be illustrated by comparing icons to symbols. Icons (which refer by resemblance, as in a portrait, or more generally by some shared topology between sign vehicle and referent, as in a diagram) are often interpretable on first inspection by relatively naive subjects, so long as the object of reference is familiar. But a good icon (e.g. a fair depiction of a familiar object) can be a complicated sign, and often takes considerable effort or talent to design, some skill to render adequately, and can suffer in detail when poorly reproduced. With most of the complexity of the referential relationship exhibited explicitly, interpretation of icons can be easy, while production can be demanding. In contrast, the sorts of token that can serve to mediate symbolic reference can be as simple as an ink mark or a brief distinctive sound. This makes symbolic sign vehicles comparatively trivial to produce. To interpret most symbol tokens, however, can take both effortful preparation to develop the necessary interpretive competence and some cognitive effort to decode. A significant amount of the complexity of the referential relationship is constructed in advance, often through explicit training to build an interpretive competence, so that the symbol token merely evokes some aspects of this existing implicit structure and need not be supported by features of the symbol tokens themselves. So interpretation of symbols is demanding, while production is easy.

During symbolic interpretive processes most of the information brought to the interpretation is provided by the interpreter and is supplied covertly. As a result, any clue in the sign vehicle that could link it to its reference can be lost without loss of reference. This is why 'arbitrariness' is often invoked as a signature feature of symbolic reference. But arbitrariness is neither a necessary nor a diagnostic feature of symbolic representation (which is why non-arbitrary features in manually signed languages and onomatopoeia in spoken languages do not make them any less symbolic). For any symbolic interpretation to be possible, a huge amount of work is necessary, in advance, to set up this interpretive system, so the gain in productive and manipulative freedom with respect to the tokens is thus paid for by a much greater demand on antecedent preparation of an interpretive system. The 'semiotic freedom' that makes it possible easily to elaborate symbolic reference by producing complex combinations of symbol tokens is achieved on the shoulders of this covert intervening interpretive system. So to understand the constraints it imposes, we must be clear about how this system is constructed.

## The Hidden Complexity of Symbolic Reference

Ultimately, anything can be a sign of any type, and the same token can serve as an icon or a symbol under different interpretations, depending on the context. What matters is not any special feature of the sign vehicle (token) but the interpretive process that takes some of its features as the basis for a particular form of representation. The competence to interpret something as symbolic of something else is not, then, a primitive psychological operation, but a consequence of a semiotic construction process (Peirce 1955 refers to this as the production of an 'interpretant'). Attending to the hidden structure implicit in the competence to interpret a given kind of sign opens a whole new window into what constitutes representation. In computational and linguistic uses of 'symbol' this interpretive infrastructure is essentially ignored. Only the features of sign vehicles, their relationships to features of reference objects, and the patterns of their arrangement are focused on, leaving the interpretive basis of this form of reference to be defined negatively: a sign lacking any configurational or causal link to its object. But when we invert this logic and focus on features of the interpretive process rather than features of the sign vehicle, symbolic relationships no longer appear simple and primitive, because this symbolic competence must be constructed from more basic non-symbolic referential relationships, including icons and indices (Deacon 1997).

Indexical reference provides the mediating link between symbols and the world. It is implicitly invoked in terms of the mapping relationship posited between between a symbol token and what it refers to. But although in hindsight it may appear as though this process of indexical 'grounding' is unconstrained, this is an illusion derived from borrowing fragments from already constructed symbol systems to create special purpose subsystems This inverts the dependency Ultimately, symbolic reference grows out of and is dependent on patterns of indexical reference. Consequently, conditions and requirements of indexical reference will constrain symbolic reference as well.

Indexicality is reference based on correlation. Pointers, samples, gauges, markers, and recognizable habits of things and people are all used for providing indexical reference. To interpret something as an index of something else one must have learned that it is invariably correlated with that something else (its referent) either as a result of noticing its physical association (part–whole, cause–effect, etc.) or merely coming to expect a correlated appearance. Indexical interpretation can be the result of conscious inference, learned habit, or even innate expectations, as in the largely innate interpretation of laughter or sobbing. Because indexical reference involves linked differences it can also enable reference to span arbitrary causal 'distances' by virtue of transitivity, e.g. pointers pointing to pointers pointing to pointers. Thus reading a meat thermometer does not burn us. But as in the case of a scientific indicator, such as the level of carbon-14 isotope measured in an ancient artefact or the extent of clotting of blood in response to a blood-type antigen, it can take considerable sophistication in understanding the causal relationships to recognize something as an index of something quite different.

This shift in analysis from the external features of sign tokens to their internal interpretive requirements helps demonstrate that iconic, indexical, and symbolic relationships are not primitive categorical alternatives of one another. The differences in the interpretive competence to 'recognize' a sign as iconic, indexical, or symbolic turns out to be hierarchically organized and of increasing complexity from icon to index to symbol. For example, an index such as the angle and orientation of a wind-sock can convey information about a wind that cannot be seen because its different behaviors are interpreted as iconic in some respects to different objects that have observed blowing in different winds. A common noun instead conveys information about a general type by bringing to mind correlated pairings (indexical uses) with different (iconically related) objects (etc.), and with different combinations of other words (token–token indexicality) in contexts that also share features. Thus indices are ultimately interpreted in terms of relationships among iconic representations, and symbols are interpreted in

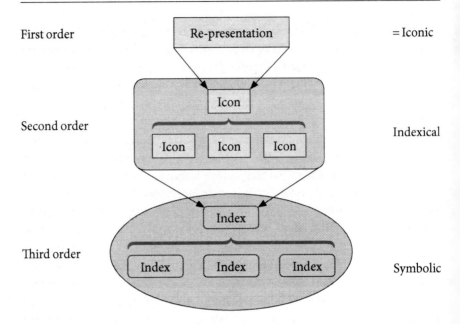

FIG. 7.1 A schematic depiction of the hierarchic compositional relationships between iconic, indexical, and symbolic relationships. Iconism (simple first-order representation) is basic, indexical referential relationships (second-order) are determined by certain patterns of relationships between icons, and symbolic referential relationships (third-order) are determined by certain systemic relationships among indices. After the analysis of Deacon (1997).

terms of relationships among indexical representations (depicted schematically in Fig. 7.1; see Deacon 1997 for a more thorough account).

So, although a symbol token can be a simple object or action that is without similarity to or regular physical correlation with its referent, it maintains its fluid and indirect referential power by virtue of its position within a *structured set of indexical relationships among symbol tokens*. Even though a symbol can, under certain circumstances, stand on its own, its representational power is dependent on being linked with other symbols in a reflexively organized system of indexical relationships. How we come to know the meaning and referential use of symbols involves two interdependent levels of correlational relationships. We individually and collectively elaborate this system by learning how each symbol token both points to objects of reference and (often implicitly) points to other symbol tokens (and *their* pointings). Symbols point to each other by virtue of patterns of replacement, alternation, cooccurrence, and so on, in context, and not just within sentences but across linguistic experiences. This is the basis of a sort of associational valency that is also a contributor to syntactic construction, though it includes semantic and referential dimensions that are far more diverse and often play no primary role in syntactic construction.

Taking advantage of the transitive ability of pointers that point to pointers, something like a closed systematically reciprocal network of sign positions and relationships is built up by the learner. The systematic regularity of these indexical relationships allows it to be used as a kind of multidimensional vector coordinate system defining a schematic space of potential reference. A system is defined by its global structure (or more generally its topology) as well as by its components. Consequently, even in the production or interpretation of isolated symbols, where no other element of the system is explicitly expressed, numerous semiotically proximate nodes in this virtual network must be implicitly referenced (and thus cognitively activated). Symbol tokens also retain a trace of their antecedent indexical links with features of the world; but whereas an index is characterized by a one-to-one association with its reference, each symbol token simultaneously marks the intersection of two planes of indexicality, one involving indexical relationships to other symbol tokens, the other involving indexical mapping to objects. Each has its own systematic regularities.

The indirect reference of symbol tokens to other symbol tokens is made explicit in such word association tools as dictionaries, thesauruses, and encyclopedias. One could, for example, conceive of a thesaurus as a vastly interconnected closed network of word pointers to other word pointers (which is essentially how its originator, Peter Mark Roget, initially conceived of it; see discussion in Crystal 1987: 104). Over the years semanticists have conceived of numerous other ways to represent this virtual space of relations among the dimensions or components of the meanings of words. But irrespective of the way it is actually represented in symbol-competent minds, this 'space' becomes the critical mediator between word–object and word–word relationships that differentiates a symbolic relationship from a simple indexical correlation of a word sound with a specific object (see Fig. 7.2). The metaphor of cognitive spaces, or schemas, with distinctive dimensions and coordinate systems, is a useful abstraction that is general enough to include attributes from a number of prior semantic relational schemes, though questions concerning how they are constructed and embodied neurologically remain unanswered (see especially Fauconnier 1985; Fauconnier and Turner 2002, and for an accessible introduction to more classic theories see Leech 1974).

One more ingredient is necessary for this systemic mediation to provide its referential function: the whole system of inter-token indexical relationships must be globally iconic of the indexical affordances implicit in the physical relations among potential referents. In other words, there must be some (highly schematic) isomorphism in the topology of the semantic network to the way the corresponding physical attributes can be used as indices capable of 'pointing' to one another. The nature of the mapping logic of

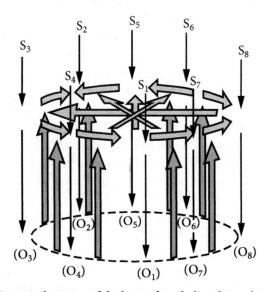

FIG. 7.2 A schematic depiction of the logic of symbolic relationships in terms of their component indexical infrastructure. Arrows depict indexical relationships and letters depict symbol tokens (Ss) and objects of reference (Os). Downward directed arrows represent the interrupted transitive indexical relationship between symbol tokens and their typically correlated objects of reference. The upper arrows point to loci in the semantic space that has been virtually constructed by token–token relationships (depicted by the horizontal system of arrows). If the symbolic reference is simple (not combinatorial), indexical reference continues from there down to a given instance of a typical object. Otherwise this referential arrow should originate from an intermediate position in the semantic space (determined by a sort of vector summation of the relationship of the combined symbols) to some other position in the object space. Upward arrows depict an extrapolated correspondence between physical relationships and semantic relationships on which this extrapolative and indirect mode of reference depends for its pragmatic fit to the world.

this global systemic iconism is anything but direct and simple (diversely related conceptions are often described as 'semantic holism', as for example represented by the classic Quine 1961) and is even subject to idiosyncratic and culture-specific variations. But the general logic of this relationship is not dependent on the degree to which this iconism is partially inchoate or idiosyncratically indirect in its mapping. The transition from indexical to symbolic reference requires only that the mapping of word to object be functionally subordinated to some systematic, implicit system-to-system relationship, though this alone is no guarantee of veridical mapping. This implicit fallibility, introduced by giving up immediate indexicality for system-mediated indexicality, changes the nature of what can be referred to from some individual object or fact to a type of object or fact. Symbol tokens therefore index positions in a semantic network and positions in a virtual space of schematic loci in the world of physical phenomena as well.

So although we initially spin this symbolic reference web from the scaffolding of discretely mapped indices, this is transformed by recognizing and using the systematicity they embody to create a 'semantic space' that affords continuously refinable gradation and evolution of reference. Since the world seldom aligns precisely with our repertoire of basic sign tokens, and continual enumeration of new tokens is limited, this systematic structure of the space that underlies symbolic reference provides a critical tool for iterative interpolation into the interstices between marked loci. In this projective role, only relatively sparsely located points ever need to be marked with basic symbol tokens, and yet still the inferred cognitive 'space' that exists between these marked 'loci' can be identified with any degree of precision. Combining symbol tokens, as in sentence construction, enables the indication of a set of loci in this space and provides additional indices for 'triangulating' from them to a specific unmarked locus. This subordination of direct singular indexicality to indirect combinatorial indexicality is what allows *the big red ball* to be picked out with respect to other big things, red things, and balls, also in the same physical context. By taking advantage of the transitive capacity of indexicality to interpose a relational system between token and referent, a many-to-one inferential condensation of reference becomes possible, while at the same time retaining the possibility of one-to-one indexicality of words as well (in which the intermediate step can be ignored).

In what follows, I hope to demonstrate that this systemic mediation of reference offers structural possibilities that are not available to indexical or iconic forms of communication, and that it also imposes some new and unavoidable demands on the composition of complex symbolic vehicles, such as sentences.

## Two Global Universals of Language Structure

I am in general agreement with arguments made in cognitive linguistic theories, in assuming that many near-universals of syntax reflect biases implicit in the structure of the semantic space (e.g. Lakoff 1987; Langacker 1987). However, the contingent nature of these—even if near-universal in expression—indicate that they are not, strictly speaking, semiotic constraints (though similar linguistic evolutionary processes may underlie their expression; see Deacon 2003). So in the remainder of this essay I will focus only on a few global linguistic consequences of the abstract structure of symbolic reference alone. I call these semiotic constraints because they are meaning-independent and derive from the structure of the referential form itself and not from any specific cognitive or discourse-related influence. They are therefore also more consistent with the kinds of content-independent universals considered by proponents of an innate UG.

To exemplify the nature of such global semiotic affordances and constraints I will focus on two ubiquitous global properties of language that I believe are derived from the structure of symbolic reference.

(a) *Non-degrading recursivity*: the principle by which representational forms can be embedded within or collapsed into other forms without any loss of referential precision, allowing the same rules to apply repeatedly across different levels of structure.

(b) *Predication–structure constraints*: requirements for elementary propositional structure regulating the composition of and relationships between unit phrases, determining the complementarity of noun- and verb-like functional elements and the hierarchic dependency constraints of phrasal and morphological construction.

## The Indirectness of Symbolic Reference Allows for Recursivity

The recognition of the general application of recursive representation across diverse domains was key to the development of mathematics and computational science, and also contributed to one of the most influential approaches in linguistics—generative theories of grammar. Words can be expanded into complex descriptive phrases, leading to the possibility of indefinite expansion and complication of embedding phrases within phrases; conversely, phrases can be replaced by words. And words in one sentence can stand in for whole phrases or sentences or even much larger chunks of text located elsewhere in an utterance or discourse (e.g. anaphoric reference). The recursive application of syntactic and morphosyntactic operations provides for the generation of indefinite numbers of forms from finite elements, with minimal expansion of expressed form, and it is critical for essentially all of the hierarchic organization of language. Without recursion, the expressive power and flexibility of modern language would be very severely limited. Descriptions would be atomistically simple and narratives would be reduced to telegraphic sequences of mentions, requiring considerable repetition or inferential analysis to sort out the ambiguities. To the extent that operations and the results of operations become interchangeable, constructions of open-ended complexity can result.

In contrast to its importance in human language and formal systems, recursion does not appear to be a principle that plays any significant role in animal communication systems. As a result, it has been considered one of the key distinguishing cognitive innovations behind the evolution of language competence. Chomsky has gone so far as to include it as a universal abstract principle of language competence—sometimes referred to as the principle of 'discrete infinity'—that must in some way be innately instantiated in human brains (Chomsky 1968; 1994; Jenkins 2000). As this phrase implies, recursion also requires that the units involved are discrete non-graded tokens.

So two questions are raised by this. (1) Why don't other animals use this powerful design principle in their communication (at least in a modest way)? and (2) What is the basis and the origin for this ability in us? I believe that we do not have to search for brain differences to answer either of these two questions. The source, I believe, is semiotic, not neurological.

Two of the most commonly recognized features of symbolization are displacement of reference and semiotic freedom: the first is the ability to embody symbols in diverse forms irrespective of their referents and the second is the ability to manipulate them in diverse combinations, respectively. These features derive from the indirectness of symbolic reference. Whatever the correlations that might exist between the kinds of object and the kinds of symbol tokens that represent them, these correlations do not directly determine the representational relationship. This is determined by the mediation of reference by a system-to-system mapping of indexical relationships. So even when a symbol token represents other symbol tokens (including such logically tangled relationships as the re-representation of the symbol itself), the specifics of that referential relationship do not influence how that token can be used. It is a kind of opacity that allows reference to be preserved without interference from manipulations of token features or relationships. This is why symbolization allows substitutions of wholes for parts, classes for members, expressions for terms, and vice versa.

In comparison to symbolic relationships, iconic and indexical relationships involve direct presentation of or involvement in what they represent, respectively. Icons are subject to blending, simplifying, and diluting effects as they are introduced into new contexts. Consider classic versions of iconic representation of recursion in such images as painters painting painters painting, self-portraits, or mirror images reflected in mirror images. In these examples something must be left out in the re-representation in the same medium, or else the re-representations must be progressively simplified or degraded in order to be 'contained' in one another. Indices suffer from a different stricture. They must maintain physical connection with and distinction from what they indicate, excluding direct recursion but also introducing the danger of circularity. I recently encountered an irritating example of indexical recursion when I followed street signs that each indicated that I should turn right around a block to find my destination, only to discover that the fourth in the series pointed to the first, leading me in a circle. Ultimately this causes a loss of the trace of reference. This vicious circular potential is the result of the necessary singularity of indexical reference. Many indices can convergently point to one referent, but no one index points to many distinct referents. Another way to conceptualize this constraint on recursion is that because indexical reference itself is based on part–whole relationships, re-representing the whole in the part is self-defeating of reference. In contrast, because a symbol token achieves reference to things in the world indirectly, any trace of direct iconism or indexicality is essentially 'encrypted' by this intermediary step. Symbolic reference can thus obscure even the quite indirect physical traces linking a symbol token and its object of reference so that the very existence of the object can remain uncertain and ambiguous. Ironically, this loss of reliable linkage with the conditions of reference is key to the possibility of fully recovering the re-represented reference in analysis.

This does not mean, however, that recursion represented linguistically must be accomplished by some sort of neural recursive function. Even fairly shallow recursive embedding of sentences or phrases can produce significant interpretive difficulties, and speech with more than three levels of recursively embedded clauses is rare. This suggests that the formal recursion that symbolization allows in the external form may nevertheless be processed in non-recursive ways neurologically—hence the resulting degradation and potential for misdirection with each step of re-represesentation (see also arguments by Christiansen and Chater 2002). But the capacity to treat symbolic constituents at different levels of token complexity and interdependency as if they are unitary, free-standing elementary units offers other benefits that more than compensate for these internal processing limitations on depth.

So an evolved innate neurally based recursive processing faculty is both unnecessary and unsupported by the evidence. The absence of anything quite like a generative grammar outside human language is consistent with the absence of symbolic reference in these forms and does not call for any special evolutionary explanation. The implicit affordance for recursion that symbolic reference provides is a *semiotic universal* that can be expected in *any* symbolic system, no matter what device implements it, whether a socially embedded in human brain or a mind rendered *in silico*. How it is neurally processed and how the symbolic medium is structured may, however, make a difference to the extent and form of its use, and so it need not be invariably expressed in any given language feature nor even severely reduced in a given language (e.g. in a pidgin or a proto-language form).

### A Symbol Token Must Be Paired With an Index in Order to Have Definite Reference

A long-recognized feature of word reference, captured formally by Gottlob Frege at the end of the nineteenth century in his sense/reference distinction, is that words on their own generally do not refer to particular concrete things in the world except when in certain combinations or contexts that determine this link (Frege 1892). Even though words have a potential 'sense'—a meaning, corresponding concept, mental image, prototypical exemplar, semantic locus, intension, and so forth—they need to be included within the special combinatorial expressions we know as sentences, propositions, predications, or ostensive occasions (e.g. pointings) in order to refer to something concretely.

There are, of course, many exceptions to this that prove the rule, so to speak: like imperatives, children's holophrastic utterances, and utterances where the reference is otherwise contextually implied, such as in the English exclamation, *Amazing!* But utterances consisting only of *Chair* or *Appear* seldom achieve concrete reference unless in a context where a previous referential frame has been explicitly established, for example by a question (what word for a piece of furniture starts with the letter 'c'?) or as a command (e.g. a magician commanding a rabbit to pop out of an apparently empty hat). Individual words outside such explicit or implicit referential frames, and which thus lack these kinds of support, seem curiously impotent and unattached to things. It is sometimes said that they refer to categories or schemas or potential objects, but this to some extent reiterates this point I wish to make. A related indefiniteness is, for example, also characteristic of badges or insignias or conventionalized icons (such as *no smoking* signs or gender icons on restroom doors). They have independent meaning, but gain their referential definiteness by being attached to something (e.g. a place or a person) to which they implicitly point. In these cases there is something more than symbolizing taking place. An additional index (even if only in the form of physical contiguity) provides definite reference. I believe that the presence of indexical operations conjoined to symbols is required in order to provide what I will call the 'grounding' of symbolic reference. This need for an indexical 'helper' when using symbols to communicate is, I believe, the result of sacrificing a direct indexical link between word and object for an indirect link mediated by a semantic network.

Philosophers of language have long argued over the role played by pointing ('ostension') in its various forms in the establishment of word reference (see e.g. Quine 1960), and the problem of defining indexical function in language has also recently become a topic of some interest (see Hanks 1992; Nunberg 1993). For my purposes it is not critical whether indexical functions are or are not sufficient for establishing initial word reference, nor is it necessary to pin down exactly what is and is not functioning indexically in a language or discourse. A number of parallels evident in the comparison of different indexical-symbolic interactions are sufficient to illustrate my point. For example, whereas a word like *chair* spoken alone and unincorporated into a sentence has a vagueness of reference and a kind of indeterminate character, merely pairing it (correlating it) with a pointing gesture to some object transforms it into a something a bit like the proposition *This is a chair* or *Here is a chair (for you to sit in)*. There often remains some ambiguity about 'what' is thereby indicated, but importantly, there is no ambiguity about the here and now of it (and perhaps even its direction, in the case of explicit pointing). Even though to a non-English speaker it would be unclear exactly what aspect of the experience was being picked out by this indication, the locus of reference would have been established with respect to a locus in some shared topology of semantic space.

Establishing a concrete locus matters. This act allows us to take advantage of the recursive powers of symbols. Though no single symbolic–indexical pairing of this sort can precisely specify what, at that locus, is being pointed to, the process need not stop there. Using the transitive power of indexical reference, one indexical–symbolic pairing can be bound to other similar 'pointings' in an extended series. Continually re-referring to an established locus in this way (e.g. with gestures or with linguistic indexicals like *that*) can constitute a linked series of coreferential indications that allows the successive incorporation of additional information from other linked symbols and indices. Add to this the fact that both discourse-external and discourse-internal loci of reference can be indicated, and also that certain classes of symbols can provide this role within a discourse (e.g. pronominal and anaphoric reference). This creates a capacity for sampling across multiple symbolic 'triangulations' of reference. The transitivity of indexicality both within and outside the symbolic system thereby provides the basis for progressively disambiguating *what* is being indicated and, to any given degree of precision, mapping it into the semantic space.

This complementary conjoining of the semiotic functions of symbols to indices is, I submit, what makes a sentence something more than merely a symbol. It is why, in order to 'do something' with words, they must be combined in such a way that these complementary semiotic roles are filled (e.g. as in a so-called 'performative' or 'speech act', such as *Not guilty!*—see Austin 1962). More precisely, I think it can be argued that it is a requirement that every symbolic operation must be complemented with an indexical operation, in order to ground the symbolic reference.

Does this mean that one could not imagine a form of symbolic communication that did not *explicitly* encode these functions in complementary classes of words and phrases? Not exactly. Imagine sitting overlooking a river watching a moonrise on some alien planet where inhabitants speak a language a lot like English, except lacking nouns. One alien might turn to the other and say *Roundly shining over flowing shimmering. Ahhh.* Even to an English speaker this could still be entirely interpretable *in context.* As in the exclamation *Whew! Hot!* the default indexicality is to the immediately enclosing physical context (part to whole) and secondly to something that is mutually salient in it. One has not escaped the need for some sort of indexical grounding. It just need not require an explicit class of special-purpose tokens or operations to specify it (on the variety and culture-specificity of such implicit indexicality, see Hanks 1992). The point is that invoking a symbol without such a bound index appears to create a kind of indexical vacuum that gets filled in with whatever is closest at hand, so to speak. But the most salient aspect of the whole physical context of which the utterance is a part is often quite ambiguous. The linguistic and cultural entrenchment of more complex, explicitly marked (symbolically encoded) indexicals both reduces this ambiguity and allows more flexible control of elaborated coreference.

So what is the semiotic basis for claiming such a requirement? The pairing of a symbolic and an indexical operation is something like an atom of definite symbolic reference, whereas a word on its own (non-indexed symbol) only points to (indicates) a locus in semantic space—a node in a virtual not actual space. As we have seen, the breaking of direct token–object indexicality is what imbues symbols with their power of collectively compressing and dissecting referential relationships, but it also undermines this referential link to things. Pairing with an additional index, however, enables the contiguous symbol token (e.g. a word) to serve a double referential role. As a physical token physically correlated with an indexical token, it transitively inherits the referential association of the correlated index; and as an indexical marker for a node in the abstract topology of the symbol system, it links this to the concrete locus of reference established by the linked index. This double mapping binds the locus in semantic space and the indicated locus of the object space together by virtue of the transitivity of indexical reference. It is this necessary semiotic synergy between symbols and indices that we recognize as determining a 'well-formed' sentence, phrase, or morphologically complex word.

This fundamental unit of grounded symbolic reference is related to what semanticists call 'predication' (see Peirce 1955; Morris 1938; 1964 for a related analysis). The essential elements that achieve predication have been well characterized in the development of symbolic logic and semantic theory. Predicate–argument structure in language or logic is typically represented notationally in the form $F(x)$ where $F$ represents some predicate function (glossed, for example, as something like *is red* or *is a dog*) and $x$ represents some possible (yet to be specified) object of reference that is being determined to exemplify (or instantiate) that function. The variable 'x' can presumably be replaced with specific objects of reference and serves as a kind of place-holder or pointer (i.e. index). Predicate functions can also be more complex, by virtue of having or requiring more than one argument variable and being subject to various required and optional kinds of modifier. For the sake of brevity and to focus on the general point, I will ignore many of the complexities of this, such as quantification, multi-place predicates, and the roles of predicate modifiers (e.g. adverbials and prepositional phrases in language), though these are quite central to a complete analysis of the semiotic infrastructure of grammar.

The role of the argument or variable component in the expression is semiotically distinct from the predicate function component, and yet the kinds of thing that can be substituted for argument variables can themselves be predicates (with their own argument variables), exemplifying the possibility for recursion implicit in symbols. Whether allowing substitution by another predicate or by some individual, the 'place' that the argument variable represents in this formalism can be understood in terms of an indexical function. The argument variable is a place-holder, or rather a pointer, or marker, for something that is not there (this is similar to an argument made on quite different—neurological—grounds by Hurford, in press).

As a sort of pictorial mnemonic, one could think of the parentheses in the form $F(x)$ as identifying a sort of logical hole that needs to be filled. Though another predicate can fill it, *that* predicate also has a hole in it that must be filled. And no hole can be left unfilled, or all predicates in the stack will inherit the equivocation of reference that remains. Any number of predicates can be embedded in this way, but the reference will fail all the way up (transitively) if all holes are not filled. And that means at some 'bottom level' there must be something other than a predicate there to stop the regress. That 'something else' must in some concrete sense be physically involved with the physical tokens of the expression. Where this becomes circular—say by being grounded in indexical 'shifters' (as in the famous *This statement is false.*)—vicious regress can result, precisely because there is no non-symbolic object at the end of the line allowing reference to extend outside the hermeneutic web of symbol–symbol relationships.

My primary hypothesis concerning predication follows from this. Predicate–argument structure expresses the semiotic dependency of symbolic reference on indexical reference. In other words 'Predicate(argument)' is a formalization of something we might analogize as 'Symbol(index)'. Binding a symbol token to an index links some locus in the mediating semantic space to some external loci embedded in the space of regularities among objects, properties, and relationships. But the semiotic freedom thereby achieved is purchased at the cost of this dependency. Being bound to an index constrains how a symbol can be placed in relationship to other symbols. Specifically, its combinatorial possibilities are now constrained by the requirements that guarantee indexicality: immediate contiguity, containment, invariant correlation, linear transitivity, and so on.

Let me just hint at a few linguistic corollaries of this requirement (which are essentially extensions from standard accounts of the role of predication in sentential semantics—see e.g. Leech 1974). Most of the world's languages explicitly distinguish two core complementary categories of word/phrase types—'noun' (subject, object, topic, etc.) and 'verb' (predicate, operation, comment, etc.)—that must (at a minimum) be paired in order to constitute a well-formed sentence. I believe that these capture the complementary indexical and symbolic functions, respectively, that are necessary to ground symbolic reference, and between them represent these semiotic roles in a symbolically encoded form. The linguistic exceptions to this simple mapping of semiotic function to word/phrase classes exemplify this. In all languages pointing or some other extralinguistic index can substitute for an explicit symbolized subject. Lone nouns or verbs bound in combination with gestures, such as pointing, or uttered in the context of explicit unmistakable stereotypical circumstances, *do* refer, and act as though they are encased in well-formed sentences, even though they stand alone. The analogues to this pairing of a linguistic symbol with a non-linguistic index are also embodied in a wide range of linguistic forms, including imperative sentences in English, so-called pro-drop languages in which subject can be regularly dropped, and discourse forms organized around a single 'topic' phrase followed by one or more 'comment' phrases that appear as sentences with a deleted subject. These exemplify the fact that the subject of a simple sentence plays an indexical role. This is also demonstrated by the way linguistic indexicals (e.g. *that*) and shifters (e.g. *you*) often take subject roles (and object roles in multi-argument predicates). It should not go unnoticed that this is consistent with arguments suggesting an evolutionary development of spoken language from ancestral forms that were entirely or partially manual.

## The Implications of Semiotic Constraints
for the Evolution of UG

Although the many implications of this cannot be explored further in this essay, I think it may turn out that many core features of UG—recursive operations, the minimally diadic structure of sentences, and many subjacency, ordering, and dependency constraints of grammar—trace their origin to this indexical requirement. These requirements apply to all symbolic linguistic relationships, and so should operate similarly in domains as diverse word morphology, phrasal relations such as movement and subjacency, and even discourse structure. In other words, these indexical constraints are implicated in most structural constraints affecting the relationships between linguistic elements in hierarchic (embedded/superordinate) and adjacency relationship to one another. The indexicality that is necessary to link a symbol to its physical context (whether to its immediate symbolic context or by transitive extension to the utterance context) imposes the more rigid constraints of indexical reference on the otherwise free symbols attached to them and combined by them. Even the symbolized re-representations of indexical relationships (e.g. via pronominalization and anaphor) are subject to the limitations of indexical reference, though often aided by agreement constraints (such as reference to the most proximate *feminine* noun).

To suggest just one rather inclusive expression of this constraint: probably the most complex and important way this indexicality constraint expresses itself syntactically is found in the constraints imposed on phrasal interdependency relationships. The condensation of many simple predications into a typical complex, multi-clause sentence is possible so long as the indexical grounding of each part is transitively passed from the most enclosed predication to the outermost enclosing predication. So there will be significant constraints affecting the extent to which symbolic reference within one unit can depend on that in others at different levels, and on the ability to retain a consistent indexical *trace* when elements are moved across phrasal boundaries. This follows from the limitations of indexicality that limit reference to immediate containment or immediate contiguity relationships, and limit transitive passing of indexical reference between elements to paths that respect this requirement. Otherwise, referential continuity and symbolic grounding will be undermined.

There are significant evolutionary implications of this. Such structural constraints have consistently been invoked as the pre-eminent examples of structural features of language that defy functional interpretation, lack semantic or cognitive motivation, and often contribute structural non-iconicity in order to maintain what appears to be a merely formal regularity. Such claims imply that only some sort of evolutionary pre-formationist argument can explain their existence. If, however, they can be shown to derive from semiotic constraints, all such protestations are rendered moot. The formalism is imposed not by fiat, but by necessity.

This is also relevant to the ancillary considerations about the learnability of grammar which are invoked to motivate innatist theories (Pinker 1994; a current overview is provided in Bertolo 2001). The semiotic approach does not require that the constraints on symbol grounding need themselves to be acquired from the corpus of language utterances one is exposed to. There is, however, agreement on many points with theories invoking innate UG. The corpus of sentences the learner encounters does not provide data for inducing grammatical rules, only helps identify local parameters with respect to universals. But semiotic constraint theory provides predictions concerning how these are discerned. Successes and failures to refer and to accomplish things with words serve as data both for discovering the universal indexical constraints and for identifying the local means to conform to them. Universals and their instantiation in the language are discovered together in the process of accomplishing things with language. To aid the process of zeroing in on the particular tricks employed to accomplish this in one's language there is a vast range of previously ignored clues that are everywhere to be discovered. These clues are neither grammatical corrections nor semantic necessities, but emerge in the form of referential consequences while trying to communicate or achieve some desired end using language.

Evidence that reference is the key arbiter, not implicit rules, is supported by the experience of most typical speakers of a language, who do not report accessing anything like rules of grammar in the construction or interpretation of well- or ill-formed sentences. When you ask a non-linguist why sentences like the following:

*John found yesterday in his shoe some candy.*
*What did John find candy and in his shoe?*
*Who do you believe that found the candy?*

are not well-formed, they usually respond with something like 'They just don't sound right'. This indeed suggests that mature language users have assimilated countless habits of expression that ultimately operate far below the level of conscious inspection. However, although most can make useful guesses (based on semantic reasonableness) what the correct form should be, their reasons sound very unlike invoking rules, such as 'X-bar' violations or 'trace-movement' errors. Instead, when you press them to explain what it is about such sentences that isn't working they often respond with something like, 'The way it is worded makes it unclear what it refers to', 'Something is left out or in the wrong place', 'It's confusing (who is doing what to whom)', or 'It can mean two different things'. These kinds of response should not, however, be taken as evidence that their knowledge of the rules they are following is innate and inaccessible (nor does it indicate the unreliability of introspection). What they are noticing is that the clarity of the reference itself has been undermined by these manipulations. To be more specific, the continuity of the transitive chain of 'pointings' from one part of the sentence to the others—which must be configured so as to end in a singular definite indication to something outside the utterance—has somehow been broken or misdirected. In other words, there is something wrong with the *indexical structure* of the sentence and so what is 'pointed to' is made somewhat uncertain.

This may help to explain why young children make good guesses about syntax as they learn a first language, as though they already know the 'rules' of grammar. If this analysis is right, they are not so much trying to figure out rules as trying to discover how their language and culture (which includes many habitual 'interaction frames' and defaults of indexical salience within which utterances are produced) implements constraints that help clearly to 'point' to a specific reference (indexically grounding the symbols). These constraints are both intuitive and well cued because they are implicit in the patterns of success and failure of reference using the tools of one's language and culture. They do not need to be inductively extracted from regularities found in the available language corpus nor do they need to be supplied by some innate template in the mind. What needs to be acquired is knowledge of the ways the tools of one's language allow these ubiquitous constraints to be respected. And for this there is plenty of feedback, even introspectively. This is a very different way of envisioning the task of language acquisition from what has been traditionally imagined. Like the pre-formationist theories, it can explain how rapid convergence on shared conventions is possible for children as well as for an incipient language community, but it does so without the baggage of innate foreknowledge.

In conclusion, I have no doubt that there is something we could ca
Universal Grammar. I also believe that it plays a powerful role in constr
ing the forms that phrases and sentences can take. And I believe the proc
of acquiring a language is made easier because of the way it vastly redu
the universe of forms that languages can take. However, I do not think i
(or can be—see Deacon 1997; 2003) an innately pre-specified neurolo
ally instantiated 'faculty'.

In this admittedly speculative essay I have instead argued that some
jor universals of grammar may come for free, so to speak, required by
nature of symbolic communication itself. Even though these semiotic c
straints still need to be discovered by learning and by communicative
periment', they are ultimately of neither biological nor cultural origin. I
guage use must be subordinated to these constraints, and language orig
and evolution in the larger sense must also have proceeded within th
constraints at every stage. With respect to the non-biological process of
guistic evolution and change, semiotic constraints can be conceived as
lection pressures, which will contribute—along with the self-organizing
namics implicit in the community of communicators—to speed evolut
toward forms that effectively communicate by not violating them.

Semiotic universals are emergent constraints, like the emergence of s
bolic reference itself. But even though they need not be biologically
specified, the evolution of human brains may also have been influenced
the presence of these semiotic selection pressures so as to make discov
of these constraints easier. Despite their abstract nature, these constrai
create 'adaptation demands' that may have selected for functional di
ences in hominid brains that ease their discovery (see Deacon 1997; 20
Ancillary biological support for arriving at highly reliable means of ach
ing these results is also plausible, such as might aid the process of auto
tization of these regularities (helping our finite brains deal with the po
tially explosive combinatorial demands that real-time symbol use crea
But the constraints themselves do not need to be redundantly built into
human brain.

The necessity of these constraints does not eliminate the demand for c
siderable cognitive effort to discover how they are reflected in the langu
nor does it does determine precisely how a given language might exemp
them. So numerous other influences and conditions need to be inclu
before the full range of universal and near-universal features of langu
can be accounted for. Furthermore, it does not render superfluous qu

major contributions to language structure by communicative, mnemonic, and neural processing constraints. Nevertheless, the origins and acquisition of language structure must be approached on very different grounds, and theoretical efforts now directed toward showing how innate UG could have evolved and how it might be instantiated in brains could well turn out to be mere intellectual exercises (e.g. Pinker 1994; Briscoe 1998; Nowak et al. 2002).

Semiotic constraints delimit the outside limits of the space of possibilities in which languages have evolved within our species, because they are the outside limits of the evolution of any symbolic form of communication. So perhaps the most astonishing implication of this hypothesis, is that we should expect that many of the core universals expressed in human languages will of necessity be embodied in any symbolic communication system, even one used by an alien race on some distant planet!

## FURTHER READING

The development of Chomsky's concept of innate Universal Grammar can be traced over three decades in such books as *Language and Mind* (1968), *Rules and Representations* (1980), and *Language and Thought* (1994). A no-holds-barred layman's-level history and defence of the concept against all critics can be found in Jenkins (2000). Darwinian variants are presented in Pinker (1994), Briscoe (1998), and Nowak et al. (2001). A co-evolutionary argument about brain evolution with respect to language competence which does *not* invoke an innate Universal Grammar is proposed in Deacon (1997), and evolutionary and neurological arguments implying the non-evolvability of innate Universal Grammar are presented in Deacon (2003). Simulation efforts have also demonstrated that at least some global attributes of UG structure can be generated without either biological evolution or an explicit linguistic evolution dynamic, but rather emerge from self-organizing effects in a population of symbolizers (e.g. Kirby 2000).

The semiotic arguments developed here, and especially this particular development of the symbol concept, were first presented in Deacon (1997: esp. chs. 2 and 3). They derive from an interpretation of the semiotic system developed by the philosopher Charles Sanders Peirce. An accessible source for the basic papers in which Peirce introduced these concepts and their hierarchic logic has been reprinted in a collected volume edited by Buchler (1955: esp. ch. 7).

# 8

# The Archaeological Evidence of Language Origins: States of Art

*Iain Davidson*

## Introduction

This chapter is principally about the archaeological evidence for the evolutionary emergence of language: how did human ancestors come to bridge the gap between humans and other animals? Over a long period of exploring the issues of language origins (Davidson and Noble 1989; Noble and Davidson 1996; 2001), Noble, a psychologist, and I, an archaeologist, have been seeking to build a mutually reinforcing argument with two main elements. One argument shows the importance of language-based interactions in defining the minded behaviour of people in our social interactions. In our view, language and mindedness are learned at our mothers' breasts through interactions which involve joint attention between mother and infant. There is a burgeoning literature on the factors that affect such joint attention (eg Langton et al. 2000; see Tomasello, Chapter 6 above). Our second argument shows how the circumstances of joint attention arose from the evolutionary emergence of bipedalism and prolonged infant dependency, leading to changed circumstances for learning and transmission of knowledge. The anatomical circumstances of bipedalism and, to a lesser extent, prolonged infant dependency can be traced in the record of physical anthropology, and the products of learned behaviour can be studied through the archaeological record. Again, issues relating to the emergence of changes in life history of hominins[1] have grown in prominence in the last few years (e.g. Alvarez 2000). Updating the evidence surrounding these arguments is beyond the scope of this chapter. Instead I will concentrate on the products of learned behaviour.

[1] The more usual word for human ancestors is 'hominid', but 'hominin' reflects more accurately the recognition that the African apes (including human ancestors) are closer to each other phylogenetically than any of them are to orangutans (Groves 1989).

## Arguments About What Language Is

Some understanding of what language is and how it relates to those products is crucial in any argument about origins and evolution of language. Without such definition it is possible to find statements such as: 'language existed in a variety of forms throughout its long evolution' (McBrearty and Brooks 2000), which allow any interpretation (or none) at all about language emergence and its relation to the evolution of behaviour.

There is a broad measure of agreement that there are two features of languages, in whatever modality they are expressed, that are generally not present among the communications of other animals: symbols and syntax (see also Tomasello, Chapter 6 above). Noble and I (1996) have argued for the priority of symbols in understanding both the relations between language and mind and in understanding the evolutionary emergence of language. Deacon (1997; see also Chapter 7 above) has called humans 'the symbolic species' and argued for coevolution between this symbolic ability and human biology—a label and a process with which it is difficult to disagree. Bickerton (1996, and Chapter 5 above) and others (e.g. Pinker 1994) would argue for the priority of syntax. The issue is whether symbols or syntax, in transforming non-human communication into human language, yield the other associations of language-mediated behaviour.

### Symbols, Syntax and Setting the Stages for the Emergence of Language

The concept of language as 'communication using symbols' is controversial only for those who do not understand that symbols are all-pervasive in human communication. Symbols are things that stand for other things, much more (and less) than the signs of religions or political ideologies.

There is a body of psychological theory about the relations between language and mind (see Noble and Davidson 1996: chs. 3–5) that seems to fit well with this definition. Two of Hockett and Ascher's (1964) 'design features' for language, productivity and displacement, depend on the characteristic that symbols stand for something other than themselves. Noble and I argue that the presence of productivity and displacement in human activity is a result of, and hence indicates, the use of symbols and language. Productivity and displacement certainly allow some archaeological interpretation.

The first element of any argument about the emergence of language is to contrast *Stage 1*, hominins and other apes communicating without language with *Stage 2*, the discovery or invention of communication using symbols. Stage 1 seems to have involved very little productivity and displacement. These are enabled by the use of symbols, so their absence implies that Stage 2 had not been reached. It is on this basis that the watercraft that carried the first people to Australia is among the earliest markers that Stage 2 had been reached (Davidson and Noble 1992). But symbols alone may not be enough, and we can postulate a *Stage 3*, working out the implications of communication using symbols, leading to *Stage 4*, language as we know it, with recognizable syntax.

There is some hint that syntax is an emergent property of communication systems with numerous 'words' acquired by observational learning. Batali (1998), Kirby (2000), and Tonkes and Wiles (2002) have independently demonstrated that complex syntactic rules emerge from quite simple systems (e.g. neural nets), which have a very small number of initial assumptions and learn from imperfect inputs. These are not empirical demonstrations of past processes, but they are very suggestive that explaining the origins of syntax may not be as big a problem as explaining the origins of symbols. That syntactic organization may be an emergent property of symbol use would be consistent with observations about the syntactic quality of communications between the ape Kanzi and his human caregivers (Greenfield and Savage-Rumbaugh 1993).

The evidence from the early acquisition of language in children also provides some support. Tomasello (2000) has suggested that infants organize their early language learning around 'concrete and particular words and phrases', and then gradually construct more syntactical utterances through imitation using general cognitive skills. He concludes that 'human language originated ultimately from a species-unique biological adaptation for symbolic communication, but the actual grammatical structures of modern languages were humanly created through grammaticalization during particular cultural histories'. The issue is how first the symbolic communication (Stage 2) and then the particular cultural histories emerged.

Bickerton (1990) postulated a stage in human evolution, which might be thought to be equivalent to my Stage 3, characterized by 'proto-language' using as an analogy modern pidgins—non-syntactic forms of speech invented by speakers of different languages to communicate without learning the other's language. There is a trap here. If the elements of the non-syntactic 'proto-language' were symbols, then, by my definition (with Noble) this is a language. If they were not symbols, then any number of other non-symbolic communication systems among apes or birds, for example, would have been under selective pressure to become syntactic. So explaining the emergence of communication using symbols is the first priority.

There has been little progress in establishing a methodology for identifying the material correlates of syntax (but see Lieberman, Chapter 14 below), except perhaps Holloway's (1969) attempt to show that, in removing flakes from a stone core, the sequence of actions is structured syntactically. Holloway's argument depended on an assumption that the stone end-products found in the archaeological record were made through an intentional sequence of hierarchically ordered flake removals. There is a strong argument that early stone tools might have been the result of much more immediate concerns about the production of flakes for use (Davidson 2002), reducing the force of any argument for a sequence of actions in hierarchic order that might resemble syntax.

While it remains the primary objective of an archaeological approach to language origins to identify the emergence of symbols, there is work still to be done to understand the relations between the earliest communication using symbols and the final stage of the evolutionary emergence of language: Stage 4, the appearance among humans of languages which are both symbolic and syntactic.

## Fossil Hominin Bones and Language

Several chapters in this volume speculate about the relationship between the timing of the emergence of language and the appearance of *Homo sapiens* in the archaeological record (e.g. Arbib, Bickerton, Dunbar, Chapters 10, 5, and 12). The identification of species from skeletal remains is made on anatomical criteria not behavioural ones. The species names given by physical anthropologists on anatomical criteria are irrelevant by themselves to arguments about the emergence of language behaviour.

Hominins with modern skeletons may or may not have been behaving like modern humans.[2] Noble and I (1996: 214) pointed out that:

- in Africa there is evidence for modern skeletons accompanied by evi-

---

[2] It has become problematic how to define 'modern human behaviour', particularly (1) its archaeological manifestations and (2) its relationship to language.

dence that suggests some had non-modern and others modern patterns of behaviour;
- in Palestine there is evidence for archaic and modern skeletons accompanied by evidence that suggests their behaviour was the same—and not modern;
- in Europe there is evidence for archaic (Neanderthals) and modern skeletons accompanied by evidence that suggests that, generally, their behaviour was different.

On the whole, this concatenation of evidence tends to suggest that modern human behaviour probably emerged earlier in Africa than elsewhere—and probably independently from the emergence of modern skeletal form, but that it was people of modern skeletal anatomy who carried modern behaviour to the world.

If we are going to try to move beyond speculation about language origins and look for evidence, from the period when the archaeological record is the only source, then we need to work out a way of looking at language that leaves some material signal. One approach, by physical anthropologists, has been to look for anatomical correlates of speech functions—an approach that has a long history, including attempts to show that, as a result of the shape of their skulls, Neanderthals could only produce three vowel sounds (Lieberman et al. 1992). All attempts to use anatomical structures to infer aspects of language evolution are relevant only in so far as we are prepared to accept that speech and language are inseparable—an area that has been addressed with some authority recently (Fitch 2000; and Chapter 9 below). Moreover, as shown in Table 8.1, many of the claims have been overstated and have not borne scrutiny. Anatomical evidence has proved a poor guide to speech abilities, and has contributed little to the understanding of the emergence of language.

## Artefacts and Language

It may seem, therefore, more appropriate to look at evidence for past behaviour by studying artefacts—the products of behaviour—in the archaeological record. Tool-making, we are led to believe, is the product of culturally determined intentions. In order to understand the issues, here, it is necessary to introduce some technicalities about stone tools.

TABLE 8.1 *Key claims about skeletal indicators of language and their equivocal nature*

|  | In favour of anatomical indication of language | Against anatomical indication of language |
| --- | --- | --- |
| Brains | Size and shape changes indicate early appearance of capacity for language (Falk 1987; Holloway 1983; Tobias 1987). | Size changes more complex and related to stature (Davidson and Noble 1998; Leigh 1992); asymmetry widespread in nature (Bradshaw and Rogers 1993); shape arguments depend on assumptions about innateness and fixed functions of cerebral regions. |
| Throats | The shape of the base of the skull indicates speech ability, and shows that Neanderthals had restricted vowel range (Lieberman et al. 1992). | The shape of the base of the skull is a product of complex developmental processes affecting many parts of the skull and is unlikely to be principally related to speech production (Lieberman and McCarthy 1999). |
| Hyoid | Dimensions of earliest surviving hyoid within range of modern humans (Arensburg et al. 1989). | Dimensions of earliest surviving hyoid within range of modern humans and many modern non-human primates (Kennedy and Faumuina 2001). |
| Hypoglossal canal | Dimensions of canal, related to innervation of tongue, within range of modern humans since 400,000 years ago (Kay et al. 1998). | All hominin specimens and most apes within range of modern humans (DeGusta et al. 1999). |
| Spinal cord | Innervation of the thoracic region significantly less in early hominins but of modern form in Neanderthals (MacLarnon and Hewitt 1999). |  |

The standard story about stone tools is well summed up in the textbook[3] by Renfrew and Bahn (2000: 319–20):

The history of stone tool technology shows a sporadically increasing degree of refinement. The first recognizable tools are simple choppers and flakes made by knocking pieces off pebbles to obtain sharp edges. The best-known examples are the so-called Oldowan tools from Olduvai Gorge, Tanzania. After hundreds of thousands of years, people progressed to flaking both surfaces of the tool, eventually producing the symmetrical Acheulian hand-axe shape, with its finely worked sharp edges. The next improvement . . . came with the introduction of the 'Levallois technique' . . . where the core was knapped in such a way that large flakes of pre-determined size and shape could be removed.

Around 35,000 years ago, with the Upper Palaeolithic period, blade technology became dominant in some parts of the world. Long, parallel-sided blades were systematically removed . . . This was a great advance, not only because it produced large numbers of blanks that could be further trimmed and retouched into a wide range of specialized tools (scrapers, burins, borers) . . .

In reality, judgements such as those concerning the 'increasing refinement' of stone tools arise from the history that much of the interpretation is determined by a progressivist view of stone tools, which was necessary when there was no other means of providing an absolute or even a relative chronology for them. More recent evidence has led to the questioning of important elements of the introductory (hence most widely understood) view of changes in stone artefacts (Davidson 2002).

### Culture and Hominin Categories

A problem arises from comparing the standard story about patterning in stone tools with the patterning in early hominin types—the clusters of variation that physical anthropologists call species. Foley (1987) argued that there was a close association between hominin species and particular stone industries; but in a more recent comparison between phylogeny and archaeology (Fig. 8.1), Foley and Lahr (1997) suggest that Mode 1 (Oldowan) industries were made by fossil species physical anthropologists call *Homo habilis*, *Homo ergaster*, and *Homo erectus*; Mode 2 (Acheulean) industries were made by *Homo erectus/Homo ergaster*, and *Homo heidelbergensis*; and

---

[3] The point here is not that I want to criticize Renfrew and Bahn for a wrongly held view. Rather, their textbook represents a widely held view, and, being popular, it is to some extent responsible for the formation of such a view.

Mode: 1 ☐ 2 ☐ 3 ☐ 4 ▨ 5 ■

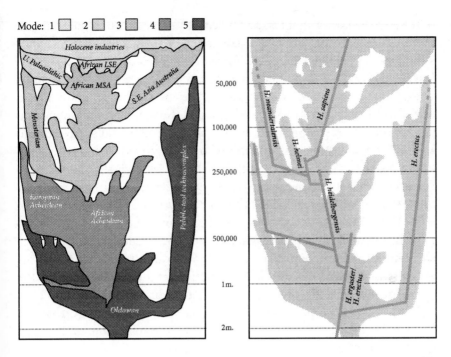

FIGURE 8.1  Foley and Lahr's phylogeny and stone industries. (Redrawn by Doug Hobbs and Mike Roach).

Mode 3 (Levallois and Mousterian) industries were made by *Homo helmei*, *Homo neandertalensis*, and *Homo sapiens*. The last of these coincidences is familiar from the record from the east Mediterranean region where modern humans and Neanderthals overlapped in time in their life and death in caves, but both engaged in Mousterian industries (Lieberman and Shea 1994).

The point (though it was not Foley and Lahr's) is that there seems to be evidence that stone industries with similar classifications (by archaeologists) were made by more than one species (as named by physical anthropologists). This feature of one stone industry being made by more than one species seems to have been quite common among early hominins.

Any definition of culture involves the traditional transmission of information *within* societies, usually involving language-based instruction. Boesch's (1991) evidence showing the beginnings of teaching of nut-cracking among chimpanzees provides an indication of how this might happen beyond humans. But in attempts to demonstrate culture in non-humans, the animals have usually fallen just short, often in the notion of traditional transmission (Boesch and Tomasello 1998; McGrew 1992; see also Davidson 1999*b*). But no definition of culture suggests that behaviours shared by *more than one* species are likely to be held in common as a result of cultural transmission.[4] In other words the techniques that appear to produce common end products of the making of stone tools were learned by some method other than traditional transmission. And it is traditional transmission, dependent on the 'ratchet effect' of language, that Boesch and Tomasello and others say is the distinctive difference between chimpanzee extrasomatic behaviour and human culture (see Tomasello, Chapter 6 above). The early stone industries, as identified by archaeologists, are a poor indicator of traditional transmission, as defined by anthropologists generally, among hominin species, as named by physical anthropologists.

The resolutions of this problem are that, *either* the classifications of the fossils are wrong, *or* the classifications of stone industries are wrong, *or* both are wrong. The first option seems highly likely, in light of the work of Collard and Wood (2000; Gibbs et al. 2000), which seems to suggest that skeletons do not offer a plausible guide to speciation.

It is also likely that the patterning of the stone industries is *not* due to processes we can identify as cultural. Villa (1981) and McPherron (2000) have suggested that raw material has an influence on the patterning of Acheulean hand axes. Some aspects of the patterning in Acheulean industries—those that contain hand axes—result from vagaries of archaeological analysis (cf. Dibble 1989) and of unplanned routine actions during knapping (Davidson 2002). This becomes more obvious when the focus moves away from the hand axes and onto the whole artefact assemblages they belong to. The impact of these few observations is to suggest that some of the patterning observed when early artefact forms, such as hand axes, are selected out of Acheulean assemblages may be due to factors other than the intentional production of hand axes—most particularly because archaeologists have already imposed their perceptions of what is or is not a hand axe before they analyse the uniformity of hand axes! Such artefacts and arguments are a poor source of evidence about the emergence of symbol use.

---

[4] Although there might ultimately be such an argument about language use in experiments with captive primates—what Savage-Rumbaugh (pers. comm.) would call the 'human–bonobo culture'.

*Stone Artefacts and the Progressive Sequence*

This observation may provide the clue to understanding the significant changes over the last fifteen years in understanding the chronology of Middle Pleistocene archaeological sites. In particular, it is now quite clear that there is no simple sequence from Oldowan to Acheulean to a late Mousterian and then the Upper Palaeolithic characterized by stone tools made on parallel-sided blades (Noble and Davidson 1997) (see Table 8.2). The new chronology requires the abandonment of the notion that there is a continuous 1.5 million-year Acheulean tradition. Instead I argue there are relatively small numbers of possible outcomes from stone-knapping such that hand axe industries occurred intermittently over much of the world.

There are two examples of the intermittent appearance of bifacial flaking. Bifacially flaked hand axes occur in the so-called 'Mousterian of Acheulean tradition' which is dated tens of thousands of years later than true Acheulean industries (Mellars 1973). Such separations in time imply there was no

TABLE 8.2 *The dating revolution*

| The standard story | The recent results |
| --- | --- |
| Acheulean (Mode 2) + Clactonian (Mode 1) 1.4 Million years (Asfaw et al. 1992) to 90,000 years (Bischoff et al. 1992). | Barnfield Acheulean and Clactonian contemporary (Ashton et al. 1994). Bifaces similar to Acheulean examples have been found in Australia which was colonized after 90,000 years ago. |
| Mousterian (Mode 3) later than Acheulean 150,000 years to 40,000 years (Bordes 1961). | High Lodge (Mousterian) and Boxgrove (Acheulean) both 500,000 years old (Ashton et al. 1992; Roberts 1986). |
| | Tabun Mousterian 275,000 years old (Mercier et al. 1995). |
| | La Cotte de St Brelade Mousterian earlier than Acheulean (Callow 1986). |
| Mousterian with bifaces is in the 'Acheulean Tradition' (Mode 2) (Bordes 1961). | Mousterian of Acheulean tradition is produced up to 50,000 years after the end of the Acheulean (Mellars 1973). |

culturally transmitted 'tradition' and that the similarity between the arte-facts from different periods did not result from cultural transmission.

The second example is the discovery of bifacially flaked hand axe-like cores in Australia (Rainey 1991) (see Fig. 8.2). The Australian region was colonised after the end of the Acheulean (Davidson and Noble 1992) from a region where, it has been said (Schick 1994), hand axes were never made. Acheulean hand axes are most often associated with the species *Homo erectus* and Australia was colonized by modern *Homo sapiens* (Davidson 1999*a*)—two species but a single stone artefact form as product of their ef-forts.

In the absence of a sequence such as that described by Renfrew and Bahn, we have to come up with a hypothesis about why hand axes (and Moust-erian artefact types) drop in and out of the prehistoric record (Noble and Davidson 1997). One possibility is that the recurrence of apparent artefact types is a result of the limited set of outcomes from flaking stone. This is why, as both Gowlett (1986) and I (Davidson and Noble 1993) have suggested, there appear to be Levallois cores at the lowest levels of Olduvai Gorge. It is not that the 'Levallois technique' was used at Olduvai, but because cores with this appearance occur as one of the possible outcomes when removing flakes from a disc-shaped core.

Finally, the progressive sequence outlined by Renfrew and Bahn (2000) emphasizes the importance in the Upper Palaeolithic of Europe of tools made on long parallel-sided flakes, called blades. Recent reviews have pointed out how widespread is the occurrence of such flakes before the Up-per Palaeolithic and outside Europe (Bar-Yosef and Kuhn 1999). Indeed,

FIGURE 8.2 Silcrete hand axe from south-west Queensland, Australia. (By permission of Dr Richard Robins and the Queensland Museum.)

one of the early industries based on such blades—the Amudian from the Palestinian site of Tabun—may have lasted longer than the Upper Palaeolithic (Jelinek 1990; Mercier et al. 1995). Interpreting this as indicating the early emergence of modern human behaviour is only plausible in the context of an interpretation that gives particular significance to the Upper Palaeolithic of Europe. If the Upper Palaeolithic is late in the story we wish to tell, then the particularities of the stone artefact sequence in Europe may be of much less significance.

### Imposed Form

Mellars (1989) tackled the question of intentionality in artefact production by introducing the phrase 'imposed form'. If archaeologists could find a way to identify that hominins or humans imposed a form on an artefact, then it would be possible to comment on the process of conceptualization among those hominins.

We have suggested previously that the first appearance of 'imposed form' was with the distinctive geometric microliths of the Middle Stone Age in southern and eastern Africa (Davidson and Noble 1993; Wurz 1999). The form was 'imposed' because the shape of the artefacts did not depend on any aspect of the mechanics of production or use, the modified edge was not that used, and the forms were standardized within a very narrow range of shapes. Artefact forms in other parts of Africa from similar periods could also be claimed to indicate 'imposed form' (Clark 1988).

There are other indicators of the emergence of modern human behaviour from the same period. There is a now well-documented presence of bone tools (Henshilwood et al. 2002) in southern Africa and elsewhere in Africa (Brooks et al. 1995). The mechanics of stone knapping impose constraints on artefact form (Davidson 1991), but these do not apply to the production of bone tools, where the whole tool was modified by scraping and grinding.

### The States of Art in Language Origins

#### Less Direct Evidence

As is to be expected in a field where there is such strong tradition of rapid oscillation between concentrations on theory and on evidence, there are many other studies of recent years—from patterns of movement, evidence

TABLE 8.3. *New evidence and its relevance or otherwise to debates about language origins*

|  | Evidence | Comment on relevance to language origins |
|---|---|---|
| Earliest migration outside Africa | Hominin fossils and stone tools at Dmanisi in Georgia by 1.7 million years ago (Gabunia et al. 2000). | Hominins have moved out of Africa many times without any sign of symbolic communication. |
| Early sea crossings | Stone artefacts claimed on the island of Flores, two sea crossings from mainland Asia, by 0.84 million years ago (O'Sullivan et al. 2001). | Short sea crossings may be possible in some environments without the use of language (Davidson 2001; Smith 2001). |
| Early spears | Pointed sticks found 0.4 million years ago in Germany (Thieme 1997). | Woodworking has been a common feature of artefact making since the earliest stone tools (Keeley and Toth 1981). |
| Evidence for the use of fire | Clear evidence for burning in pre-modern sites in Palestine (Albert et al. 2000), South Africa (Goldberg 2000), and France (Rigaud et al. 1995). | Hominin making and control of fire much more difficult to demonstrate. |
| Changing patterns of foraging | Chimpanzee hunting more complex than previously known (Stanford 1996). Early hunting may be more common than was thought (Boëda et al. 1999; Marean and Kim 1998), but interpretation needs care (Stiner 1998). Small animals may have contributed significantly to diet (Stiner et al. 2000). | Interpreting hominin ability to plan hunting activity (e.g. Gibson 1996) from the evidence of bones at archaeological sites is no easier than it was, but more urgent. |

of foraging activities and technology, and the use of fire—that appear to provide new insights into the abilities of our hominin ancestors (Table 8.3). Nevertheless, I do not believe that these amount to an overwhelming case, yet, for adjusting expectations about the date of emergence of language.

*The Berekhat Ram Pebble—Is It Art?*

A different problem concerns the 230,000-year-old scratched pebble from Berekhat Ram (Goren-Inbar 1986) in Syria, about which I have previously shown scepticism (Noble and Davidson 1996). It is one of a number of early objects that have been claimed to indicate early symbolic marking—art?— among pre-modern hominins (see eg Davidson 1990; 1991), most of which are not susceptible to interpretation as symbolic when discussed critically (e.g. d'Errico and Villa 1997).

I suggested (e.g. Davidson 1990) a *Convention Criterion* for identifying a crucial aspect of symbols:

*Convention Criterion:*
We can infer a convention or code through which meaning might be recognized if we find depictive or non-depictive marks on objects in repeated patterns, restricted in time and distribution.

Marshack (1997) and d'Errico and Nowell (2000) have separately shown that the Berekhat Ram pebble is scratched in at least two places, which Marshack describes as the 'neck' and the right 'arm'. These anatomical identifications and orientations were made prior to the identification that this object really does represent a creature that has a head, a face, a neck and an arm, and a single breast. It is only because there are scratch marks on the pebble that it becomes plausible to produce an argument about what different parts of the pebble represent if the marks were made with the deliberate intent of creating or enhancing a perceived representation.

We cannot be seduced by the promise of iconicity into accepting the plausibility of the original identification. Even an accumulation of implausible claims does not strengthen the claim for any one of them. There is still no evidence for early materials that satisfy my Convention Criterion. Finds displaying such conventions occur in Europe only in the Upper Palaeolithic (e.g. Bolus and Conard 2001), but they are earlier in southern Africa and possibly in Australia (Davidson and Noble 1998).

*Burial*

Burial of human cadavers can be taken as an indicator of the sorts of displacement and reflectivity made possible by the use of symbols. The question of burial earlier than modern humans can now be answered definitively.

There does not seem to be any good evidence that any Neanderthals buried their dead. Although Gargett (1999) is more cautious, he showed that good taphonomic information, recently recovered from well-excavated Neandertal skeletons, demonstrates the taphonomic histories of the bodies. Among the claimed burials, two processes seem to have operated. On the one hand are bodies crushed by rockfall, which tend to be complete but broken collections of bones. On the other hand are bodies that had lain in natural depressions in the sediment. Natural processes of sediment formation had generally covered these bodies slowly because the typical absence of significant limb segments strongly suggested that the flesh had rotted before interment of the bodies. Neither of these contexts is persuasive that burial took place. None of the Neandertal skeletal remains have been found in situations that could be interpreted as open-air burial (Stringer and Gamble 1993).

Among fully modern humans there are some early burials in caves (see (Aldhouse-Green and Pettitt 1998), and some human remains that probably resulted from accidents of deposition similar to those suffered by Neanderthals (Svoboda 2000). Most importantly, some of the earliest burials are in the open-air: Lake Mungo (Bowler et al. 1970; Bowler and Thorne 1976) in Australia (perhaps as old as 43,000) (Gillespie 1998); Dolni Vestonice (Formicola et al. 2001) in the Czech Republic (about 26,000 years ago); Sungir (Pettitt and Bader 2000) in Russia (between 24,000 and 21,000 years). Both in Europe and Australia, these burials are accompanied by clear signs of ritual—particularly through the use of ochre. In Europe, the burials are also accompanied by personal decorations such as beads and bracelets.

### Early Ochre in Africa

Among the early evidence in Australia is the discovery of ochre as powder surrounding the 43,000-year-old burial at Lake Mungo (Bowler and Thorne 1976), and with ground facets in the oldest layers at Malakunanja 2 and Nauwalabila (Jones and Johnson 1985), perhaps over 50 thousand years old (Roberts et al. 1993). Ochre use seems to be earlier in Africa, with well-documented finds at Blombos Cave earlier than 70,000 years ago (Henshilwood et al. 2002) and finds from several sites, such as Klasies River of similar antiquity. It is more difficult to tell what the ochre was used for, but one possibility is that it was used for body decoration, which may be one of the earliest uses of symbolism marking modern human behaviour (Gorman 2000). This would be an important stage in the emergence of self-awareness

(Davidson 1997) (see Dunbar, Chapter 12 below, for markers of group membership)—a major aspect of reflectivity in a language-based society (Davidson and Noble 1989).

### Early Art

Beads and bracelets appear among the early manifestations of personal ornaments in Asia by 40,000 years ago (Kuhn et al. 2001) as well as in Africa by similar dates (Ambrose 1998), Australia by 32,000 (Morse 1993), and Europe by the same date (White 1989).

The question of the earliest appearance of art has generally been addressed in terms of the Upper Palaeolithic of Europe, particularly the remarkable ivory sculptures from Germany (Hahn 1986) and the engraved limestone blocks from France (Delluc and Delluc 1978). Now cave paintings have been dated from the same period, older than 30,000 years ago, at Chauvet Cave (Clottes 2001). But the signs of art seem to be of the same order of age in Australia, Africa, and Europe (Davidson 1997), with some probability that they are earlier outside Europe. The recent documentation of ochre and bone objects engraved with deliberate marks, at the South African site of Blombos Cave older than 70,000 years ago (Henshilwood et al. 2002), confirms earlier, less well-documented, findings from Klasies River (Singer and Wymer 1982) which are marked in repeated and deliberate patterns (personal observation), meeting my Convention Criterion. In addition, recent re-dating of Apollo XI Cave in Namibia (Miller et al. 1999) opens the possibility that the painted slabs of rock generally said to be 26,000 years old (Wendt 1976) may actually be dated to 40,000 or more.

## The State of the Art of the Empirical Evidence for the origins of language

It is not enough in an argument about language origins to point simply to an accumulation of archaeological evidence. Any argument about language origins depends absolutely on the development of theory. Since the publication of a review of Middle Palaeolithic symbolism (Chase and Dibble 1987), there have been new developments in theory, new discoveries of clear documents relevant to the use of symbols, and many reassessments of previous claims. The evidence for symbol use goes back 70,000 years, but there are hints that it may first appear slowly and not pervade all aspects of behaviour. The state of the art in the archaeology of language origins still depends very much—but not only—on the state of the archaeology of art.

But many questions remain unresolved. Among these questions are:

- How could we establish whether hominins had a rich symbolic life that never manifested itself in the surviving material record?
- Early hominins seem to have carried stone tools with them, which might indicate displacement, but can this occur without a greater degree of productivity?
- Did symbols appear before syntax?
- Does any behaviour of early hominins imply sequencing of actions that required syntactic thought before the evidence for symbols?
- How would we theorize a communication system that disconnected syntax and symbols, without an analogy with modern communications that derive from language-based communication?
- Can the archaeological record be interpreted to reveal syntactic thought or other features that are less than 'whole' modern languages?
- Was the emergence of language simply a behavioural change, or are there features of language use that demonstrate that a genetic change was necessary?

Given recent refinement in our understanding of the emergence of symbols, it seems likely that it will be possible in future to tease out more of the story of the first emergence of language, and with it the complex interaction of material objects, symbols, and syntax.

## FURTHER READING

Anyone beginning to study language origins needs background information on a wide range of subjects. The best general introduction is Lock and Peters (1996), a massive compilation of specialist chapters on almost all aspects of the topic, a little dated now because of long publishing history, but an excellent starting point on any topic.

One of the key requirements is some understanding of the evidence from modern humans, including psychology and linguistics. There are various compilations of specialist studies from a variety of disciplines, mostly the result of biennial conferences on the Evolution of Language. The latest of these is Wray (2002) with chapters by several leading thinkers in different aspects of language. Probably the current definitive work of synthesis on language origins is by Deacon (1997), which features a very dense account of neurological studies. The weakness of this synthesis is its rather uncritical approach to the archaeological record. One approach that seems promising, but has not yet been operationalized in relation to other data, is that by Jackendoff (1999), who 'decomposes' modern language to suggest nine stages of emergence.

One other key source of evidence and ideas is studies of modern non-human primates, including their communicative abilities. McGrew (1992) is a landmark study synthesizing evidence for chimpanzee behaviour which shows how much these animals can achieve without language, particularly in the production and use of material tools. Savage-Rumbaugh and Lewin (1994) is a popular account of the remarkable studies at the Language Research Centre which made the break-through to getting apes to communicate with humans in controlled circumstances. More detailed studies have been published elsewhere, but this volume is also valu-able for its frank summary of the successes and failures of other language experi-ments with apes. The symposium volume by King (1999) collects important papers by a number of major players (including primatologists, linguists, and anthropolo-gists) in the debates about the contributions of primate studies to understanding of language origins.

The only evidence of what actually happened in the evolutionary emergence of language is from archaeology. But to make such a contribution, the archaeological record needs to be critically assessed, particularly in light of the theoretical as-sumptions of the various interpretations. Noble and Davidson (1996) attempted to combine psychological theory with a reinterpretation of the archaeological record to produce an integrated account of the origins of language and its implications for the evolution of human behaviour. Finally, the most recent book looking at the archaeological evidence is Nowell (2001), a collection of papers from leading authorities on the contribution to studies of the evolution of mind from different aspects of archaeology and physical anthropology.

# 9

## What Are the Uniquely Human Components of the Language Faculty?

*Marc D. Hauser and W. Tecumseh Fitch*

### Introduction

From a biologist's perspective, language has its own particular design features. It is present in virtually all humans, appears to be mediated by dedicated neural circuitry, exhibits a characteristic pattern of development, and is grounded in a suite of constraints that can be characterized by formal parameters. Thus, language has all the earmarks of an adaptation, suggesting that it could be fruitfully studied from a biological and evolutionary perspective (Bickerton 1990; Deacon 1997; Jackendoff 1999; Lenneberg 1967; Lieberman 1984; Pinker and Bloom 1990). On the other hand, linguistic behaviour leaves no fossils, and many characteristics of language appear unique to our species. This suggests both that the phylogenetic approach (constructing adaptive narrative that captures the timing and functionality of language evolution in our species) and the comparative approach (using data from other species to gain perspective on characteristics of our own) will be fraught with difficulty.

We tend to agree that the construction of historical narratives of language evolution is too unconstrained by the available data to be profitable at present, especially since many plausible scenarios have already been exhaustively explored (see e.g. Harris 1996). At best, this practice provides a constrained source of new hypotheses to be tested; at worst it degenerates into fanciful storytelling. In contrast, we claim that the comparative approach to language has been and will continue to be a powerful approach to understanding both the evolution and current function of the language faculty. Our purpose in this chapter is to review the current state of the art in comparative studies of the faculty of language, focusing specifically on the sensory-motor system involved in the production and perception of acoustic signals. We show that comparative research is an extremely valuable source of information in biolinguistics, allowing us to isolate and study those components of the language faculty inherited from our non-human ancestors. We also point out that such work is logically necessary before claims of human uniqueness and/or language-specific evolution can be made.

Despite the wide variety of theoretical and empirical perspectives in the study of human language, everyone agrees that language is a complex entity, incorporating a variety of interacting subsystems, from neurobiological and cognitive to social and pragmatic. As a research strategy, especially one aimed at uncovering similarities and differences across species, it appears necessary first to fractionate the 'language faculty' into a set of relevant subsystems and then explore which are uniquely part of the human capacity to acquire language. It is clear that some aspects of the language faculty must be unique to our species as no other theoretical stance can account for the fact that chimpanzees raised identically to humans cannot attain full language competence, despite their impressive achievements (Savage-Rumbaugh et al. 1993). At a minimum, all the subcomponents necessary for language might be present in chimpanzees, without being adequately interconnected. More likely, we think, a number of critical systems are not fully present in chimpanzees, though a significant proportion of the necessary machinery is in place. It is also possible that most of the subsystems of language have been modified to suit language in the course of human evolution, and thus are unique in a more limited sense. From this viewpoint, the lexicon would be an exception that proves the rule. Despite its obvious homology with memory systems in non-humans, the huge number of words that every child learns dwarfs the capabilities of the most sophisticated non-humans. This suggests that, despite a broadly shared neural basis, even the lexicon has undergone some special modifications in humans. In view of the currently available data, all of these possibilities seem at least reasonable, and a priori commitments to one viewpoint or the other seem premature.

We hope that the points made thus far are rather obvious and uncontroversial. Somewhat more controversially, we argue that the study of language must proceed by a detailed, comparative approach to each of the mechanisms contributing to language, and to their interactions and interface conditions. The most obvious reason concerns empirical research into the neural systems that contribute to language, much of which is limited to animals due to both practical and ethical considerations. To the extent that component subsystems are shared with animals, only there can we study them

with the full panoply of modern neuroscientific techniques. A second reason concerns communicative function. Language functions (among other
things) as a system of communication for members of our species, although
some have argued that important aspects of language did not initially evolve
for communicative purposes (Hauser et al. 2002). During human evolution,
it was minimally necessary for the faculty of language to coexist with the
extant vocal communication system. Further, the evolving language faculty probably co-opted certain aspects of the pre-existing communication
system (e.g. parts of the phonetic system) for its own use. This constraint
makes it important to have a clear sense of what the pre-linguistic hominid communication system was like, an understanding that can only come
from the comparative study of communication. Finally, and most critically,
it is logically necessary to study animals before making claims about human
uniqueness, or positing that a particular subsystem or mechanism evolved
'for' language. Although this point also seems obvious to us, it has traditionally been ignored or misunderstood. In summary, animal studies are a necessary component of the biological study of the language faculty, allowing
us to discover by a process of elimination those mechanisms that are unique
to human language, as well as to study exhaustively those language-related
mechanisms that are shared.

In this chapter we will concisely review comparative research that has
specifically focused on mechanisms believed important for human language, many of which were or are posited to be uniquely human and/or
specific to language. Because much of this work has focused on speech production and perception, we will focus on these areas. It is, of course, clear to
us that speech and language are logically and sometimes practically separa-
ble (written and signed languages providing two clear examples) and might
well have had independent evolutionary trajectories (Fitch 2000a). Nonetheless, all human societies use speech as the primary input/output system
for language, and the constraints of the phonetic interface surely played a
role in the evolution of the language faculty in its broad sense. Furthermore,
the empirical grounding of speech science, combined with its measurable
behavioural manifestations, has allowed the comparative study of speech
and vocal production/perception to progress far beyond that of other aspects of comparative language research (e.g. syntax or semantics). This
should not be taken as an indication that the comparative study of these
latter topics is impossible, but simply that other research areas have lagged
behind that of animal speech research. We expect that many of the same

theoretical perspectives and empirical tools that have been applied in the arena of speech can be fruitfully extended to other aspects of language as well (Hauser et al. 2002).

## Speech Production

A basic understanding of the physics and physiology of speech was attained more than fifty years ago (Chiba and Kajiyama 1941; Fant 1960; Stevens and House 1955), providing a necessary first step for adequate analysis and synthesis of speech, and thus paving the way for the major advances in speech perception that followed (Liberman 1996). Surprisingly, the ethological study of mammalian vocal communication proceeded in the opposite direction, with an almost exclusive focus on the perception of signals and little understanding of their production. Extremely basic questions concerning animal sound production have only recently become the focus of concerted research (Fitch 1997; 2000a; Goller and Larsen 1997; Nowicki and Capranica 1986; Owren and Bernacki 1988; Suthers et al. 1988). Such information is important both for the practical reason that adequate analysis and synthesis of animal signals requires a solid understanding of how they are produced, and because the evolution of communication systems is characterized by a constant interplay between signal production and perception (Bradbury and Vehrencamp 1997; Hauser 1996). As we describe below, the physics of the production mechanisms can lead to certain types of information being available in signals, providing the conditions for subsequent selection of perceptual mechanisms to access this information. New perceptual mechanisms can then create selection pressures on production mechanisms, to conceal, enhance, or exaggerate such acoustic cues. Thus, an adequate understanding of the evolution of acoustic communication systems requires mature theories of both production and perception.

### *Source–Filter Theory of Animal Vocal Production*

A central insight of modern speech science is that human vocal production can be broken down into two components, termed source and filter (Titze 1994). This source–filter model has more recently been generalized to vocal production in other terrestrial vertebrates (Fitch and Hauser 2002) and holds true for virtually all vertebrates whose production mechanisms are well understood. The *source* (typically the larynx) converts the flow of air from the lungs into an acoustic signal. Typically this source signal is periodic (it has a fundamental frequency which determines its perceived pitch) and has energy at many higher frequencies (harmonics). This sound then propagates through the vocal tract (including the oral and nasal cavities). The air contained in the vocal tract, like any tube of air, possesses multiple resonances at which it can oscillate, termed *formants*. Formants act as bandpass filters, allowing energy to pass through at their centre frequency, and suppressing frequencies higher or lower than this. Together, all the formants of the vocal tract create a complex, multi-peaked *filter* function (a formant pattern), which forms the acoustic basis for much of speech perception. Thus vocal production is typically a two-part process: the production of an acoustic signal by the source and the subsequent spectral shaping by the formants constituting the vocal tract filter. In humans, and the other mammals studied thus far, these two components are independent. Thus, properties of the source (such as pitch) can be varied independently of the filter (formants), and vice versa.

### The sound-producing source

From a comparative perspective, the larynx is a conservative structure (Harrison 1995; Negus 1949). The original function of the larynx was to protect the airway by acting as a gatekeeper to the respiratory system. This function is preserved in all tetrapods. The larynx in mammals also typically produces sounds: it contains paired vocal cords which are set into vibration by air flowing through the glottis. The rate of vibration, which can be modified voluntarily via changes in vocal fold tension, determines the voice pitch. This sound-generating function evolved later than the more basic gatekeeper function of the larynx, and must coexist within the constraints imposed by it. Nonetheless, interesting modifications of the mammalian larynx exist, including such phenomena as vocal membranes for high-pitch phonation and laryngeal air sacs (Fitch and Hauser 1995; Gautier 1971; Kelemen 1969; Mergell et al. 1999). However, the cartilaginous framework, innervation, vasculature, and musculature of the larynx are essentially invariant in mammals. It seems likely that the 'dual-use' constraint following from a single organ serving the dual functions of airway protection and vocalization is one reason for the relative conservatism of the mammalian larynx (Fitch and Hauser 2002). Conveniently, the conservatism of the larynx allows us to apply the insights of the theory of human vocal fold vibration, termed the myoelastic-aerodynamic theory (Titze 1994), to laryngeal function in other mammals.

One interesting type of variation in laryngeal anatomy, extreme among mammals, is in its relative size: howler monkeys have a massive larynx and hyoid complex, nearly the size of their head (Schön Ybarra 1988), while the hugely enlarged larynx of male hammerhead bats fills the entire chest (Schneider et al. 1967). Less extreme, but nonetheless impressive, hypertrophy appears to have evolved independently in many mammalian lineages, particularly among males (Frey and Hofmann 2000; Hill and Booth 1957). Because the lowest frequency producible by the vocal cords is determined by their length (Titze 1994), the primary function of a large larynx is the production of loud, low-pitched calls. Any force selecting for low voices (e.g. female mate choice or aggressive encounters with competitors) will select for lengthening of the vocal folds and accompanying enlargement of the larynx. This is nicely illustrated in humans: at puberty the male voice drops in pitch due to a testosterone-dependent enlargement of the larynx and concomitant lengthening of the vocal folds (Kahane 1978); this doubling of vocal cord length leads explains why men's voices are lower than women's (Titze 1994). In addition to providing an excellent example of convergent evolution between humans and animals, laryngeal dimorphism in our species also shows that laryngeal size has no critical impact on speech production.

Thus, the presently available comparative data indicate that the human larynx does not differ from that of most other mammals in ways obviously relevant to speech production. The aspects of human speech that involve the larynx (control of voicing and pitch) are almost certainly built on a phylogenetically ancient set of mechanisms shared with other mammals (although the possibility remains that humans have finer control over these functions than other mammals: Lieberman 1968*a*). In contrast, the human vocal tract is strikingly different from that of other mammals.

### The vocal tract filter

The vocal tract includes the pharyngeal, oral, and nasal cavities. Their size and shape determine the complex formant pattern of the emitted sound. A central puzzle in the study of speech evolution revolves around the fact that human vocal tract anatomy differs from that of other primates: the human larynx rests much lower in the throat. This fact was recognized more than a century ago (Bowles 1889). In most mammals, the larynx can be engaged into the nasal passages, enabling simultaneous breathing and swallowing of fluids (Crompton et al. 1997). This is also true of human infants, who can suckle (orally) and breathe (nasally) simultaneously (Laitman and Reidenberg 1988). At about the age of 3 months, the larynx begins a slow descent to its lower adult position, which it reaches after 3–4 years (Lieberman et al. 2001; Sasaki et al. 1977; Senecail 1979). A second, smaller descent occurs in human males at puberty (Fitch and Giedd 1999). A comparable 'descent of the larynx' occurred over the course of human evolution.

The acoustic significance of the descended larynx was first recognized by Lieberman and his colleagues (Lieberman and Crelin 1971; Lieberman et al. 1969), who realized that the lowered larynx allows humans to produce a wider range of formant patterns than other mammals. The change in larynx position allows us to independently vary the area of the oral and pharyngeal tubes and to create a broad variety of vocal tract shapes and formant patterns, thus expanding our phonetic repertoire. In contrast, the standard mammalian tongue rests flat in the long oral cavity, making vowels such as the /i/ in 'beet' or the /u/ in 'boot' difficult or impossible to produce because they require extreme constriction in some vocal tract regions and dilation in others. These vowels are highly distinctive, found in virtually all languages (Maddieson 1984), and play an important role in allowing rapid, efficient speech communication to take place (Lieberman 1984).

Until recently, the descended larynx was believed unique to humans. Considerable debate has centred on when in the course of human evolution the larynx descended (reviewed in Fitch 2000a), a debate which remains unresolved because the tongue and larynx do not fossilize. Attempts to reconstruct the vocal tract of extinct hominids must thus rely on skeletal remains, combined with tenuous assumptions about the relationship between the anatomy of the skull or hyoid bone and the position of the larynx. Recent studies of the dynamics of vocal production in non-human mammals raise serious doubts about the reliability of such reconstructions. This work, involving X-ray video of vocalizing dogs, pigs, goats and monkeys (Fitch 2000b), shows that the mammalian larynx is surprisingly mobile, flexibly moving up and down during vocalization. This flexibility suggests that attempts to estimate the resting position of the larynx based on skull anatomy are superfluous, since the larynx typically moves far from its resting position during mammalian vocalization. These data also indicate the existence of a gradualistic evolutionary path to laryngeal descent: speech in early hominids might have been accompanied by a temporary laryngeal retraction (as seen in other mammals during vocalization). Finally, recent

data demonstrate that humans are not unique in having a permanently descended larynx. Male red deer show laryngeal descent exceeding our own,
and further lower the larynx to the sternum while vocalizing (Fitch and
Reby 2001). These comparative data indicate that a descended larynx is not
necessarily indicative of speech. We conclude that questions of historical
timing and attempts at fossil reconstruction have been overemphasized in
the literature concerning speech evolution, at the expense of detailed consideration of robust comparative data available from living animals.

### *Communication With Formants: The Comparative Perspective*

Although the phonetic, and hence communicative, importance of formants
is axiomatic in speech science, there has until recently been little discussion of formants in animal communication, and one might easily conclude
(incorrectly) that formants play little role in non-human communication.
Here we briefly review the literature on the communicative function of
formants in animals, to provide a richer perspective on the evolution of
human speech. This research not only reveals that formants are present in
vocalizations and perceived by animals but suggests that communicative
uses of formants have a rich evolutionary history, long preceding human
evolution (Fitch 1997; Owren and Bernacki 1988; Rendall et al. 1998).

Although researchers have recognized the existence of formants in
the vocalizations of non-human primates for many years (Andrew 1976;
Lieberman 1968; Richman 1976), little attention has been paid until recently to the information they might convey. Two types of information
that might theoretically be conveyed via formants are individual identity
and body size. Because the detailed shape of the oral and nasal vocal tracts
vary, individuals should have slightly different formant patterns that would
allow listeners to determine the identity of a vocalizer. For example, individual differences in the sizes and locations of the nasal sinuses lead directly
to individual differences in the speech output spectrum (Dang and Honda
1996), and discriminant function analysis of rhesus macaque calls suggests
that similar phenomena may apply in monkeys (Rendall 1996). This has led
Owren and Rendall (1997) to suggest that formants could provide important cues to individual identity in primates. While plausible, this suggestion
has yet to be rigorously tested, and the flexibility of the mammalian vocal
tract during calling (Fitch 2000*b*) suggests caution in interpreting individual vocal tract morphology as 'fixed'.

The idea that formants in animal vocalizations convey body size information has received more empirical support. Formant frequencies are strongly
influenced by the length of the vocal tract (Fitch 1997; Titze 1994). Vocal
tract length, in turn, is largely determined by the size and shape of the skull,
which is strongly correlated with total body size (Fitch 2000c). Thus vocal
tract length and formant frequencies are both closely tied to body size in
the species examined so far (Fitch and Giedd 1999; Fitch 1997; Riede and
Fitch 1999). This linkage should hold true for most mammals (Fitch 2000c).
Thus, a listener that can perceive formants could gain accurate information
about the body size of the vocalizer. A variety of birds and mammals can be
easily trained to perceive formants (Hienz et al. 1981; Sommers et al. 1992),
or perceive them spontaneously (Fitch and Kelley 2000), suggesting that the
ability to perceive formants was present in the reptilian common ancestor
of birds and mammals. Because body size is highly relevant to social behaviour and reproductive success in most terrestrial vertebrates, it seems likely
that an initial function of formant perception was to help judge the body
size of a vocalizer. Particularly in dense forest environments or in dark-
ness, an ability to perceive body size based on acoustic cues would be highly
adaptive. These data provide strong support for the notion that communication via formants has a long evolutionary history in terrestrial vertebrates.
Though more comparative data are necessary to exclude the possibility
that formant perception in birds and mammals is a convergent adaptation
(homoplasy), the most parsimonious interpretation of current data is that
formant perception represents a homologous character, present in the common ancestor of birds and mammals that lived during the Palaeozoic several hundred million years ago.

### Convergent Evolution: The Descent of the Larynx in Non-Humans

As mentioned earlier, the permanently descended larynx in humans represents an important difference between humans and our primate relatives,
highly relevant to speech production. For many years researchers believed
that this trait was uniquely human (Lieberman 1984; Negus 1949). Recent
comparative studies demonstrate otherwise: at least two deer species have
a descended larynx (red and fallow deer: Fitch and Reby 2001), which is
pulled down to its physiological limit during vocalizations, substantially
surpassing laryngeal descent in our own species. Other species have simi-

larly lowered and/or lowerable larynges, including lions, tigers, and other members of the genus *Panthera* (Peters and Hast 1994; Pocock 1916; Weissengrüber et al. 2002), as well as koalas (Sonntag 1921); we confine our discussion mainly to deer. The descent of the larynx in deer is clearly not an adaptation to articulate speech, but its dynamic retraction during vocalization strongly suggests that it serves a vocal function. Why does the larynx descend in these species?

Detailed audio-video analysis of deer vocalizations demonstrates that laryngeal retraction lowers formant frequencies, as predicted by acoustic theory (Fitch and Reby 2002). One possible function of this formant lowering might be to increase the propagation of sounds through the environment, since atmospheric absorption is more pronounced for high frequencies. However, when a sound source is close to the ground (less than a metre, as with deer), the interference with reflections from the ground can actually weaken low-frequency transmission. This, along with behavioural data, suggests that formant lowering does not aid sound propagation in deer.

A more likely hypothesis is that laryngeal retraction serves to exaggerate the size of the vocalizer, a form of 'bluffing' that would be valuable in animals that often vocalize at night and in dense foliage, as do deer. Once perceivers use formants as a cue to size, the stage is set for deception: any anatomical mechanism that allows a vocalizer to evade the normal constraint linking body size and vocal tract length enables a smaller animal to duplicate the formant pattern of a larger individual by elongating its vocal tract, thus exaggerating its apparent size. Male red deer have partially evaded the constraint linking skull size to vocal tract length by evolving a highly elastic thyrohyoid ligament (which binds the larynx tightly to the hyoid skeleton in most mammals). Combined with powerful laryngeal retractor muscles, this allows stags to extend their vocal tracts far below the normal position, to about a third of their body length. The impressive roars thus produced have very low formants, serving to intimidate rivals (Clutton-Brock and Albon 1979) and attract females (McComb 1991) and creating an 'arms race' where all males without the trait will be out-competed. Finally, this sets the stage for the next round of perceptual evolution. This 'size exaggeration' hypothesis for laryngeal descent in deer is consistent with the available behavioural and acoustic data, and with data on vocal tract elongation in other taxa (Fitch 1999).

Because the common ancestor of deer and humans did not have a descended larynx, laryngeal descent in these species represents an example of convergent evolution. There is obviously no guarantee that the descent of the larynx in each lineage occurred for the same reasons. However, since the size exaggeration hypothesis is based on physical and physiological principles that are common to all mammals it also provides a plausible alternative explanation for the initial descent of the larynx in our own species. By this argument, the permanently descended human larynx might have evolved early in the hominid lineage (e.g. in australopithecines), serving a size exaggeration function, long before the advent of language. The increased phonetic potential allowed by this arrangement may have lain dormant for millennia (as it still does in red deer) before being exapted for use in spoken language by later hominids. Consistent with this hypothesis, the initial descent of the human larynx, which happens in infants, is followed by a second descent which occurs at puberty, but only in males (Fitch and Giedd 1999). This second descent does not increase the phonetic abilities of teenaged boys, but probably serves a function similar to that in deer: increasing the impressiveness of the adult male voice via size exaggeration (Fitch and Giedd 1999; Ohala 1984).

### Speech Production: Conclusions and Future Directions

The data reviewed in this section indicate that researchers interested in the mechanisms underlying human speech production can gain important insights from the study of vocal production in other animals. Far from being unique to humans, communication via formant frequencies appears to be an ancient characteristic antedating the origin of humans. Communication via formants originally functioned for size perception or individual identification, not for transmitting sophisticated linguistic messages. Most non-human mammals lower the larynx during vocalization, suggesting that the unusual descended larynx in our species probably evolved gradually, based on a pre-adaptive flexibility in larynx position in mammals. Furthermore, several non-human species show a permanent descent of the larynx which evolved convergently with humans, and a likely explanation for descent in these species might apply to humans as well. Thus, certain key aspects of speech are likely built upon an ancient foundation that can be fruitfully studied from a comparative perspective.

A broader conclusion that is that the importance of changes in the hominid vocal periphery has historically been overemphasized. Future work should focus on changes in the neural mechanisms underlying speech production (e.g. Deacon 1997; Fitch 2000a; Lieberman 2000; MacNeilage 1998). There are at least two important candidate mechanisms for the role of critical adaptations in the evolution of spoken language: vocal imitation and hierarchical composition. Although vocal imitation is not uniquely human (it is seen in most songbirds and a number of marine mammals), it is obviously critical for the acquisition of large open-ended vocabularies, and is not shared with other non-human primates (Janik and Slater 1997). Thus, the neural basis and evolutionary history of vocal imitation should be a focus of future research (Studdert-Kennedy 1983; Hauser et al. 2002). Second, speech requires a flexible and powerful ability to recombine small acoustic units (phonemes and syllables) into larger composites (words and phrases); this is the only way that an open-ended vocabulary of readily discriminable vocalizations can be created. Again, recombination of small units into larger units is seen in other animals (Hauser 1996), but not to the same degree as in human speech (MacNeilage and Davis 2000). The comparative study of the neural bases of these abilities, both cortical (Deacon 1997; MacNeilage 1998) and subcortical (Lieberman 2000), will be an important source of new information relevant to the evolution of spoken language.

## Speech Perception

Our ears are bombarded with sound. However, when we hear spoken language, as opposed to sounds associated with either human emotion (e.g. laughter, crying) or music, different neural circuits appear to be engaged. The fact that specialized and even dedicated neural circuitry is recruited for speech perception is certainly not surprising, especially when one considers the evolution of other systems of communication. Exploration of this comparative database reveals that the rule in nature is one of special design, whereby natural selection builds, blindly of course, adaptations suited to past and current environmental pressures. Thus, by looking at the communicative problems that each organism faces, we find signs of special design, including the dance of the honey bee, electric signalling of mormyrid fishes, the song of passerine birds, and the foot drumming of kangaroo rats. The question of interest in any comparative analysis then becomes which aspects of the communicative system are uniquely designed for the species of interest, and which are conserved. In the case of speech perception, we know that the peripheral mechanisms (ear, cochlea, and brainstem) have been largely conserved in mammals (Stebbins 1983). The focus of this section is to inquire which components of speech perception are mediated by a specialized phonetic mode, and which by a more general mammalian auditory mode. Evidence that non-human animals parse speech signals in the same way that humans do provides evidence against the claim that such capacities evolved for speech perception, arguing instead that they evolved for more general auditory functions, and were subsequently coopted by the speech system.

### Categorical Perception and the History of the 'Speech Is Special' Debate

In the 1960s, Liberman and his colleagues (reviewed in Liberman 1996) began to explore in detail the mechanisms underlying human speech perception. Much of this work was aimed at identifying particular signatures of an underlying, specialized mechanism. An important early candidate mechanism was highlighted by the discovery of categorical perception.

When we perceive speech, we divide a continuously variable range of speech sounds into discrete categories. Listening to an artificially created acoustic continuum running from /ba/ to /pa/, human adults show excellent discrimination of between-category exemplars, and poor discrimination of within-category exemplars, a phenomenon termed 'categorical perception'. When first discovered, this phenomenon seemed both highly useful in speech perception and specifically tailored to the speech signal. This fact led Liberman and colleagues to posit (before any comparative work was done) that categorical perception was uniquely human and special to speech. To determine whether the mechanism underlying categorical perception is specialized for speech and uniquely human, new methods were required, including subjects other than human adults. In response to this demand, the phenomenon of categorical perception was soon explored in (1) adult humans using non-speech acoustic signals as well as visual signals, (2) human infants using a habituation procedure with the presentation of speech stimuli, and (3) animals using operant techniques and the precise speech stimuli used to first demonstrate the phenomenon in adult humans (Harnad 1987). Results showed that categorical perception could be demonstrated for non-speech stimuli in adults, and for speech stimuli in both human infants and non-human animals (reviewed in Kuhl 1989). Although the earliest work on animals was restricted to mammals (i.e. chinchilla, macaques), subsequent studies provided comparable evidence in birds (reviewed in Hauser 1996). This suggests that the mechanism underlying categorical perception in humans is shared with other animals, and may have evolved at least as far back as the divergence point with birds. Although this finding does not rule out the importance of categorical perception in speech processing, it strongly suggests that the underlying mechanism is unlikely to have evolved for speech. In other words, the capacity to treat an acoustic continuum as comprising discrete acoustic categories is a general auditory mechanism that evolved before humans began producing and perceiving the sounds of speech.

### Beyond Categorical Perception

The history of work on categorical speech perception provides both a cautionary tale and an elegant example of the power of the comparative method. If you want to know whether a mechanism has evolved specifically for a particular function, in a particular species, then the only way to address this question is by running experiments on a broad array of species. With respect to categorical perception, at least, it appears that the underlying mechanism did not evolve for processing speech. We cannot currently be absolutely confident that the underlying neurobiological mechanisms are the same across species, despite identical functional capacity (Trout 2000). Nonetheless, a question arises from such work: What, if anything, is special about speech, especially with respect to processing mechanisms? Until the early 1990s, animal scientists pursued this problem, focusing on different phonemic contrasts as well as formant perception (reviewed in Trout 2000; Hauser 2002); most of this work suggested common mechanisms, shared by humans and non-human primates (for a recent exception, see Sinnott and Williamson 1999). In the early 1990s, however, Kuhl and colleagues (1991; 2000) published intriguing comparative results showing that human adults and infants, but not rhesus monkeys, perceive a distinction between so-to-speak *good* and *bad* exemplars of a phonemic class. The good exemplars or *prototypes*, functioned like perceptual magnets, anchoring the category, and making it more difficult to distinguish the prototype from sounds that are acoustically similar; non-prototypes function in a different way, and are readily distinguished from more prototypical exemplars. In the same way that robins and sparrows, but not penguins or flamingos, are prototypical birds because they carry the most common or salient visual features (e.g. wings for flying, small beaks) within the category bird, prototypical phonemes consist of the most common or salient acoustical features. Although there is controversy in the literature concerning the validity of this work in thinking about the perceptual organization and development of speech (Kluender et al. 1998; Lotto et al. 1998), our concern here is with the comparative claim. Because Kuhl failed to find evidence that rhesus monkeys distinguish prototypical from non-prototypical instances of a phonetic category, she argued that the perceptual magnet effect represents a uniquely human mechanism, specialized for processing speech. Moreover, because prototypes are formed on the basis of experience with the language environment, Kuhl (2000) further argued that each linguistic community will have prototypical exemplars tuned to the particular morphology of their natural language. We consider this work to be a an elegant example of the comparative method, especially with respect to testing animals before claiming a uniquely human speech processing mechanism.

To further investigate the comparative claim, Kluender and colleagues (1998) attempted a replication of Kuhl's original findings, using European starlings and the stimuli used in Kuhl's original work: the English vowels /i/ and /I/, as well as the Swedish vowels /y/ and /ʉ/. These vowels have distinctive prototypes that are, acoustically, non-overlapping. Once starlings were trained to respond to exemplars from these vowel categories, they readily generalized to novel exemplars. More importantly, the extent to which they classified a novel exemplar as a member of one vowel category or another was almost completely predicted by the prototypical acoustic signatures of each vowel, as well as by the exemplar's distance from the prototype or centroid of the vowel sound. Because the starlings' responses were graded, and matched human adult listeners' ratings of *goodness* for a particular vowel class, Kluender and colleagues concluded, *contra* Kuhl, that the perceptual magnet effect is not uniquely human, and can be better explained by general auditory mechanisms.

In contrast to the extensive comparative work on categorical perception, we have only two studies of the perceptual magnet effect in animals. One study of macaques claims that animals lack such capacities, whereas a second study of starlings claims that animals have such capacities. If starlings perceive vowel prototypes but macaques do not, then this provides evidence of an analogy or homoplasy. Future work on this problem must focus on whether the failure with macaques is due to methodological issues (e.g. differences in exposure to speech prior to training) or to differences in sensory-motor capacities that are indirectly (e.g. starlings are vocal mimics whereas macaques show no such evidence) or directly linked to recognizing prototypical vowels. If macaques lack this capacity while starlings have it, then our evolutionary account must reject the claim concerning uniqueness, but attempt to explain why the capacity evolved at least twice, once in the group leading to songbirds and once in the group leading to modern humans; again, we must leave open the possibility of a difference in the actual neurobiological mechanisms underlying the perceptual magnet effect in starlings and humans.

*Spontaneously Available Mechanisms*
*for Speech Perception in Animals*

To date, when a claim has been made that a particular mechanism X is special to speech, animal studies have generally shown that the claim is false. Speech scientists might argue, however, that these studies are based on extensive training regimes, and thus fail to show what animals spontaneously perceive or, more appropriately, *how* they actually perceive the stimuli. They might also argue that the range of phenomena explored is narrow, and thus fails to capture the essential design features of spoken language (Trout 2000). In parallel with work on other cognitive abilities (e.g. number, tool use, food: Hauser 1997; Hauser et al. 2000; Santos et al. 2002), we have been pushing the development of methodological tools that involve no training and can be used with animals or human infants, thereby providing a more direct route to understanding which mechanisms are spontaneously available to animals for processing speech, and which are uniquely human. Next, we describe several recent experiments designed to explore which of the many mechanisms employed by human infants and children during the acquisition of spoken language are spontaneously available to other animals.

A powerful technique for exploring spontaneous perceptual distinctions is the habituation/dishabituation procedure. Given the variety of conditions in which our animals live, each situation demands a slightly different use of this technique. The logic underlying our use of the procedure for exploring the mechanisms of speech perception is, however, the same. In general, we start by habituating a subject to different exemplars from within an acoustic class. A response is scored if the subject turns and orients in the direction of the speaker. Once habituated, as evidenced by a failure to orient, we present test trials consisting of exemplars that deviate in some specified way from the training set. A response to the test stimuli constitutes evidence for perceptual discrimination, while the failure to respond (i.e. transfer of habituation) constitutes evidence for perceptual clustering or grouping across habituation and test stimuli.

*The role of rhythm in discriminating human languages*

The first comparative habituation/dishabituation experiment on speech perception (Ramus et al. 2000) explored whether the capacity of human infants both to discriminate between, and subsequently acquire two natural languages is based on a mechanism that is uniquely human or shared with other species. Though animals clearly lack the capacity to produce most of the sounds of our natural languages (see previous section, 'Speech production'), and are never faced with the natural problem of discriminating different human languages, their hearing system is such (at least for most primates: Stebbins 1983) that they may be able to hear some of the critical acoustic features that distinguish one language from another. To explore this problem, we asked whether French-born human neonates and cotton-top tamarin monkeys can discriminate sentences of Dutch from sentences of Japanese, and whether the capacity to discriminate these two languages depends on whether they are played in a forward (i.e. normal) or backwards direction; given the fact that adult humans process backwards speech quite differently from forward speech, we expected to find some differences, though not necessarily in both species. For neonates we used a non-nutritive sucking response, whereas for tamarins we used a head orienting response.

Neonates failed to discriminate the two languages played forward.[1] Rather than run the backwards condition with natural speech, we decided to synthesize the sentences and run the experiment again, with new subjects. One explanation for the failure with natural speech was that discrimination was impaired by the significant acoustic variability imposed by the different speakers. Consequently, synthetic speech provides a tool for looking at language discrimination, while eliminating speaker variability. When

---

[1] Strictly speaking, when subjects fail to dishabituate in the test trial following habituation, one cannot conclude that subjects have failed to discriminate. Specifically, and in contrast to psychophysical experiments that uncover just noticeable differences (JNDs) the habituation/dishabituation technique only reveals meaningful or salient differences (JMDs); even though two stimuli may not be considered meaningfully different, they may nonetheless be discriminable under different testing conditions.

synthetic speech was used, neonates showed discrimination of the two languages, but only if the sentences were played in the normal, forward direction. In contrast to the neonates, tamarins showed discrimination of the two languages played in a forward direction, for both natural and synthetic sentences. Like neonates, they also failed to discriminate Dutch from Japanese when the sentences were played backwards. More recent work (Tincoff et al. in prep) shows that tamarins can discriminate two other languages differing in rhythmic class (Polish and Japanese), but not two languages from the same rhythmic class (English and Dutch).

These results allow us to make five points with respect to studying the 'speech is special' problem. First, the same method can be used with human infants and non-human animals. Specifically, the habituation/dishabituation paradigm provides a powerful tool to explore similarities and differences in perceptual mechanisms, and avoids the potential interpretive problems associated with training. Second, animals such as cotton-top tamarins not only attend to isolated syllables as previously demonstrated in studies of categorical perception, but also attend to strings of continuous speech. Third, given the fact that tamarins discriminate sentences of Dutch from sentences of Japanese in the face of speaker variability, they are clearly able to extract acoustic equivalence classes, a capacity that comes online a few months after birth in humans (Jusczyk 1997; Oller 2000). Fourth, because tamarins fail to discriminate sentences of Dutch from sentences of Japanese when played backwards, their capacity to discriminate such sentences when played forward shows that they must be using specific properties of speech as opposed to low-level cues; the capacity to discriminate languages falling between rhythmic classes, but not within, adds support to this claim. Fifth, because the tamarins' capacity to discriminate Dutch from Japanese was weaker with synthetic speech, it is possible that newborns and tamarins are responding to somewhat different acoustic cues during this task. In particular, newborns may be more sensitive to prosodic differences (e.g. rhythm), while tamarins may be more sensitive to phonetic contrasts. Future research will explore this possibility.

*Speech segmentation and the implementation*
*of statistical learning mechanisms*

A real-world problem facing the human infant is how to segment the continuous acoustic stream of speech into functional units, such as words and phrases. How, more specifically, does the infant know where one word ends and another begins? Since periods of silence occur within and between words, and since stress patterns might only help with nouns (*Look at the ball!*'), what cues are available to the child?

A recent attempt to tackle this problem builds on early intuitions from computational linguistics, and in particular the possibility that infants extract words from the acoustic stream by paying attention to the statistical properties of a given language (Harris 1955). For example, when we hear the consonant string *st* there are many phonemes that we might expect to follow (e.g. *ork, ing*), but some that we explicitly would not expect (e.g. *kro, gni*). Saffran et al. (1996) tested the hypothesis that infants are equipped with mechanisms that enable them to extract such statistical regularities from a particular language. Eight-month old infants were familiarized for two minutes with a continuous string of synthetically created syllables (e.g. *tibu-dopabikudaropigolatupabiku* ...), with no pauses between syllables. Within this continuous acoustic stream, some three-syllable sequences always clustered together, whereas other syllable pairs occurred only occasionally. To determine whether infants would extract such statistics, they were presented with three types of test items following familiarization: *words* consisting of syllables with a transitional probability of 1.0, *part-words* where the first two syllables had a transitional probability of 1.0 while the third syllable had a transitional probability of 0.33, and *non-words* where the three syllables were never associated (transitional probability of 0.0) in the familiarization corpus. Based on dozens of comparable studies on human infants, Saffran et al. predicted that if the infants have computed the appropriate statistics, and extracted the functional words from this artificial language, then they should show little to no orienting response to familiar words, but should show interest and an orienting response to both the part-words and the non-words. Results provided strong support for this hypothesis. They further show that infants are equipped with the capacity to compute conditional statistics. And it is precisely these kinds of computation, together with others, that might help put the child on the path to acquiring a language. Is the capacity to compute such statistics uniquely human and, equally important, special to language?

Saffran and collegues have excluded the 'special to language' hypothesis by showing that, at least for transitional probabilities, the same kinds of result hold for melodies, patterns of light, and motor routines (Hunt and Aslin 1998; Saffran et al. 1999). A different approach comes from testing non-human animals.

Several studies of pigeons, capuchin monkeys, and rhesus monkeys demonstrate that, under operant testing conditions, individuals can learn to respond to the serial order of a set of approximately eight to ten visual or auditory items (Orlov et al. 2000; Terrace et al. 1995; Wright and Rivera 1997). These results show that at least some animals, and especially some primates, have the capacity to attend to strings of items, extract the relevant order or relationship between items, and use their memory of prior responses to guide future responses. In addition to these data, observations and experiments on foraging behaviour and vocal communication suggest that non-human animals also engage in statistical computations. For example, results from optimal foraging experiments indicate that animals calculate rates of return, sometimes using Bayesian statistics, and some animals produce strings of vocalizations such that the function of the signal is determined by the order of elements (Hailman and Ficken 1987; Zuberbühler 2002). Recently, studies by Savage-Rumbaugh and colleagues (1993) suggest that at least some human-reared bonobos have some comprehension of speech and, specifically, attend to the order in which words are put together in a spoken utterance. Together, these studies suggest that, like human adults and infants, non-human animals are equipped with statistical learning mechanisms.

Hauser et al. (2001) used the original Saffran et al. (1996) material in order to attempt a replication with cotton-top tamarins of the statistical learning effects observed with human infants. The procedure was the same as that used with human infants, with two exceptions. Unlike human infants, who were exposed to the familiarization material for two minutes and then presented with the test items (in association with a flashing light), we exposed the tamarins in their home room to twenty-one minutes of the familiarization material on day 1 and then, on day 2, presented individuals located in a soundproof chamber with one minute of the familiarization material followed by a randomly presented set of test items.

Like human infants, tamarins oriented to playbacks of non-words and part-words more often than to words. This result is powerful, not only because tamarins show the same kind of response as human infants, but because the methods and stimuli were largely the same, and involved no training.

In terms of comparative inferences, our results on statistical learning should be treated somewhat cautiously because of subtle differences in methods between species, the lack of information on where in the brain such statistics are being computed, and the degree to which such computations can operate over any kind of input (i.e. visual, motoric, melodic). Methodologically, the tamarins received more experience of the familiarization material than did the infants. We provided the tamarins with more input because we were unsure at the time that they would even listen to such synthetic speech, much less orient to it. Nonetheless, future work must establish how much experience is necessary in order to derive the appropriate statistics, and how the properties of certain statistics are either learnable or unlearnable by both humans and non-humans. For example, recent work (Newport et al. in preparation) suggests that both human adults and adult tamarins can learn about non-adjacent statistical relationships, but that the relevant perceptual units may differ between species; during these tasks, humans apparently extract units at the level of the phonemic tier (consonants and vowels), while tamarins extract at both the syllabic and phonemic tier, with the latter restricted to vowels as opposed to consonants. It is now important to ascertain whether human infants are more like tamarins or human adults, and the extent to which different kinds of statistical computation may or may not play a significant role in language acquisition. It is, of course, also important to ascertain which of these computational abilities are uniquely human and uniquely evolved for the purpose of language processing as opposed to other cognitive problems.

## The Future of Comparative Studies

We have argued that a crucial component for discovering how the subsystems underlying speech production and perception evolved is to explore whether such mechanisms operate in other species. Our results show that many of the subsystems that mediate speech production and perception are present either in our closest living relatives or in other, more distantly related species; the work on speech perception also integrates nicely with work on computational issues, including statistical mechanisms for extracting the relationships between abstract variables in a sequence (Hauser et al. 2001; Hauser et al. 2002). As a result, we argue, such mechanisms did not evolve for speech production or perception, but for other communicative or cognitive functions. We conclude here with a few comments about the connection between the neurosciences and behavioural studies of speech and language.

Are our verbal abilities unique or not? If we had to place a wager, we would bet that humans share with other animals the core mechanisms for speech perception. More precisely, we inherited from animals a suite of perceptual mechanisms for listening to speech—ones that are quite general, and did not evolve for speech. Whether the similarities across species represent cases of homology or convergence (homoplasy) cannot be answered at present and will require additional neuroanatomical work, tracing circuitry and establishing functional connectivity. What is perhaps uniquely human, however, is our capacity to take the units that constitute spoken and signed language, and recombine them into an infinite variety of meaningful expression (Hauser et al. 2002). Although many questions remain, we suspect that animals will lack the capacity for recursion, and their capacity for statistical inference will be restricted to items that are in close temporal proximity. With the ability to test animals and human infants with the same tasks, with the same material, we will soon be in a strong position to pinpoint when, during evolution and ontogeny, we acquired our specially designed system for spoken language.

One direction that is likely to be extremely productive, in terms both of our basic understanding of how human infants acquire a language and of how the brain's representational structure changes over time, is to use nonhuman animals as models for exploring the specific effects of experience on acoustic processing. A major revolution within the neurosciences over the last ten or so years has been the discovery of remarkable plasticity in the adult brain, influenced by experience (Recanzone 2000). This revolution actually started earlier, driven in part by the magnificent findings on some songbird species and their capacity to learn new songs each season (reviewed in Nottebohm 1999). More recent work on mammals (rats and primates) has shown that when an individual engages in repetitive motor routines, or is repeatedly presented with sounds falling within a particular frequency range, the relevant cortical representations are dramatically altered. Similar kinds of effect have been suggested for language acquisition in human infants (Kuhl 2000), as well as for patients suffering from phantom limb (Ramachandran and Blakeslee 1998).

This evidence for cortical plasticity suggests experiments providing animal subjects with specific 'linguistic' experience and then testing for reorganization of perceptual sensitivity. For example, consider the results on tamarins showing a capacity to distinguish two different languages from two different rhythmic groups (i.e. Dutch and Japanese). Studies of human infants suggest that whereas natives of one rhythmic group (e.g. French) can discriminate sentences of their own language from sentences of another language within the same rhythmic group (e.g. Spanish), infants exposed to a language that falls outside this rhythmic group can not discriminate French from Spanish. To test whether this rapidly developing selectivity follows from general auditory principles or from a specialized speech mechanism that is uniquely human, we can passively expose animals to one language over a period of weeks or months, and then explore whether such experience influences their capacity to discriminate this 'native' language with other languages, or the capacity to make fine-grained discriminations within the exposed language. Similarly, it is possible to selectively expose captive primate infants at different stages of development, and thereby determine whether there are critical periods for responding to such exposure. These results can then form the basis for further studies exploring the neurophysiology underlying behavioural or perceptual changes.

It is apparent to us (Hauser et al. 2002), and many other scientists (Nowak et al. 2002), that the comparative approach will be a critical branch of empirical research into the nature of the human language faculty. At a minimum, comparative research will play the necessary if somewhat negative role of determining, by process of elimination, which components of language are *not* uniquely human or specific to language. More positively, we can expect that the comparative study of brain function, evolution, and development will provide the basis for a future theory of the neural implementation of the language faculty. Such research will combine with detailed behavioural study of animal capabilities to provide insights into the neural and behavioural mechanisms that were present at the evolutionary divergence between chimps and humans, which the evolving language faculty incorporated and elaborated. We foresee an iterative process in which studies on animals help to fractionate the language faculty naturally, 'cleaving nature at its joints', thus providing insight into how brains produce and process sounds, into how genes build brains, and eventually into the specific genetic changes that were necessary for the evolution of the language faculty.

## FURTHER READING

Classics in the evolution of speech that provide a starting point for all further discussion include Lenneberg (1967) and Lieberman (1975; 1984). Lieberman's early work on non-human primate vocal production also provides an early example of the value of comparative work in understanding the evolution of speech—work that was far ahead of its time (Lieberman 1968; Lieberman et al. 1969). A rarely quoted gem is Nottebohm (1976). For a more broad-ranging and revealing comparative analysis of the parallels between birdsong and speech, a nice introductory article is by Doupe and Kuhl (1999).

At a higher level, important ongoing work in the evolution of phonology is provided by MacNeilage's work (1998), which provides a Darwinian framework for understanding basic phonological distinctions (e.g. consonant and vowel, place of articulation) based upon the basic function and motor control of the jaw. This work provides a nice bridge between the low-level aspects of speech considered in this chapter and more theoretical issues at the heart of linguistics.

# 10

# The Evolving Mirror System:
# A Neural Basis for Language Readiness

*Michael A. Arbib*

## Language Readiness

When we say 'Humans have families' we refer to a basic biological inherit-ance of the human species, albeit one whose form varies greatly from soci-ety to society. When we say 'Humans have cities' or 'Humans have writing' we refer to human cultural achievements with a history of at most 10,000 years. What, then, are we to make of the claim that 'Humans have language'? Certainly, the human brain and body evolved in such a way that we have hands, larynx, and facial mobility suited for generating gestures that can be used in language, and the brain mechanisms needed to rapidly perceive and generate sequences of such gestures. However, I seriously question Chom-sky's hypothesis that our genetic constitution further includes the basic ground plan of grammar. The immense diversity of the Indo-European lan-guages, encompassing languages as different as Hindi, German, Italian, and English, took about 6,000 years. How can we imagine what has changed in 'deep time' since the emergence of *Homo sapiens* some 200,000 years ago, or in 5,000,000 years of prior hominid evolution? At the minimum, we must treat with suspicion any claim that the commonality of present-day human languages implies that the first *Homo sapiens* had communication systems of similar complexity. I shall instead argue that the human brain and body are 'language-ready' in the sense that the first *Homo sapiens* used a form of vocal communication which was but a pale approximation of the richness of language as we know it today, and that language evolved *culturally* as a more or less cumulative set of 'inventions' that exploited the pre-adaptation of a brain that was 'language ready' but did not genetically encode general properties of, for example, grammar.

Bickerton (1995) defines a *protolanguage* as any system—like infant language, pidgins, and the 'language' taught to apes—made up of utterances comprising a few words (in the current sense) without syntactic structure.[1] To keep the argument clear, we introduce the term *prelanguage* for any system of utterances used by a particular hominid or prehominid community which was a precursor to human language in the modern sense. Bickerton assumes that the prelanguage used by *Homo erectus* was a protolanguage in his sense. Language just 'added syntax' through the evolution of Universal Grammar. I offer the counter-view that the prelanguage of early *Homo sapiens* was composed of 'unitary utterances' naming events as well as a few salient actors, objects and actions, and that this preceded the discovery of *words* in the modern sense of units for compositional formation of utterances. On this view, words in the modern sense co-evolved with syntax through *fractionation*, a process of discovery and diffusion quite distinct from the formation of a genetic module for grammar. Unfortunately, we have no traces of any hominid prelanguage, but I hope to show that the views of Bickerton and Chomsky should not be accepted unquestioningly.

## Criteria for Language Readiness and Language

I next list properties of 'language readiness' and then list properties that set language apart from 'simpler' forms of communication.

### Language Readiness

Under language readiness, we have the properties:

### Symbolization

The ability to associate an arbitrary symbol with a class of episodes, objects or actions. This might also be called 'naming', but I want here to stress especially that these 'symbols' need not, in early *Homo sapiens* and other hominids, have been words in the modern sense. My hypothesis is that the early forms of 'prelanguage' generated arbitrary novel utterances to convey complex but frequently important situations, and it was a major later discovery

---

[1] By contrast, Dixon (1997) (as do other writers on historical changes in languages) views a *protolanguage* as a human language ancestral to a specific family of human languages.

en route to language as we now understand it that one could gain expressive power by fractionating such utterances into shorter utterances conveying components of the scene or command.

### Intentionality

The communicative gestures are intended. This is hard to characterize, but the contrast here is with 'communicative gestures' that are involuntary: seeing a leopard, the vervet monkey has no choice but to utter the 'leopard call' and the monkey who hears this call has no choice but to execute the biologically mandated 'leopard escape response' such as running up a tree.

### Parity (the 'mirror property')

What counts for the speaker must count in much the same way for the listener. Indeed, the search for brain mechanisms which can account for this property will provide a major impetus for the approach to language evolution given here, in which (following Rizzolatti and Arbib 1998 and their precursors) I will argue that the homology (via a complex process of evolutionary change) between the 'mirror system for grasping'—which is active both when the monkey grasps something with his own hand and when he observes another monkey or human carry out a similar action—and Broca's area in the human, a major component of the human language system, provides the key to parity in language.

### Temporal ordering

The temporal order of gestures encodes 'hierarchical structures of the mind'.

### Beyond the here-and-now 1

Utterances are not just related to currently perceptible events, actions, or objects. The utility of prelanguage may well have been related to the ability to imagine future events and to make (and understand) utterances to announce these events, or to issue commands that might bring about these events. In any case, it seems clear that the human brain provides the basic cognitive capacity to recall past episodes and to imagine possible future outcomes of our actions.

### Paedomorphy and Sociality

The social structure of the hominids so evolved that infants would remain helpless longer than in other species (this phenomenon is called paedomorphy), and that adults would provide the care and instruction to enable infants to adapt to diverse environments and social structures.

### Language

With this we can move on to the list of features that may properly be said to characterize languages rather than language readiness:

### Symbolization 2

Whereas utterances in prelanguage may simply be strings of gestures corresponding to unanalysed wholes, the symbols become words in the modern sense, interchangeable and composable in the expression of meaning.

I have already suggested that it was a major *discovery* en route to language as we now understand it that one could gain expressive power by *fractionating* the holophrastic utterances of prelanguage into shorter utterances conveying components of the scene or command. Indeed, I suggest that many ways of combining these shorter utterances—grammatical structures like adjectives, conjunctions such as *but*, *and*, or *or* and *that*, *unless*, or *because*—were the discovery of *Homo sapiens*, i.e. these might well have been 'postbiological' in their origin. Two examples:

(a) If one starts with holophrastic utterances then statements like *baby want milk*, *brother change baby* and so on, each must be important enough or occur often enough for the tribe to agree on a symbol (e.g. arbitrary string of phonemes) for each one. Discovering that separate names could be assigned to each actor, object and action, would require five words instead of two to recode the above two utterances. However, once the number of utterances with overlap reaches a critical level, economies of word learning would accrue from building utterances from 'reusable' components. Separating verbs from nouns lets one learn $m+n+p$ words (or less if the same noun can fill two roles) to be able to form $m*n*p$ of the most basic utterances. Of course, not all of these combinations will be useful, but the advantage is that new utterances can now be coined 'on the fly' rather than each acquiring group mastery of a novel utterance.

(b) Similarly, discovery of the one word *ripe* halves the number of fruit names to be learned—one no longer needs separate words for 'apple to be left on tree' and 'apple ready to be eaten', and so on. Note, though, that a particular society might slowly accumulate such 'inventions' over many generations before some genius realized the common semantic pattern and devised a single syntactic device—such as NP → Adj N—to accommodate it.

Put differently, the utterances of prelanguage were similar to the 'calls' of modern primates—such as the 'leopard call' of the vervet monkey—in the sense that the community could only communicate about situations that occurred so frequently that a specific utterance was available to 'name' it. However, they *differed radically* from the primate calls—and thus exhibit a dimension of language readiness absent in non-human primates—in that new utterances could be invented and acquired through learning within a community, rather than emerging through many generations of biological evolution. The use of more finely differentiated words that can be combined to express novel meanings sets us on the path to genuine language, as we shall now see.

### Syntax and semantics

The matching of syntactic to semantic structures co-evolves with the fractionation of utterances. This enables the meaning of complex utterances to be at least approximately inferred using a compositional semantics.

### Recursive structuring

The utterances of language are recursively structured. Specifically, language use requires the production and recognition of utterances as composed of components with specific relations between them, with these components potentially themselves composites. (I will use 'components' for the verbal expression of semantic subunits of the utterance; this leaves open whether or not, in a particular language, their expression takes the form of 'constituents' in the sense employed by syntacticians.)

### Beyond the here-and-now 2

Utterances can be marked (e.g. by the use of verb tenses) as referring to the past, present or future, and can also express negations or counterfactuals. This clearly relates to the neural machinery (language readiness) to recall past events or imagine future ones.

### Learnability

To qualify as a human language, a set of symbolic structures must be learnable by most human children. This is not as uncontroversial as it seems, for it suggests that language is (at least in part) something to be learned rather than something encoded in the genome. However, even if one accepts the notion of an innate Universal Grammar, one must still require mechanisms which enable the child's brain to 'set parameters' and acquire the lexicon of a particular language, and we may call these learning mechanisms for the purpose of the present discussion. I add that, whereas a child of 5 may have acquired much of the basic syntax of its language, it will in general be another ten years or more before the young adult's use of language deploys anything like a typical adult lexicon or range of syntactic subtlety.

## The Human Language System

It is beyond the scope of this chapter to characterize 'the human language system', i.e. those parts of the human brain that are most directly involved in language for a human who has grown up within a particular language community. Here I simply want to introduce two key components of that system, Broca's area and Wernicke's area, that were first studied in the latter half of the nineteenth century (see Benson and Ardila 1996; Arbib et al. 1982). Broca (1861) described a patient with an anterior lesion of the brain (including what is now called Broca's area). The deficit seemed essentially motoric in that the patient was able to comprehend language, but could speak only with effort and then only 'telegramatically' in short utterances, omitting most of the grammatical markers. The patient had non-fluent speech with many pauses; words came with great effort and the melodic intonation was flat and monotone in pitch. However, comprehension of speech was relatively intact. By contrast, Wernicke (1874) described a patient with a posterior lesion (in the so-called Wernicke's area) whose deficit was deemed essentially sensory, in that the patient seemed not to comprehend speech (but was not deaf to other auditory stimuli) and would speak a fluent but meaningless stream of syllables devoid of any content but with free use of verb tenses, clauses, and subordinates. Sometimes even single words were not comprehended. The specific observations Broca and Wernicke made of their patients were later generalized to define the syndromes of Broca's aphasia and Wernicke's aphasia, respectively, but the localization of brain lesions which yield given aphasic symptoms is highly variable.

## The Mirror System Hypothesis

The time has come to introduce the 'mirror system' for grasping in the monkey mirror brain, for this system plays a key role in the approach to the evolution of language readiness presented in these pages. But first we need a quick appreciation of the brain mechanisms which enable a monkey (and a human) to look at an object and determine how to grasp it.

### The Mirror System for Grasping in the Monkey Brain

Two areas in the monkey brain ground our study of brain mechanisms for the visual control of grasping. F5 is a part of premotor cortex (the fifth area of frontal cortex of monkey in an arbitrary numbering system) which elaborates motor grasp (the relevant data are from the group of Giacomo Rizzolatti); while AIP, the anterior intraparietal area (an area in parietal cortex), computes grasp affordances, i.e. processes visual input from an object to extract data on how to grasp it, rather than on the identity of the object (the relevant data are from the group of Hideo Sakata). At present, the fullest interpretation of these and related findings is given by the FARS (Fagg–Arbib–Rizzolatti–Sakata) model (Fagg and Arbib 1998). AIP extracts the set of affordances for an attended object. These affordances highlight the features of the object relevant to physical interaction with it. It is especially worth emphasising that although the FARS model hypothesizes that it is the path from visual cortex to AIP and thence to F5 that provides these visually based affordances, the actual selection by F5 of which affordance to act upon depends crucially on input from prefrontal cortex (PFC), and that this PFC input in turn may depend heavily upon processes in inferotemporal cortex (IT) which mediate recognition or classification of the object.

Rizzolatti et al. (1995) found a subset of the grasp-related premotor neurones of F5 which discharge not only, as other F5 neurones do, when the monkey executes a certain class of actions but also when the monkey observes more or less similar meaningful hand movements made by the experimenter (or another monkey). They call these 'mirror neurones', and there was always a link between the effective observed movement and the effective executed movement. Thus F5 in the monkey is endowed with an 'observation/execution matching system', and we refer to this system in the monkey brain as the *mirror system for grasping*. We refer to those F5 neurones which are active only during the monkey's own movements as *canonical* F5 neurones.[2]

The monkey mirror system, as observed neurophysiologically by Rizzolatti et al. is concerned with observation of a single action that is already in the monkey's repertoire. Possible roles for such a mirror system may include:

(1) Self-correction: based on the discrepancy between intended and observed self action.
(2) Learning by imitation at the level of a single action
(3) Social interaction. By anticipating what action another monkey has begun, a monkey can determine how best to compete or cooperate with the other monkey.

To date, analysis of monkey behaviour seems consistent with roles (1) and (3), but monkeys seem poor at imitation in any extended sense. In fact, a major part of our hypothesis on the evolution of language readiness is that evolution of the hominid brain equipped it to support richer and richer forms of imitation.

Where our understanding of the mirror system for grasping in the monkey brain rests on neurophysiological recording of the activity of single cells in selected brain regions, the discovery of a mirror system for grasping in the human brain rests on techniques for measuring rCBF (regional cerebral blood flow) to get a measure of how activity in various brain regions differs from task to task. The two main methods are PET (positron emission tomography) and fMRI (functional magnetic resonance imaging). Rizzolatti et al. (1996) used PET imaging to show that Broca's region is activated by observation of hand gestures. They compared three experimental conditions: object observation (control condition); grasping observation (the subject observed someone else grasp an object); and object prehension (the subject grasped the object). The most striking result was highly significant activation in the rostral part of Broca's area for both the execution and observation of grasping. Thus, a key area in the human language system is a possible mirror system for grasping! Other PET data (Bonda et al. 1994) showed that

---

[2] Most current papers seem to view each mirror neurone as coding a specific (more or less broadly tuned) class of actions. However, the available data seem more consistent with the view that mirror neurones encode "components" of the actions of interest. I suggest that, rather than seek "the" neurone for each action, we should look at sets of neurones during related movements to try to find the differences and similarities to extract the ensemble code. See 'Discussion' below for the adaptability of the mirror neurone repertoire.

there was a highly significant activation of Broca's area during execution of a sequence of self-ordered hand movements. Moreover, Massimo Matelli (as reported by Rizzolatti and Arbib 1998) provides an anatomical argument that F5 in the monkey is homologous to (i.e. shares an evolutionary history as a distinguishable brain region with) Area 45 in the human (Broca's area = Areas 44 + 45).

### The Mirror System Hypothesis

Rizzolatti et al. (1995) thus hypothesized that the functional specialization of human Broca's area derives from an ancient mechanism related to the production and understanding of motor acts. The 'generativity' which some see as the hallmark of language is present in manual behaviour which can thus supply the evolutionary substrate for its appearance in language. As we shall elaborate later, this lies at the root of our explanation for the evolutionary prevalence of the lateral motor system over the medial (emotion-related) primate call system in becoming the main communication channel in humans. Rizzolatti and Arbib (1998) further developed these ideas in their paper 'Language Within Our Grasp'. Their theory is within the tradition that roots speech in a prior system for communication based on manual gesture—and views the present human capacity for language as based not on speech alone but rather on communication integrating vocal, facial and manual gestures. What it adds to the manual origins hypothesis is that the mirror system provides a possible neural 'missing link' in the evolution of human language readiness:

*The Mirror System Hypothesis (MSH)*: Broca's area in humans evolved from a basic mechanism *not* originally related to communication—the *mirror system for grasping in the common ancestor of monkey and human*. The mirror system's capacity to generate *and* recognize a set of actions provides the evolutionary basis for *language parity*, in which an utterance means roughly the same for both speaker and hearer.

It is important to also note what MSH does *not* say:

(1)  It does not say that having a mirror system is equivalent to having language. Monkeys have mirror systems but do not have language, and we expect that many species have mirror systems for varied socially relevant behaviours. As noted, the monkey mirror system may have adaptive value in supporting both self-correction of movements and forms

of social interaction based on recognition of the actions of others. Hurford (2002) reviews a wide range of animal behaviours—schooling fish, flocking birds, escape behaviour, the way in which cuttlefish change colour to match the background, human yawning and laughter, and more—that 'probably involve arrangements more or less like mirror neurones'. But whether or not mirror systems are widespread is irrelevant to assessment of MSH. What MSH does say is that a *specific* mirror system, the mirror system for grasping, is the shared heritage of man and monkey, and that in humans it evolved to provide core components for the language system.

(2) It does not say that the ability to match the perception and production of *single* gestures is sufficient for language.

(3) It does not say that language evolution can be studied in isolation from cognitive evolution more generally. In using language, we may use, for example, negation, counterfactuals, and verb tenses or other devices to indicate the past and future, but these linguistic structures are of no value unless we can understand that the facts contradict an utterance, or can recall past events and imagine future possibilities.

## Beyond the Mirror: Imitation is the Key

The monkey mirror system, as observed neurophysiologically by Rizzolatti et al. is concerned with observation of a single action that is already in the monkey's repertoire. 'Beyond the mirror' (as so conceived) lies the imitation of complex behaviours, 'parsing' them into variations of familiar elements and then being able to repeat the observed structure composed from those elements. In this section (based on Arbib 2003), I expand MSH by adding two imitation stages to those discussed by Rizzolatti and Arbib (1998) to define seven stages[3] of evolution, from manual grasping through imitation to language:

(1) grasping;

(2) a mirror system for grasping (i.e. a system that matches observation and execution);

---

[3] I do not mean to suggest a sharp transition from one stage to the next; rather, I suggest a process of cumulative changes whereby creatures with many of the skills of Stages 1 to n+1 gradually emerged from creatures who possessed only the skills of Stages 1 to n.

(3) a simple imitation system for grasping;

(4) a complex imitation system for grasping;

(5) 'protosign': a manual-based communication system, breaking through the fixed repertoire of primate vocalizations to yield a combinatorially open repertoire;

(6) 'protospeech': a vocal-based communication system which breaks through the closed nature of primate vocalizations as a result of the 'invasion' of the vocal apparatus by collaterals from the communication system based on F5/Broca's area.

(7) language: the change from action–object frames to verb–argument structures to syntax and semantics: co-evolution of cognitive and linguistic complexity.

This list expands the four stages given by Rizzolatti and Arbib (1998): (1) grasping; (2) a mirror system for grasping; (3) a manual-based communication system; and (4) speech as a result of the 'invasion' of the vocal apparatus by collaterals from the manual communication system. Hurford (2002) is thus mistaken when he asserts that Rizzolatti and Arbib claim that the mirror neurone system constitutes the final key evolutionary development leading to human language.

We have already given a full account of stages 1 and 2, and now move on to the later stages. Note the careful use of the term 'protospeech' in Stage 6, rather than the term 'speech'. This is because 'speech' may be taken to mean 'spoken *language*', and I want to emphasize the possibility of an evolving capacity for *protospeech* as the open-ended production and perception of sequences of vocal gestures, without these sequences *at first* constituting utterances of a language.

### Imitation

Imitation involves, in part, seeing another's performance as a set of familiar movements. But to imitate successfully (consider imitating a dance or a song), one must not only observe *actions* and their *composition*, but also novelties in the constituents and their *variations*. One must also perceive the overlapping and sequencing of all these basic actions and then remember the 'coordinated control program' (Arbib 1981) so constructed. Each successive approximation to the overall performance provides the framework in which attention can be shifted to specific components which can then be *tuned* and/or *fractionated* appropriately, or better *coordinated* with other components of the skill. This process is *iterative*, yielding both the mastery of ever finer details and increasing grace and accuracy of the overall performance.

It is widely held that the capacity for imitation is limited in monkeys, though much further study is needed to settle the matter. In particular, I shall assume that the monkey mirror system for grasping is part of a system for recognition of specific hand movements, not for the imitation of novel patterns of hand movements. However, chimpanzees use and make tools (Beck 1974; Goodall 1986), and the fact that different tool traditions are apparent in geographically isolated groups of chimpanzees suggests that this may involve some form of imitation. But the form of imitation reported for chimpanzees is a long and laborious process compared to the rapidity with which humans can acquire novel sequences. Myowa-Yamakoshi and Matsuzawa (1999) observed that chimpanzees typically took twelve trials to learn to 'imitate' a behaviour in laboratory studies, and in doing so paid more attention to where the manipulated object was being directed, rather than the actual movements of the demonstrator. This involves the ability to learn novel actions which may involve using one or both hands to bring two objects into relationship, or to bring an object into relationship with the body.

If we assume (i) that the common ancestor of monkeys and apes had no greater imitative ability than present-day monkeys, and (ii) that the ability for simple imitation shared by chimps and humans was also possessed by their common ancestor, but that (iii) only humans possess a talent for 'complex' imitation, then we have established a case for the following claim.

> *Stage 3*: a simple imitation system—imitation of short novel sequences of object-directed actions through repeated exposure—for grasping developed in the 15 million year evolution from the common ancestor of monkeys and apes to the common ancestor of apes and humans; and

> *Stage 4*: a complex imitation system—acquiring (longer) novel sequences of more abstract actions in a single trial—developed in the 5-million-year evolution from the common ancestor of apes and humans along the hominid line that led, in particular, to *Homo sapiens*.[4]

---

[4] See fig. 4.2 of Gamble (1994), and accompanying discussion of the molecular clock of evolution, for these timeline estimates.

What marks humans as distinct from their common ancestors with chimpanzees is that whereas the chimpanzee can imitate short novel sequences through repeated exposure, humans can acquire (longer) novel sequences *in a single trial* if the sequences are not too long and the components are relatively familiar. The very structure of these sequences can serve as the basis for imitation or for the immediate construction of an appropriate response, as well as contributing to the longer-term enrichment of experience.

We hypothesize that the F5 mirror system would at any time bridge the execution and observation of the unitary actions currently in the animal's repertoire, but that it would require a superordinate system to perceive or execute the relationships between these units that define a compound action. My view, then, is that *extension of the mirror system from a system for recognition of single actions to a system for recognition and imitation of compound actions* was one of the key innovations in the brains of hominids relevant to language readiness. With this, let us look in more detail at sequential behaviour.

### From Vocalization to Manual Gesture and Back Again: The Path to Protospeech is Indirect

As background for language readiness, we accept that biological evolution equipped the hominid brain-body complex with bipedality, manual dexterity, and a larynx well suited for vocal production

MSH stresses the ability to relate the actions of others to one's own actions. The preceding discussion stresses key abilities *beyond the mirror*: the ability rapidly to acquire a vast array of flexible strategies for pragmatic and communicative action; and the ability to generate and comprehend hierarchical structures 'on the fly'.

Let us try to match some of these ideas to what is known about hominid evolution. Unfortunately, brain tissue does not fossilize. However, endocasts made of the interior of the cranial cavity of hominid fossils, which can only give a rough impression of the overall external shape of the brain, with no information about its finer subdivisions indicate[5] that 'speech areas'—i.e. the areas whose enlargement we view, on the basis of aphasia studies, as related to speech and language—were already present in early hominids such as *H. habilis* long before the larynx reached the modern 'speech-opti-

[5] This is controversial. Wilkins and Wakefield (1995) cite evidence in support of this view, but the claim is strongly contested by certain of the commentators on their target article.

mal' configuration. This may fit in with our view that the evolution of language readiness was a matter of gestural communication as much as vocal communication. There is a debate over whether such areas were already enlarged in australopithecines, but we shall make here the hypothesis that the transition from australopithecines to early *Homo* coincided with the transition from a mirror system used only for action recognition to a human-like mirror system also used for intentional communication.

With this, we can now offer our view of the evolution from a complex imitation system for grasping along the hominid line to protosign, a manual-based communication system, breaking through the fixed repertoire of primate vocalizations to yield a combinatorially open repertoire. We thus need to understand how hand movements could evolve from (without superseding) praxis (e.g. manipulation of physical objects) to communication. Our hypothetical sequence for manual gesture starts from praxis (action directed towards a goal object) and then develops in two substages:

*Stage 5a*: The key transition to *intentional* communication is, we hypothesize, *pantomime*, in which similar actions are produced away from the goal object. Imitation is the generic attempt to reproduce movements performed by another, whether to master a skill or simply as part of a social interaction. By contrast, pantomime is performed with the *intention* of getting the observer to think of a specific action or event. It is essentially communicative in its nature. The imitator observes; the pantomimic *intends to be* observed.

*Stage 5b*: However, pantomime would at first be closely linked to pragmatic action. In the next posited stage, abstract gestures divorced from their pragmatic origins (if such existed) become available as elements for the formation of compounds which can be paired with meanings in more or less arbitrary fashion. In pantomime it might be hard to distinguish a grasping movement signifying 'grasping' from one meaning 'a [graspable] raisin', thus providing an 'incentive' for coming up with an arbitrary gesture to distinguish the two meanings.

Together these define *Stage 5*: a *distinct* manuo-brachial communication system evolved to complement the primate calls/orofacial communication system. At this stage the '(not yet) speech' area of early hominids (i.e. the area somewhat homologous to monkey F5 and human Broca's area) mediated, I hypothesize, orofacial and manuo-brachial communication but not protospeech.

But how did such a system of manual gesture lead to protospeech? For want of better data, I will assume that our common human–monkey ancestors shared with monkeys a finite set of discrete vocal calls coupled with orofacial gestures for expressing emotion and social states. Note the linkage between the two systems—communication is inherently multi-modal. Combinatorial properties for the openness of communication are virtually absent in basic primate calls and orofacial communication, though individual calls may be graded. The neural substrate for primate calls is in a region of cingulate cortex *distinct from* F5, the monkey homologue of human Broca's area. Yet for most humans we think of speech (a structure of vocalizations) as the primary form of language. Why then is F5, rather than the cingulate area already involved in monkey vocalization, homologous to the Broca's area's substrate for language? We hypothesize that this is because the evolution from gestural communication to protospeech built upon the combinatorial properties inherent in the manuo-brachial system.

*Stage 6*: The manual-orofacial symbolic system 'recruited' vocalization. Association of vocalization with manual gestures allowed them to assume a more open referential character, yielding protospeech.

It should be noted that the evolution of language readiness involved changes in body as well as brain. Humans have a different vocal apparatus from that seen in the apes. Lieberman (1991) has done valuable research on the supralaryngeal vocal tract of humans and hominids. He places particular emphasis on the descent of the larynx in the human which enhances speech production but makes swallowing liquids and solid food more risky. He thus argues that the restructuring of the human supralaryngeal vocal tract to enhance the perceptibility of speech would not have contributed to biological fitness unless speech and language were already present in the hominid species ancestral to modern *Homo sapiens*. However, Fitch and Reby (2001) showed that lowering of the larynx has occurred in other species. For example, lowering of the larynx in the red deer may have been selected to deepen the animal's roar so that the animal would seem larger than it was. Thus the lowering of the larynx in humans or pre-human hominids might have served a similar purpose—without denying that further selection could have exploited the resultant increase in degrees of freedom to increase the flexibility of speech production. Moreover, this selective advantage would hold even for a species that employed holophrastic utterances devoid of syntax. Thus, I cannot agree with Lieberman (2000) that speech

and language must have already been present in *Homo erectus* and in Neanderthals—a core of protospeech would have been sufficient to provide pressures for larynx evolution. This is why I view the evolution of protospeech
as occurring en route to *Homo sapiens* (and, quite probably, other now extinct hominids) to yield a brain that was language-ready and, in particular
speech-ready—but not necessarily possessed of language from the start.

The hypothesis is that *Homo habilis* and even more so *Homo erectus* had
a 'proto-Broca's area' based on an F5-like precursor mediating communication by manual and orofacial gesture. This made possible a process of collat-
eralization whereby this 'proto-Broca's area' gained primitive control of the
vocal machinery, thus yielding increased skill and openness in vocalization.
Larynx and brain regions could then co-evolve to yield the configuration
seen in *Homo sapiens*. However, where Rizzolatti and Arbib (1998) stressed
that this showed why speech did not evolve 'simply' by extending the classic primate vocalization system, I would now stress the *co-evolution* of the
two systems. Lesions centred in the anterior cingulate cortex and supple-
mentary motor areas of the brain can cause mutism in humans, similar to
the effects produced in muting monkey vocalizations (Benson and Ardila
1966). I hypothesize cooperative computation between cingulate cortex and
Broca's area, with cingulate cortex involved in breath groups *and emotional
shading* (and imprecations!), and Broca's area providing the motor control
for rapid production and interweaving of elements of an utterance.

With Stage 6, we have defined a highly socialized hominid with a variety
of more or less conventionalized manual, orofacial, and vocal gestures which
refer to *something else*—we have come a long way from the original mirror
system for grasping in which a hand movement signifies nothing other than
the praxic action that it is performing. Hurford has shown this distinction
much as in Fig. 10.1. However, the bottom row of the figure takes us into

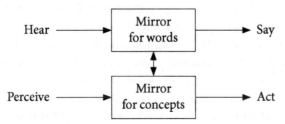

Fig. 10.1 The Saussurean sign: linking word and meaning.
(Adapted from Hurford 2002.)

territory that MSH has remained silent on. It makes the claim that there is a mirror system for concepts, which links perception and action related to each concept. However, I do not believe that there is a mirror system for all concepts, and do not regard this notion as a necessary part of MSH. Recognizing an object (an apple, say) may be linked to many different courses of action (to place the apple in one's shopping basket; to place the apple in the bowl at home; to pick up the apple; to peel the apple; to cook with the apple; to eat the apple; to discard a rotten apple, etc.). Of course, once one has decided on a particular course of action then specific perceptual and motor subschemas may be invoked. But note that, in the list just given, some items are apple-specific whereas other invoke generic schemas for reaching and grasping. It was considerations like this that led me to separate perceptual and motor schemas (Arbib 2003*b*): a given action may be invoked in a wide variety of circumstances; a given perception may, as part of a larger assemblage, precede many courses of action. Putting it another way, there is no one 'grand apple schema' which links all 'apple perception strategies' to 'every act that involves an apple'. Thus I reject the notion of a mirror system for concepts (or a view of one schema, one concept). Instead, I visualize the brain as encoding a varied network of perceptual and motor schemas. Only rarely (as in the case of certain basic actions) will the perceptual and motor schemas be integrated into a 'mirror schema'. In general, a word may be linked to many schemas, with varying context-dependent activation strengths. On this view, I do not see a 'concept' as corresponding to one word, but rather as being a graded set of activations of the schema network.

## Discussion

Many papers on the mirror system for grasping in the monkey focus on grasps that seem so basic that it is tempting to think of them as prewired. However, observation of human infants shows that little more than the sketchiest coordination of reaching with the location of a visual target plus the palmar grasp reflex is present in the early months of life, and that many months pass before the child has in its motor repertoire the basic grasps (such as the precision pinch) for which mirror neurones have been observed in the monkey. Oztop et al. (to appear) thus argue that, in monkey as well as human, the basic repertoire of grasps is attained through sensorimotor feedback. They also provide the Infant Learning to Grasp Model (ILGM) that explains this process of grasp acquisition; a complementary model, the MNS1 model of Oztop and Arbib, explains how mirror neurones may organize themselves to recognize grasps as they become added to the repertoire. Future modelling will address the issue of how the infant may eventually learn through observation, with mirror neurones and grasping circuitry developing in a synergistic manner (see Zukow-Goldring et al. to appear, for a study of infant skill acquisition and further discussion).

Rizzolatti and Arbib (1998) asserted: 'Manual gestures progressively lost their dominance, while in contrast, vocalization acquired autonomy, until the relation between gestural and vocal communication inverted and speech took off'. But speech did not 'take over' communication.

- McNeill (1992) has used videotape analysis to show the crucial use that people make of gestures synchronised with speech.
- Even blind people use manual gestures when speaking.
- Sign languages are full human languages rich in lexicon, syntax, and semantics.
- It is not only deaf people who use sign language, so do some aboriginal Australian tribes, and some native populations in North America.

All this suggests that we locate phonology in a speech–manual–orofacial gesture complex, consistent with the observation that language acquisition takes various forms: A normal person shifts the major information load of language—but by no means all of it, since hand movements and facial gesture shade and punctuate speech—into the speech domain, whereas for a deaf person the major information load is removed from speech and taken over by hand and orofacial gestures. We can thus say there are not three separate systems for manual, facial, and vocal communication, but a single system operating in multiple motor and sensory modalities.

This completes my analysis (as far as this chapter is concerned) of the seven stages from grasping via imitation to language. My current hypothesis is that Stages 4, a complex imitation system for grasping, and 5, an open manual-based communication system, and a rudimentary (pre-syntactic) form of protospeech (Stage 6) were present in pre-human hominids; but that the 'explosive' development of protospeech may have been coupled with further evolution of imitation abilities, and that what we know as language (Stage 7) depended on 'cultural evolution' well after biological evolution had formed modern *Homo sapiens*, as suggested by my discussion of fractionation changing unitary utterances to composites of words with the progress from prelanguage to language. This remains speculative, and one should note that biological evolution may have continued to reshape the genome for the human brain even after the skeletal form of *Homo sapiens* was essentially stabilized, as it certainly has done for skin pigmentation and other physical characteristics. However, the fact that people can master any language equally well, irrespective of their genetic community, shows that these changes are not causal with respect to the structure of language.

## FURTHER READING

The original statement of the mirror system hypothesis appears in Rizzolatti and Arbib (1998) and is most fully developed to date in Arbib (2003*a*).

Since the present chapter is more related to neuroscience than others in this book, readers may wish to turn to Benson and Ardila (1996) for a thorough example of one way to view the brain's language mechanisms through the spectrum of language disorders. Lieberman (2000) provides another view of human brain mechanisms for language with excellent reviews of material on speech production and subcortical mechanisms. The stress is on evolution, but there is little attempt to delineate the specific steps, biological or cultural, that lead to language. Readers wanting a more general perspective on the structure and function of the brain may turn to Arbib et al. (1997) for a comprehensive overview of the brain from the perspectives of neuroanatomy, dynamical systems, and cognitive and computational neuroscience. Arbib (2003*a*), on the other hand, provides a comprehensive guide both to realistic biological models of the brain and to connectionist models of linguistic and psychological performance. It has 285 original articles, as well as 'road maps' to help the reader navigate through twenty-two different themes—including the road map on 'Linguistics and Speech Processing', which contains articles on imaging of human brain mechanisms as well as connectionist models.

Cangelosi and Parisi (2001) may be seen as a companion to the present volume, providing a useful compendium of computational approaches to the evolution of language and communication, but with very little about the brain.

Finally, four books that provide very different framing perspectives for the present article: Dixon (1997) provides a somewhat controversial view of historical linguistics, but this is certainly a good corrective to too innatist or uniformitarian a view of human language capabilities. Gamble (1994) gives an accessible view of human evolution which emphasizes the twofold emergence of hominids and then humans out of Africa. Finally, McNeill (1992) is a good place to start learning about the role of gesture in human cognition, while Corballis (2002) offers a view on the role of the hands in the evolution of language that complements the theory presented here.

# 11

# From Hand to Mouth:
# The Gestural Origins of Language

*Michael C. Corballis*

## Introduction

Speech is so much a part of our lives that it may seem obvious that it must have always been that way—at least as long as we have had language. Primates are noisy creatures, so it must seem equally obvious that speech, and indeed language itself, evolved from their persistent vocalizations. Yet there are good reasons to believe that animal calls have nothing to do with the evolution of speech, or even of language itself. Noam Chomsky certainly believed this to be so. In his 1966 book *Cartesian Linguistics*, he wrote:

The unboundedness of human speech, as an expression of limitless thought, is an entirely different matter [from animal communication], because of the freedom from stimulus control and the appropriateness to new situations . . . . Modern studies of animal communication so far offer no counter-evidence to the Cartesian assumption that human language is based on an entirely different principle.  (pp. 77–8)

Chomsky was vague as to how language did come about in our species, and in this chapter I want to try to persuade you that it emerged from manual gestures, rather than from primate calls.

## Precursors to Language in Primates

### Primate Vocalization

Primates certainly *can* vocalize. Gelada baboons, for example, are said to have at least twenty-two different vocal calls, which have been given labels such as *moan, grunt, yelp, snarl, scream, pant*, and so on (Dunbar and Dunbar 1975). A similar list of chimpanzee calls was compiled by Goodall

(1986), who also gave names to them, such as *pant-hoot, food grunt, wraah,* and many others. Unlike human speech, though, these calls exist as wholes and cannot be broken down into interchangeable parts, and they are not combined into sequences.

For the most part, primate calls seem to be involuntary, and even our closest relative among the primates, the chimpanzee, seems to have difficulty either producing vocalizations in the absence of a triggering situation or suppressing vocalizations induced by automatic triggers. Jane Goodall, who has spent a good part of her life observing chimpanzees in the wild at Gombe in Tanzania, records an instance of a chimpanzee who found a cache of bananas, and evidently wished to keep them for himself. He was unable to suppress the excited pant-hoot signalling the discovery of food, but attempted as best he could to muffle it by placing his hand over his mouth. She also remarks that, conversely, 'The production of sound in the *absence* of the appropriate emotional state seems to be an almost impossible task for a chimpanzee' (Goodall 1986: 125).

Since most primate calls signal warning, or relate to territory or mating, it may be that a *lack* of voluntary control is adaptive, because it makes them hard to fake (Knight 1998). Warning signals must be reliable, and not subjected to the whim of the animal who might 'cry wolf'. It is true that some birds, such as the parrot or mocking bird, are skilled imitators, but to my knowledge primates have not made a practice of vocal deception. But it is precisely because primate calls cannot be faked that they are ill-suited to exaptation for intentional communication. A much better candidate lies in the forelimbs.

### Manipulative Primates

Tens of millions of years of adaptation to life spent mostly in the trees have given primates, humans included, intentional control over the limbs, and especially the hands and arms. The hands, and the brain areas that control them, are specialized for activities such as grasping, manipulating, catching insects, and grooming. Primates are also visual creatures, unique among non-human mammals in possessing sophisticated colour vision, and it has been estimated that about half of the primate brain is involved, in one way or another, with the analysis of the visual world (Blakemore 1991).

There are also specialized brain mechanisms for mapping perceptions onto actions. Rizzolatti and his colleagues have discovered neurons in the premotor cortex of the monkey that respond when animals make particular grasping movements or gestures with their arms and hands. Some of these neurons, dubbed 'mirror neurons', also respond when the animal observes the same movement made by another. These neurons therefore perform the function of mapping gestures perceived onto gestures produced—exactly the sort of mapping one would expect if gestures are to serve to communicate. This is not to say that these monkeys possess gestural language: the mirror neurons may serve communicative functions, but there is no evidence that monkeys have anything resembling gestural language itself. Nevertheless, mirror neurons may well have formed the basis for the subsequent evolution of a gestural language (Rizzolatti and Arbib 1998).

It is becoming clear that great apes, at least, make extensive use of gestures to communicate. Tanner and Byrne (1996) counted some thirty different gestures made by lowland gorillas in the San Francisco Zoo, where the animals are enclosed in a large, naturalistic area. Similarly, Tomasello and his colleagues have extensively studied the gestures made by free-ranging captive chimpanzees. In one study, thirty gestures were selected for study; these did not exhaust the full range of gestures, but were chosen because they were readily observed and counted by human observers (Tomasello et al. 1997). Nearly all of them included reference to another individual, usually in a way that invited reciprocation, indicating that they were *dyadic*. Gestures were produced more often when the recipient was looking, indicating that the gesturers were sensitive to when others were watching them. The dyadic nature of these gestures distinguishes them from chimpanzee *vocalizations*, which are generally not directed to specific others.

One exception, perhaps, might be 'duetting', in which chimpanzees, gorillas, and baboons, among others, exchange calls with one another when they are out of visual contact. Detailed analyses of the sequences of chimpanzee calls during such vocal exchanges show, however, that they have none of the properties of conversation. When people converse, they tend to choose words different from those they have just heard—the response to a question is not the same as the question itself. Even from an acoustic point of view, human conversation consists of the alternation of sounds that are in general dramatically different from one another, whereas chimpanzees produce sequences that tend to be similar to what they have heard. These exchanges probably have to do simply with maintaining contact (Arcadi 2000).

### Teaching Language to Apes

Given these considerations, it is not surprising that attempts to teach chimpanzees to actually talk have failed dismally (e.g. Hayes 1952). In contrast, each of the species of great ape has been taught to communicate quite well using visual and manual signals. Chimpanzees, gorillas, and an orang-utan have been taught a simple form of sign language, and both chimpanzees and bonobos have learned to use a keyboard containing symbols, which they point to in sequence to deliver messages. At least one of these animals, the bonobo Kanzi, has invented gestures to add to the repertoire, and can understand spoken sentences uttered by humans—although he cannot himself speak (Savage-Rumbaugh et al. 1998).

Although these animals can produce and understand short sequences of signs or symbols, their accomplishments cannot be described as true language. The systems they use typically consist of symbols for objects and actions, usually combined to form requests. There is no way of representing different tenses, such as past, future, or conditional, and no way of distinguishing between requests, statements, questions, commands, negations, and so on. There is no recursion, whereas in human speech we readily embed phrases within other phrases in a recursive manner to convey complex propositions, as in *I suspect that she knows that I'm watching her talking to him.* The level of language reached by the so-called linguistic apes is roughly that of a 2-year-old child, and has been called protolanguage rather than true language (Bickerton 1995).

Just as the 2-year-old must await the next stage of development for syntax to emerge, so the common ancestor of ourselves and the chimpanzee was not yet ready for true language.

### Hominin Evolution

If we accept that this common ancestor did not possess true language, it follows that language must have evolved at some point in the hominin branch, which split from the branch leading to modern chimpanzees and bononbos some six million years ago (Waddell and Penny 1996). The hominins were distinguished chiefly from the other great apes by being bipedal: they habitually walked upright, although the earliest hominins probably retained some adaptations to living in the trees. Bipedalism would have freed the hands and opened up a more confrontational stance, perhaps leading to a wider range of communicative gestures. However there is no evidence to suggest that anything approaching syntax would have evolved until the emergence of the genus *Homo* around 2.5 million years ago.

### The Emergence of Homo, and a Cognitive Advance

Stone tools first emerge in the archaeological record at around the same time as the first known species of the genus *Homo*, *Homo rudolfensis*. This also marks the beginnings of an increase in brain size—the earlier hominins had brains no larger, when corrected for body size, than that of the chimpanzee. *Homo ergaster* and its Asian cousin, *Homo erectus*, emerged a little later, and had still larger brains, while both the Neanderthals and modern *Homo sapiens*, whose common ancestor dates from around 500,000 years ago, have brains that are about three times the size of that predicted for an ape of the same body size. For example, the gorilla is the largest of the great apes, with an average body weight of about 62 kg., and its average brain volume is around 400 cc. Humans have an average body weight of around 68 kg., but a brain volume averaging around 1300 cc. And nearly two million years ago, *Homo erectus* began what appears to be a series of migrations from Africa to Asia.

These events all suggest an advance in cognitive capacity—or the ability to think and plan. Migrations and manufacture also suggest that communication may have become more effective. It therefore seems reasonable to suppose that language developed beyond protolanguage, probably gradually (Pinker and Bloom 1990), over the past two million years. I shall argue, though, that language developed first as a primarily gestural system, involving movements of the body, and more especially the hands, arms and face. Nevertheless there was probably increasing vocal accompaniment, with speech finally becoming the dominant mode only following the emergence of our own species, *Homo sapiens*, within the last 170,000 years.

### Why Speech Arrived Late

One reason to believe that speech evolved late is that the vocal apparatus and the brain mechanisms controlling it had to undergo considerable change before speech was possible. One change relates to the control of the tongue, which of course is critically involved in speech—that's why languages are sometimes called 'tongues'. The muscles of the tongue are innervated by the hypoglossal nerve, which in mammals passes through the hypoglossal canal at the base of the skull. The hypoglossal canal is much larger relative to the size of the oral cavity in humans than it is in the great apes, presumably because there are relatively more motor units required to produce speech. Fossil evidence shows that the size of the hypoglossal canal in early australopithecines, and perhaps in *Homo habilis*, was within the range of that in modern great apes. From this it has been concluded that the capacity for human-like speech was present by at least 300,000 years ago, which is the approximate date of the earliest Neanderthal skull (Kay et al. 1998). It is generally recognized that the Neanderthals were distinct from *Homo sapiens*, but share a common ancestor dating from some 500,000 years ago (e.g. Ward and Stringer 1997), and it might also be reasonable to conclude that this common ancestor also possessed sufficient control of the tongue for articulate speech.

Philip Lieberman (1998) has long argued, though, that the changes that resulted in the modern human vocal tract were not complete until the emergence of our own species around 170,000 years ago, and that they were also incomplete in the Neanderthal even as recently as 30,000 years ago. In human children, the lowering of the larynx in the first few years of life is accompanied by a flattening of the face, so that, relative to chimpanzees and other primates, we humans have short mouths (D. Lieberman 1998). The length of the mouth more or less matches the length of the pharynx—the other arm of the right-angled vocal tract. Philip Lieberman argues that the two arms of the tube must be of approximately equal length to enable us to produce the range of vowel sounds we use in normal speech. The fossil evidence shows that our Neanderthal cousins did not have flattened faces like ours, but had long mouths more like those of apes. Since the flattening of the face had apparently not occurred in the Neanderthal, it is a reasonable assumption that the lowering of the larynx had also not taken place, or was at least incomplete.

Moreover, for the length of the pharynx to match the length of the mouth the larynx would have to have been placed in the chest. This would surely have prevented the poor creatures from swallowing! While this might perhaps explain why the Neanderthals became extinct, it is more plausible to suppose that the changes to the face and vocal tract that have given us the power of articulate speech had not yet occurred, or were incomplete, in the

Neanderthal (P. Lieberman 2000). If Lieberman's argument is correct, the fully formed human vocal tract must have emerged *since* the parting of the ways between the Neanderthals and the line leading to *Homo sapiens*. Indeed, it might be considered a critical part of the 'speciation event' that gave rise to our own species some 170,000 years ago, although I shall argue below that fully autonomous speech may have merged even later.

Lieberman's views are controversial (e.g. Gibson and Jessee 1999), but it is unlikely that speech itself arrived suddenly. Even Lieberman has acknowledged that the Neanderthals could probably speak, but without the full range of articulation possessed by *Homo sapiens*. Presumably, if Lieberman is correct, they would have the vocal range of modern human infants. The alterations to the vocal tract, along with other changes to be reviewed below, must surely have occurred gradually, perhaps reaching their present level of elaboration with the emergence of our species.

Speech, unlike breathing for air, involves the muscles of the thorax and abdomen, which are innervated through the thoracic region of the spinal cord. This region is considerably enlarged in modern humans relative to that in non-human primates, including the great apes, presumably reflecting the extra demands placed on these muscles by speech. Studies of hominin fossils also show that the enlargement was not present in the early hominins or even in *Homo ergaster*, dating from about 1.6 million years ago. However, it is clearly present in several Neanderthal fossils (MacLarnon and Hewitt 1999). This again suggests that at least a rudimentary form of speech may have evolved in the common ancestor of ourselves and the Neanderthals by some 500,000 years ago.

## Clues from Present-Day Humans

### Gesticulation

People habitually gesture as they speak. This is often considered to be nonverbal communication, unrelated to speech itself. McNeill (1985) has shown, on the contrary, that the gestures we use when we speak are in fact precisely synchronized with the speech, suggesting that speech and gesture together form a single, integrated system. Goldin-Meadow and McNeill (1999) propose that speech carries the syntactic component, while iconic gestures help convey the actual content, especially if it includes spatial or emotional components difficult to put in words. If people are prevented from speaking while they explain something, their gestures then begin to take on the syntactic component as well (Goldin-Meadow et al. 1996). On-line language is therefore really a blend of sound and action. It is readily pushed in either direction, toward total gesture when we try to get a message across to those who speak a different tongue, or stripped of all gesture when it reaches us via telephone or radio. To be sure, speech dominates, but gesture is not far below the surface. And people still gesture when they talk on the telephone, or on radio.

The gestures that people use when they talk are often called *gesticulation*, to distinguish them from those that occur in sign languages.

### Sign Languages

Some communities communicate entirely in manual gestures, or have developed gestural systems for occasions when speech is impossible or for some reason not permitted. These include religious communities sworn to silence, people working in extremely noisy environments, indigenous peoples involved in rituals of silence, and, above all, deaf people. The sign languages invented by deaf people all over the world provide perhaps the strongest evidence that language at least *could* have been purely gestural, without any loss of expressivity or complexity.

The most extensively studied of the sign languages of the deaf is American Sign Language (ASL), and it is now clear that it has a syntax as complex as that of any spoken language (Neidle et al. 2000). It incorporates such features as tense and mood. It is the official language of Gallaudet University, said to be the world's only university that caters exclusively for the deaf, and the students learn all the usual subjects, even poetry, without a word being spoken (Sacks 1991). Many of the signs used by ASL signers are iconic, which is to say they mime or mimic the objects or actions they represent, but with extended use they become abstract. Charles Darwin (1965: 62) quotes a passage from a book published in 1870:

[The] contracting of natural gestures into much shorter gestures than the natural expression requires, is very common amongst the deaf and dumb. This contracted gesture is frequently so shortened as nearly to lose all resemblance of the natural one, but to the deaf and dumb who use it, it still has the force of the original expression.[1]

---

[1]  Quoted from W. R. Scott's 1870 book, *The Deaf and the Dumb* (2nd edn., 1870, p. 12).

This tendency for signs to shorten and become more abstract is a feature of virtually all communication systems, and is called *conventionalization*.

Children exposed only to sign language go through the same developmental stages as those who learn to speak, including a stage when they actually babble in sign, making repetitive movements analogous to the repeated syllables of vocal babbling (Pettito and Marentette 1991). Indeed there is some evidence that deaf children learn sign language more quickly than hearing children learn to speak (Meir and Newport 1990). Chomsky (2000: 121) has recently remarked: 'Though highly specialized, the language faculty is not tied to specific sensory modalities, contrary to what was assumed not long ago. Thus, the sign language of the deaf is structurally very much like spoken language, and the course of acquisition is very similar'.

Browman and Goldstein (1991), among others, have argued that speech itself is best considered as made up of gestures of the articulators, rather than as combinations of phonemes. Their model of how speech works was based on a theory previously developed to describe skilled motor actions *in general*, and an early version of the model was 'exactly the model used for controlling arm movements, with the articulators of the vocal tract simply substituted for those of the arm' (p. 314). Viewed in this light, one might suppose that language is in fact a combination of oral and manual gestures, with oral gestures gradually but not completely replacing manual ones. Even today, we often naturally include manual gestures in a sentence to substitute for spoken words, as in *He's completely [crazy]*, where a spiralling gesture in the region of the head substitutes for the word *crazy*.

Many of the gestures of sign language involve expressions of the face, including the mouth and even the tongue. Voicing might initially have been a device for making some of these gestures more accessible, simply by rendering them audible as well as, or instead of, visible. Voicing can increase the range of gestures in another way too, by creating distinctions between voiced and unvoiced sounds—examples include the distinctions between /s/ and /z/, /t/ and /d/, or /k/ and /g/.

## The McGurk Effect

If you dub a sound such as *ga* onto a video of a mouth that is actually saying *ba*, then you *hear* the syllable *da*, which is a sort of compromise between the sound itself and what the lips seem to be saying. This phenomenon is known as the McGurk effect (McGurk and MacDonald 1976). Ventriloquists are also able to trick us into thinking the dummy is talking by keeping their own lips as still as possible, while moving the dummy's mouth in time with their own utterances. One might infer from this that speech itself has not escaped from its gestural origins.

### Handedness and Cerebral Dominance

One characteristic that distinguishes humans from the other primates is the fact that about 90 per cent of us are right-handed. We prefer the right hand for skilled actions that involve a single hand, like writing or throwing, and it is the right hand that usually has the dominant role in bimanual actions, such as unscrewing the lid of a jar or uncorking a bottle of wine. Although there is some evidence that captive chimpanzees may show a population bias for some activities, such as feeding (Hopkins 1996), there is little evidence for such a bias among chimpanzees in the wild (McGrew and Marchant 1997), and the bias is in any event much less pronounced than it is in humans (Corballis 1997).

In marked contrast, there is evidence that a left-hemispheric dominance for both the production and perception of vocalizations is widespread among animals and birds (see Vallortigara et al. 1999 for a review). The left hemisphere controls vocalization even in the frog, suggesting that this asymmetry may go back to the very origins of vocal behaviour (Bauer 1993). We have also seen that control of vocalization in primates is largely subcortical, though it clearly came under cortical control at some point in hominin evolution.

A clue to this transition is provided by recent evidence on the so-called 'mirror neurons', discussed earlier. In the monkey, these neurons respond to both the production and perception of grasping movements. This region of the monkey premotor cortex in which these neurons have been found is homologous to Broca's area in humans—an area critically involved in the production of articulate speech. There is also evidence that Broca's area is involved in a gestural mirror-neuron system in humans (e.g. Nishitari and Hari 2000), as well as in the programming of ASL (Neville et al. 1997).

In monkeys, mirror neurons have been recorded on both sides of the brain, and there is no evidence of cerebral asymmetry. In humans, though, Broca's area is predominantly left-hemispheric, whether involved in speech or in the programming of gesture. In the course of evolution, therefore, a system that was at one stage bilateral has become lateralized in humans. This suggests that gestural language may have been initially bilateral, but increasingly left-hemispheric as the vocal component increased. That is, the mirror-neuron system gradually began to control vocalizations, previously subcortical and left-sided, and this had the effect of introducing a left-sided bias into manual as well as vocal action. The link between gesture and speech may therefore explain why right-handedness, a consequence of dominant left-hemispheric control, emerged in humans. There is some evidence that a population-level right-handedness was present in early *Homo*, suggesting that vocalization may have already begun to exert an influence (Toth 1985), although it is not clear whether it was more extreme than that observed in modern captive chimpanzees.

## Why Did Autonomous Speech Emerge?

The evidence thus far suggests that language evolved from manual and facial gestures. These were increasingly accompanied by vocalizations, perhaps initially in the form of grunts accompanying actions (Diamond 1959), but increasingly refined as they came under the influence of Broca's area, so that they could eventually act as vocal 'signs' for the representation of real-world objects and events. At some point in hominin evolution, language became autonomously vocal, at least to the point that communication can be fully intelligible over the phone or radio—although we still embellish speech with gestures in normal conversation.

Given that speech is now the dominant form of language, there must have been compelling reasons for the selection of the changes necessary for its evolution. Indeed, the modifications to the vocal tract were not achieved without a risk to survival. The lowering of the larynx means that breathing and swallowing must share the same passage, so that people, unlike other mammals, cannot breathe and swallow at the same time. They are therefore especially vulnerable to choking. Why then have we been endowed with a risky form of communication when a gestural system seems to have provided a natural and flexible way to communicate?

### A Linguistic Advantage?

One advantage may have to do with linguistic considerations. As we have seen, the signs of sign language tend to be iconic, which is to say that they resemble physically what they represent. This gives them an advantage over abstract signs in that they are easier to learn, and can often be understood even by those who do not know the particular sign language. But iconic signs are a disadvantage as the number of signs increases, especially since it becomes increasingly difficult to distinguish between objects or actions that are physically similar. It would be difficult, for example, to make iconic signs that would distinguish ducks from drakes, or even cats from dogs. Moreover, many of the words we use in language refer to abstract concepts, like love or irony, that cannot be conveyed iconically—except indirectly, perhaps, through metaphor.

Spoken words provide a ready-made solution to this problem, since most of the things we communicate about have to do with the spatial world. With the exception of a few onomatopoeic words like *screech* or *miaow*, spoken words cannot be iconic representations of real-world objects or events. They can therefore be calibrated to minimize confusion between physically similar objects—the names of similar animals, such as *cats, lions, tigers, cheetahs, lynxes*, and *leopards*, all *sound* rather different. Spoken words can also be calibrated so that common words tend to be shorter than uncommon ones, an association noted by Zipf (1935), who related it to a principle of 'least effort'.

Nevertheless there must be strong reservations about any claim that speech is *linguistically* superior to signed language. After all, students at Gallaudet University seem pretty unrestricted in what they can learn, suggesting that signed language can function well right through to university level. We have seen that signs do become conventionalized over time, and there are many signs representing abstract concepts. It is nevertheless true that many signs remain iconic, or at least partially so, and are therefore somewhat tethered with respect to modifications that might enhance clarity or efficiency of expression. But there may well be a trade-off here. Sign languages may be easier to learn than spoken ones, especially in the initial stages of understanding what language is about, and linking objects and actions with their linguistic representations. But spoken languages, once acquired, may result in messages being relayed more accurately, since spoken words are better calibrated to minimize confusion.

### Practical Advantages of Vocal Language

There are a number of practical reasons why speech might have gradually replaced a gestural language. One is obvious—we can use it in the dark, or when obstacles intervene. Sound is also less demanding attentionally: you do not have to look at people to understand what they are trying to tell you.

By much the same token, vocal sounds can be used to attract or command attention, as when Mark Antony cried *Friends, Romans, countrymen, lend me your ears.* Sometimes, of course, sound is a disadvantage, as when we want to convey information secretly, or organize a hunting party to stalk prey. These difficulties can be overcome by whispering, or by reverting to gestural language! The !Kung San, who are present-day hunter-gatherers, sometimes use bird calls to communicate with each other while seeking prey, and then revert to silent signals as they advance on unsuspecting victims (Lee 1979).

But perhaps the main advantage of vocal language was that it freed the hands. Charles Darwin, who seems to have thought of almost everything, wrote:

We might have used our fingers as efficient instruments, for a person with practice can report to a deaf man every word of a speech rapidly delivered at a public meeting; but the loss of our hands, while thus employed, would have been a serious inconvenience.   (Darwin 1896: 89)

It would clearly be difficult to communicate manually while holding an infant, or driving a car, or carrying the shopping, yet we can and do talk while doing these things. But a more positive consequence of the freeing of the hands is that it allowed our forebears to speak while they carried out manual operations, and so explain what they were doing. This may well have led to advances in manufacturing techniques, and the ability to teach techniques to others.

If vocal language freed the hands for other activities, and extended the range of pedagogy, this might explain why our own species is unique among the primates for its extraordinary ability to manufacture objects and exploit and extend the natural environment. The process was cumulative, as invention piled upon invention, and clearly depended on the transmission of information between generations. If this argument is correct, it may be possible to infer *when* speech emerged from what is known about the prehistory of material culture.

## *When* Did Language Become Predominantly Vocal?

Crow (1998) has suggested that there was a 'speciation event' around 170,000 years ago that gave rise to language, cerebral asymmetry, and other uniquely human attributes. This new species was *Homo sapiens*. The idea that language itself might have emerged suddenly with our own species is also supported by authors such as Bickerton (1995), and might be termed the 'big bang' theory of language evolution.

But it may not have been language *per se* that characterized our species, but rather the emergence of autonomous speech. Prior to that, language was a combination of speech and gesture, but the accumulation of changes to the vocal tract and the cortical control of vocalization and breathing allowed speech to take over. The switch to autonomous speech would have freed the hands from any crucial role in language, which in turn might explain the extraordinary developments in manufacturing and manipulation of the environment that characterize our species, and clearly distinguish us from the other great apes. The separation of language from manufacture would also have led to the development of pedagogy, whereby techniques could be explained at the same time as they are demonstrated 'online'. This is demonstrated by any TV cooking show, where the chef is seldom at a loss for either words or ingredients.

One difficulty with this scenario is that a fully 'modern' behaviour does not seem to have emerged until the Upper Palaeolithic, beginning about 40,000 years ago, or perhaps earlier in Africa. The sudden flowering of art and technology in Europe around 30,000–40,000 years ago has been termed an 'evolutionary explosion' (Pfeiffer 1985). It includes a dramatic expansion of tools to include projectiles, harpoons, awls, buttons, needles, and ornaments (Ambrose 2001), as well as cave drawings in France and northern Italy depicting a menagerie of horses, rhinos, bears, lions, and horses (Balter 1999). There is also widespread evidence across Russia, France, and Germany for the weaving of fibres into clothing, nets, bags, and ropes, dating from at least 29,000 years ago (Soffer et al. 2000). These apparently sudden developments are usually attributed to the arrival of *Homo sapiens* in Europe and Russia, eventually replacing the Neanderthals who were already established there. The evolutionary explosion was evident at a slightly earlier date in Africa and the Levant (Ambrose 2001), and probably began in Africa around 50,000 years ago. It seems, in fact, that there was a major dispersal of artists and artisans from Africa from about 50,000 years ago: the archaeologist Richard Klein is quoted as saying 'There was a kind of behavioural revolution [in Africa] 50,000 years ago. Nobody made art before 50,000 years ago; everybody did afterwards' (Appenzeller 1998: 1451).

Even a date of 50,000 years ago leaves unexplained the gap of around 100,000 years between the emergence of *Homo sapiens* and evidence of modern behaviour. One possibility is that sophisticated technology had been accumulating gradually, and reached some kind of critical mass around 50,000 years ago. For example a bone industry, which probably included the manufacture of harpoons to catch fish, has been discovered in the republic of Congo, and dates from about 90,000 years ago (Yellen et al. 1995). Even so, there seems to have been a clear quickening of pace around 50,000 years ago.

### A Genetic Mutation?

One possibility is that this was due to a genetic mutation within the last 100,000 years, presumably in Africa, that may have led finally to speech becoming autonomous. Evidence for this stems from studies of an extended family, known as the KE family, with a genetic disorder of speech and language. The disorder is transmitted as an autosomal-dominant monogenic trait, encoded by a mutation on a gene on chromosome 7 known as *FOXP2* (Lai et al. 2001). It has been argued that this gene is a grammar gene (e.g. Pinker 1994), but although those affected have difficulties with both receptive and expressive grammar, the core deficit is more likely one of articulation (Watkins et al. 2002). The *FOXP2* gene underwent changes in hominins at some point after the split between hominin and chimpanzee lines, and probably within the past 100,000 years (Enard et al. 2002). Enard et al. write that their discovery 'is compatible with a model in which the expansion of modern humans was driven by the appearance of a more-proficient spoken language' (p. 871).

Evidence from the analysis of mitochondrial DNA suggests indeed that although *Homo sapiens* probably emerged in Africa some 170,000 years ago, modern non-Africans have a common African ancestor dating from perhaps as recently as 52,000 years ago (Ingman et al. 2000), implying that migrations from that time eventually replaced all other hominins outside Africa. Perhaps the mutation of the *FOXP2* gene was the final adjustment that allowed speech to become autonomous, freeing the hands for the development of technologies. Armed with vocal language and a new-found capacity for manufacture, including the manufacture of weapons, our forebears may have migrated from Africa and effectively conquered the world. The Neanderthals in Europe persisted until as recently as 30,000 years ago, and there is also evidence that *Homo erectus* in South-East Asia may have survived until about the same time (Swisher et al. 1996), but evidently could not withstand the invasive presence of *Homo sapiens*. These other species of

*Homo* may have lacked the biological requisites to match the communicative and tool-making capacities of *sapiens*.

The evidence from mitochondrial DNA effectively establishes a common *female* ancestor, sometimes dubbed 'African Eve', since mitochondrial DNA is passed down the female line. It is of interest that evidence based on variations in the Y chromosome, which is passed down the male line, has suggested a much more recent date, estimated to be between 35,000 and 89,000 years ago (Underhill et al. 2000), for 'African Adam'. Further, mitochondrial DNA taken from a fossil male discovered at Lake Mungo in Australia has been dated at around 62,000 years ago, but its structure appears to be different in fundamental ways from that of any modern human (Adcock et al. 2001). Yet this fossil, dubbed Mungo Man, was clearly a member of our own species. These claims are controversial, but they are consistent with a more recent wave of migrants to Australia having replaced earlier members of our species there.

If this scenario is correct, then one might ask why the estimate of the date of our most recent common male ancestor is so much more recent than that of our most recent common female ancestor. These estimates are subject to possible artefacts, and may be unreliable, but there is one rather unsavoury possibility that may explain the difference. It may be that the migrating bands out of Africa comprised men, who killed the indigenous men and abducted the women—a scenario controversially documented and endorsed by Pinker (1997).

### Conclusions

The overall evolutionary scenario may have gone something like this. With the emergence of bipedalism, the early hominins evolved more sophisticated ways to gesture to one another than their immediate primate ancestors. But these gestures may still have consisted of relatively isolated signs until around two million years ago, when brain size increased and migrations out of Africa began. This may have led to the combining of gestures to form new meanings, and perhaps the beginnings of narrative. Thus, eventually, was syntax born.

It is also likely that the face became increasingly involved in gesturing, especially as the manufacture and use of tools became more sophisticated, and increasingly occupied the hands. No doubt, too, gestures were punctuated with vocalizations and other sounds, such as chimpanzee-like lip-smacking and teeth chattering, but perhaps increasingly associated with facial gestures. As we have seen, speech itself can be regarded as a system of gestures, formed by the addition of sound to movements of the mouth and tongue. Some speech sounds retain a visual aspect, as demonstrated by the McGurk effect and ventriloquism. The addition of sound to facial gesture, however, would have allowed many gestures to retreat inside the mouth, where they became largely invisible but could still be distinguished acoustically. The addition of voicing would also have allowed the distinction between voiced and unvoiced sounds, such as /t/ vs. /d/, or /s/ vs. /z/, thereby increasing the range of possible signals.

Nevertheless, the addition of voicing and articulatory movements internal to the mouth would have required extensive modifications of the tongue and vocal tract, as well as of the brain mechanisms for vocal control. This would have taken considerable time to achieve through the mechanisms of natural selection, and may not have been complete until the emergence of *H. sapiens* some 170,000 years ago. The gestural theory that I have outlined in this chapter implies that the switch from visual gestures to vocal ones was gradual, and that for much of our evolutionary history language was both visual and vocal—at first, manual and facial gestures would have been accompanied by grunts, but in modern humans vocal language is embellished by manual gestures, although the purely visual mode remains an option for the deaf. My surmise is that autonomous vocal language, with a largely nonessential visual component, may have arisen from a genetic mutation some time between 100,000 and 50,000 years ago in Africa.

Some of the most profound cultural revolutions have come about through changes in the ways in which we communicate with one another. Consider the impact of the inventions of writing, of printing, of the telephone, of the computer, of the internet. No doubt it was the development of syntax, perhaps in the context of gesture, that helped lift human culture to a level of complexity well beyond that of our primate forebears, and it may well have been the shift to autonomous speech that led to the cultural upheavals of 50,000 years ago.

## FURTHER READING

The gestural theory of the origins of language was proposed by the philosopher Condillac in the eighteenth century—see Condillac (1947). As an ordained priest, Condillac was forbidden to speculate as to the origins of language, considered by the Church to be a gift from God. The theory is therefore presented in the form of a fable, which makes interesting reading.

Gordon W. Hewes deserves much of the credit for reviving the theory, and hi 1973 article in *Current Anthropology* provides the platform for modern account including my own, and the commentary following that article provides a useful cr tique. Armstrong et al.'s (1995) book *Gesture and the Nature of Language* explore the implications of sign language for an understanding of language in terms of ge ture, and also explains how syntax might have evolved from simple manual ge tures. This theme is further developed in Armstrong (1999).

The article by Rizzolatti and Arbib (1998) provides an accessible account of th nature of 'mirror neurons' and their possible role in the evolution of language, an my own recent book *From Hand to Mouth* (Corballis 2002) expands on all of th themes developed in this chapter.

# 12

# The Origin and Subsequent Evolution of Language

*Robin I. M. Dunbar*

Language has two remarkable properties. First, it allows us to communicate ideas with each other; second, languages evolve and diversify with a speed and facility that is quite unique within biological evolution. The first has been the focus of much of the research on language over the past half century: ever since Chomsky's formative statement of the programme for linguistics, the mechanism that makes it possible for language to transmit information (i.e. grammar) has been the focus of a concerted attempt to unravel the processes that allow language to encode information. This emphasis on the formal aspects of grammatical structure has, however, tended to obscure a number of interesting questions that lie at the very roots of language as a biological phenomenon, namely whence it came and why it mutates so readily, spawning dialects and ultimately new languages in profusion.

The reasons why language evolved exclusively in the hominid (and possibly only human) lineage have been dealt with at length elsewhere: the burden of these considerations has been to suggest that language evolved for an essentially social function to facilitate the bonding of large social groups (see e.g. Dunbar 1996; Deacon 1997). In this chapter, I want to explore the other two issues (the precursors of language and the ease with which dialects form) in a little more detail. My concern here is to highlight questions rather than to provide answers. First, however, let me sketch out briefly the arguments for why language evolved.

## Why Did Language Evolve?

Perhaps the fundamental problem associated with the evolution of language is why it evolved at all. Conventional wisdom has always assumed that it did so to facilitate the exchange of information—after all, that is what grammar is basically designed to do. This claim is, I guess, not in serious contention. A more exacting issue is what kind of information language was intended to convey. It would, I think, be fair to say that most people would, at least until very recently, have supposed that this was related to information about hunting or the manufacture of tools. 'There were bison down at the lake yesterday when I was passing there' or 'If you want to make an arrowhead, you need to hit the flint nodule right here to strike off a suitable flake'.

What is unsatisfactory about such claims is that (a) these kinds of technological activities take up a relatively small proportion of our time and (b) when we do engage in them, we actually rarely use language when doing so. Hunting is often best done in silence, and tool-making is best done by demonstration rather than instruction. In addition, the size of human social groups gives rise to a serious problem: grooming is the mechanism that is used to bond social groups among primates, but human groups are so large that it would be impossible to invest enough time in grooming to bond groups of this size effectively. The alternative suggestion, then, is that language evolved as a device for bonding large social groups—in other words, as a form of grooming-at-a-distance. The kind of information that language was designed to carry was not about the physical world, but rather about the social world. Note that the issue here is not the evolution of grammar as such, but the evolution of language. Grammar would have been equally useful whether language evolved to subserve a social or a technological function.

Testing between alternative hypotheses is not easy when, as in the case of language, there is only a single instance of it. However, the issue essentially boils down to a choice between two possible pathways to language as we have it. Both assume that, in the here and now, language can be used for both social and technological information exchange: the real question is which came first. The traditional view is that language evolved to subserve a technological function and, once we had it in place, its intrinsic capacities allowed it to be used for other more trivial purposes (such as social information exchange); the alternative view is that language evolved to subserve the social function, but its intrinsic properties allowed it eventually to be used for other purposes (e.g. technology) that were in themselves valuable. The issue, then, is which was cause and which consequence?

Two lines of evidence have been adduced (see Dunbar 1993; 1996) to support the second view. One is the factual observation that about two thirds of speaking time in informal conversations is devoted to social top-ics (comments about myself (as speaker), the listener or a third party, plan-ning social activities and the like). Only a very small proportion is devoted to technical topics or descriptions of the physical world. Indeed, we tend to find the latter topics rather boring and hard going, even when we are explic-itly engaged in learning about them. In other words, most of what we devote our conversations to is servicing our social relationships. In some cases, this involves direct investment in a particular relationship (by spending time talking to that person), in others it involves finding out what has been hap-pening in the community to which I belong. Language allows us to keep track of events in the community in a way that our monkey and ape cousins simply cannot do: what they do not see, they can never know about.

The second line of evidence is a purely logical one: if language does not act as bonding agent, what does? We cannot ignore that question, because the claim that language's primary function is the exchange of technical in-formation assumes that large, stable, coherent social groups actually exist. Dismissing that question as irrelevant to language leaves the technological version of language evolution founded on sand, because the conventional primate bonding devices (grooming) are not good enough to maintain groups of the size typical of modern humans.

Finally, we may note that there are at present three versions of the so-cial language hypothesis: Dunbar's (1993) social gossip (or social bonding) hypothesis, Deacon's (1997) social contract hypothesis and Miller's (1999) Scheherazade Effect hypothesis. Deacon's argument, in a nutshell, is that language became necessary in order to allow us to make the contracts on which sociality depends (in particular, the contracts that prevent us from running off with someone else's spouse when wives get left behind while husbands are off hunting). Miller's argument is that language evolved to al-low us to advertise our suitability as mates and/or to allow mates to keep each other entertained (and so interested in each other) once a relation-ship has been established. In fact, these three hypotheses can be unified fairly easily, since the social contract and Scheherazade effect hypotheses both require the pre-existence of large bonded groups (otherwise they run into the same problem as above) while at the same time providing valuable additional benefits to languages in different domains once language (in its gossip/bonding form) is in place. In effect, we can see them as valuable con-sequences of language that reinforce its further rapid evolution rather than its primary selective advantage as such.

Having sketched in the background, let me now turn to the two issues that I want to discuss in more detail: how language evolved and why languages diversify so readily.

## The Precursors of Language

The origins of language (and speech) are inevitably shrouded in mystery, buried in an inaccessible fossil past that leaves only the most indirect traces of behaviour. Although there is strong anatomical (Kay et al. 1998; Mac-Larnon and Hewitt 1999) and comparative (Aiello and Dunbar 1993) evidence to suggest that speech first appeared with the earliest *Homo sapiens* at around half a million years ago, the process by which speech evolved out of an organism with an essentially primate communication system remains uncertain. There have been two general positions: one claims that speech evolved out of an essentially primate vocalization system (Dunbar 1993; Aiello and Dunbar 1993; Burling 1993; MacNeilage 1998), the other that it evolved via a stage of gestural communication (presumably from an origin in primate gestural communication) (Hewes 1973; Calvin 1983; Corballis 1983; Kimura 1993).

Although given very wide credence, the latter position has been based on what is essentially indirect evidence. Typical examples include the observation that apes appear to be better at gestural languages like ASL than vocal languages, or that deaf and dumb human children seem to develop signing languages complete with grammar spontaneously (i.e. without need of teaching), or that we use gestures as a part of everyday speech-based communication. The claim that aimed throwing would have provided a natural substrate off which to build fine motor control for hand signals has been a particularly potent symbol of this view.

Most of these arguments do not really provide more than circumstantial (and far from conclusive) evidence in support of a gestural origin for language. If aimed throwing provided fine motor control for gestures, for example, it might well have provided a template for the genuinely fine motor control that is crucial for speech, but this does not imply that language necessarily passed through a gestural phase before becoming based in speech. Similarly, gestural languages may develop effortlessly in the deaf and dumb children in today's environment, but in the past the deaf and dumb were notoriously regarded as imbeciles *precisely* because they were unable to communicate *in any form* with the rest of social community; the gestural languages of the deaf and dumb seem to develop naturally off the back of *spoken* language (or at least the cognitive substrates required for spoken language—the so-called 'language of thought') when members of the speaking community encourage the deaf and dumb to communicate using a gestural form of transmission. Gestural languages do not seem to develop spontaneously all that easily.

Gestural languages, however, have two genuinely serious disadvantages. One is the fact that they are entirely useless at night. Second, they require line-of-sight contact between communicators (and thus gain no advantage over the social grooming of primates: see Dunbar 1993)—a matter of very serious concern in a natural environment where a great deal of time has to be devoted to foraging. This is likely to be especially problematic if (as would seem more likely) the ancestral environment was more like scrub woodland (with its dense under-storey and poor visibility) than open, short-grass savannah. The merit of speech in this latter context is precisely that it allows time-sharing with other activities without reducing the efficiency with which the latter can be carried out. In other words, we can talk while we eat, dig, and walk in a way that we simply cannot while gesturing—in the last case, not least because gestural communication needs to be done more or less face to face, and walking face to face is not the most efficient way of progressing.

More recently, Rizzolatti and Arbib (1998) have attempted to provide a neural underpinning for the gestural origins theory. They point to the fact that watching an experimenter pick up and manipulate an object triggers responses in two key areas in the human neocortex, the superior temporal sulcus and Broca's area. The latter turns out to be anatomically homologous with area F5 in the macaque brain, where so-called 'mirror neurons' (neurons that fire both when performing an action and when watching another individual perform the same action) were first discovered in the macaque. Rizzolatti and Arbib suggested that mirror neurons might have provided a bridge between performing actions and communicating with others. They suggested that the very slight muscle movements that occur when we observe the actions of others might have provided the basis for a gestural phase during language evolution.

Interesting though this suggestion is, it does not provide unequivocal evidence of a gestural origin for speech: if primitive language principally involves statements about actions, then we might well expect mirror neurons to be involved, irrespective of whether communication is gestural or vocal. An even stronger claim might be made if it is the case (as has been argued) that the real function of Broca's area is the management of fine breath control for speech: in this event, it is hardly surprising that its neurons should fire when upper limb action is involved, since bracing the chest is necessarily involved in any such movements. There is, however, another reason why the Rizzolatti and Arbib claim may be misleading: there is evidence that mirror neurons may also be involved in the cognitive phenomenon known as 'theory of mind', the process by which we simulate the mind states of others (Gallese and Goldman 1998). Theory of mind is probably essential for language, not so much because it is involved in the production of speech *per se* but because it provides the mechanism that both enables speakers to ensure that their message has got through and allows hearers to figure out just what the speaker's message actually is (subtext and all) (Worden 1998; Dunbar 1996). In short, the evidence for mirror neurons does not allow us to differentiate clearly between the two alternative hypotheses.

All this notwithstanding, the weakness of the gestural theory for language origins is perhaps that it is difficult to see what real advantage a gestural phase would have offered. Much hangs here on what is envisaged in any such gestural stage. If the gestural stage is assumed to involve full conceptual language as we have it now (or even some more modest precursor to what we have now), then the hypothesis raises some puzzles. The gestures that punctuate everyday speech do not seem to have the same illocutionary force as the speech acts they accompany. Nonverbal aspects of speech seem to have more to do with emphasis and command. They elaborate on what is said, but they almost never replace what is said. In this respect, human gestural communication does not seem to differ from the gestural communication systems of our primate cousins. Because of this, it is difficult to see what would have been gained by moving from one already efficient communication channel (sound) to another (gesture) and then back again, when a direct route from primate vocal communication to human verbal communication is clearly possible. In addition, one must ask why, if gestural languages were so well developed as to allow the exchange of conceptual information, we should have lost so efficient a mode of communication since the rise of speech. Atrophy is not an adequate answer in biological terms: if gestures can supplement emphasis and instruction as effectively as they do, then it does not make sense to suggest that they cannot do the same for the conceptual content of speech. Conversely, if the gestural stage of language involved gestural communication as we have it now (i.e. essentially emotional rather than conceptual in content), then there is not much to choose between the vocal and the visual channels. Nonverbal vocal cues can be (and are) just as effective as gestural cues. Moreover, that channel already exists and is used to good effect by all primates (and, indeed, by most other mammals too).

The burden of the argument, then, is that a vocal origin to human language seems a more plausible starting point than a gestural one. Although Owren and Rendell (2001) have argued vehemently against the claim that any aspect of non-human primate vocal communication pre-shadows human speech, their argument seems to focus on the assumption that what is being claimed is that non-human primates exhibit some (or even all) of the psycho-acoustic foundations of language (e.g. the ability to form consonants and vowel sounds). This is not the issue at all: such abilities probably depend on some uniquely human features, namely the fine control of breathing (to enable the phrasing of sound elements) and the peculiar physical structure of the human laryngeal and palatal spaces. The substantive issue is whether non-human primate vocal communication shows those features of human speech whereby *meaning* is attached to sounds. In this respect, the answer must be yes if any credence is to be given to the work of Cheney and Seyfarth (1990). As with the original chimpanzee language projects, the issue is not whether monkeys and apes have the physical apparatus to produce human speech sounds (after all, if they did, then they surely would be able to speak), but rather whether they have the cognitive components that allow meaning to be attached to arbitrary signals in order to transfer information from one mind to another. In other words, is the non-human primate mind a kind of paralinguistic mind waiting in the wings for the acquisition of an appropriate speech apparatus? Hovering behind this is the more fundamental question of whether other primates possess the kind of *social* mind that seems to be essential for human language: although there are dissenters (Povinelli 1999), many would argue that they do (Call 2001; Sudendorff and Whiten 2001). The question of whether or not non-human primates can produce vowel sounds is beside the point: that is merely a consequence of minor adjustments to the size and shape of the vocal apparatus.

The point here, of course, is that there is a world of difference between speech per se and language. The one is a (relatively) uninteresting channel of communication; the other is what the whole process of human communication is all about—the mental gymnastics involved in formulating and, subsequently, decoding utterances that have real meaning over and above the sign value of their basic constituents (words).

The issue, then, is whether non-human primates (or, if it comes to that, any other animals) share with us the ability to express conceptual information in a vocal channel. If the answer is in the affirmative, then the business of getting language off the ground is very easy. Indeed, Nowak et al. (1999; Nowak and Krakauer 1999; Nowak and Komarova 2001) have used a modelling approach to demonstrate that if the initial function of language is to allow a speaker to name 'objects' (things or events) and both speaker and listener gain a fitness pay-off if the information is successfully communicated between them (i.e. they cooperate in the exchange of information such that both gain in fitness as a result), language can easily evolve out of naming.

The key to their model is the risk of misunderstanding (i.e. the listener being mistaken about which object or event is being named), with the risk of misunderstanding being a function of how acoustically similar any two sounds (names) actually are. Nowak et al. (1999) showed that, if a language's fitness (indexed by its capacity to transfer information) is maximized by having a small number of simple sounds to describe a few valuable concepts or objects, then increasing the number of signals does not necessarily increase the fitness of the language. This suggests that non-human communication systems may be error-limited. However, this error limit can be overcome by combining sounds (phonemes) into words, since this has the effect of reducing the number of sounds that can be mistaken for each other. Since fitness increases exponentially with the length of the words used, word formation must have been a crucial step in the evolution of human language (the equivalent of moving from an analogue to a digital channel). Note that there have been claims that at least some non-human primates form utterances that are structured in this way (though these regularities in phonemic structure have usually been interpreted as syntactical: Cleveland and Snowdon 1981).

Nowak et al. (2000) extended this approach to the evolution of grammatical structure. They considered a language that uses *noun+verb* complexes to describe events and showed that, when there is a large number of relevant events, syntactic organization of the stream of words conveys a significant fitness advantage (less risk of misunderstanding). The point at which syntax becomes advantageous in this respect depends on (a) the difficulty of memorizing the particular signals and (b) the fraction of all word combinations that describe real meaningful events (contrast *pigs fly* with *pigs walk*). When a noun+verb phrase is twice as difficult to memorize as a non-syntactic utterance (e.g. a noun on its own) and one third of all possible noun–verb combinations are real possible events, then the minimum object–event combination that will make syntax advantageous is surprisingly small: a matrix of eighteen nouns by eighteen verbs. The dynamics of this system is, however, complex: in smaller systems (say, a six by six matrix), the number of learning opportunities becomes crucial: the fitness of a syntactic language only exceeds that of a non-syntactic one when the number of exposures (i.e. learning opportunities) is below about 400. In other words, when the number of possible noun–verb combinations is small but the frequency of exposure is very large (> 400 occasions), a non-syntactic system is actually more efficient.

In other words, so long as there is a modest selective advantage in terms of more effective cooperation between individuals, both language (in the simple sense of naming objects) and, more importantly, grammar (the ability to describe events) can evolve quite easily by a conventional Darwinian process. This makes it all a great deal easier to move from a simple non-human primate vocal system with referential meaning (like that of the Cheney–Seyfarth vervets) to a more sophisticated human-like language with grammar. Essentially, all that is required is an increase in the number of relevant events that it is useful to be able to describe to another individual. It is difficult to see how anything in the physical world of the later hominids could have been so different from that of their forebears (or indeed, most other primates) to fulfil that condition, but the demands of an increasingly large social world would seem to do so admirably.

One other point of interest: Nowak and Komarova (2001) show that what is important here is the accuracy of language acquisition. When this is low, a variety of different grammars can emerge in a population, but when it is moderate to high the population rapidly converges on one standard grammar, and a coherent communication system emerges. The point is that any cognitive mechanism that facilitates accuracy of acquisition (such as a Universal Grammar) necessarily pushes the system towards a single language form. This would suggest that there would be a strong selection pressure favouring some kind of neural hardwiring for a predisposition to attend to and learn language—any language—which, of course, is exactly what we see in human infants.

While it is feasible to argue that language evolved directly out of non-human primate vocalizations as a device for exchanging information, there remains an alternative possibility that might suggest a more long-drawn-out process. This would view information exchange as a later phase responsible for the explosive evolution of the language capacity but not, perhaps, for its initial appearance. Aiello and Dunbar (1993) noted that we might see language evolution as just such a multi-stage process. We might envisage these as (1) an initial form much like conventional non-human primate contact calls that serve to keep track of other group members (or, perhaps, as in the case of the gelada, to act as a form of grooming-at-a-distance between pairs of friends), (2) a more developed form of this (chorusing?) designed to overcome the physical constraints on grooming that limit group sizes to around sixty individuals, (3) a more fully fledged language that uses grammatical structures to convey social information and, finally, (4) a fully developed modern human language capable of deeply abstract symbolic representation of concepts (with the last of these perhaps involving changes more at a software than a hardware level). Aiello and Dunbar (1993) envisaged these stages kicking in successively over the course of hominid evolution as the need to evolve increasingly large social groups to cope with key ecological problems became more intrusive.

I want to focus here on the second step—the development of a more communal form of contact calling as an intermediate step between conventional non-human primate contact calling and fully fledged grammatical (but still exclusively social) language. The issue is about bonding social groups. Among Old World monkeys and apes at least, this is done mainly through social grooming, a deeply intense activity that facilitates the release of endogenous opioids (see e.g. Keverne et al. 1989). If the effectiveness of social bonding (and a willingness to cooperate and exchange support) is a function of the opioid high generated by social grooming, and if the size of the group that can be supported off the back of such a form of social interaction is limited by the amount of time that animals can afford to devote to social interaction, then the limiting size of social groups that we should expect to see among primates is about sixty—much as seems to be the case. Part of the problem here is created by the fact that social grooming is very much a one-on-one activity: primates (including humans) do not groom more than one individual at a time (even though several animals may occasionally groom the same recipient). So intense and personalized is this activity that failure to groom with sufficient commitment and intensity is often the precursor to the end of a grooming bout.

If the time available for social interaction is limited, then the only way that group size can be increased beyond the limit created by this time constraint is for the available social time to be used more efficiently. There are, in principle, two ways of doing this: one is cognitively expensive, the other cognitively cheap. The expensive way would seem to be to introduce a quantitative change in the kind of information that can be transmitted during interactions: language does this very effectively by allowing us to seek out information on who is doing what with whom and how often. This, however, requires the computational devices capable of supporting language (minimally, the capacity to encode and process grammar, plus advanced theory of mind). An alternative, and cheaper, option is to increase the number of individuals that can be 'groomed' simultaneously. Language actually allows us to do this, because we can talk to several individuals at once in a way that is just not possible with physical grooming (see Dunbar et al. 1995). But we also have a way of achieving the same effect without language, namely by means of communal singing (and possibly dance). Singing seems to be an exceptionally powerful stimulant of endogenous opioids, as evidenced by our enthusiasm for it and the way it makes us feel generally relaxed and positive towards those with whom we have been singing.

The significance of singing in this context is that it does not require words (language as we know it), merely rhythmically synchronized vocalizations—in fact, just the kind of thing we already find widely among other primates. Although it is clear that communal singing must increase the interaction group size (the number of people simultaneously subjected to an opioid release effect), it is not clear what the magnitude of this effect is likely to be or by how much it would increase social group size above the limiting size that can be sustained by conventional social grooming. What it might at least do is bridge the gap between this limiting size (somewhere around sixty individuals) and the size of social groups that eventually emerged off the back of conventional language (around 150) (see Dunbar 1993). An intermediate step in the process of language evolution that involved an intense communal activity like chorusing would provide a strong natural pressure in the direction of a verbal means of communication.

## Why Do Languages Diversify?

In the previous section I focused on the question of how language came about. But given that language evolved to allow information to be exchanged (everyone agrees, for example, that this is what grammar is well designed to do), why on earth do languages diversify so rapidly that they very quickly become mutually unintelligible? After all, there are now some 6,000 living languages, plus an untold number that have already become extinct. Within languages, dialects are equally diverse: until as late as the 1970s, a native English speaker's dialect was sufficient to identify the speaker's natal location to within 40 km. of his or her place of birth (Trudgill 1999). Languages spawn dialects with unseemly speed, and dialects in turn eventually give rise to new languages. Ancestral Indo-European, the language spoken by a small group of (probably Anatolian) agro-pastoralists around 6000 BC, has given rise to around 150 descendent languages from Gaelic in the west to Bengali in the East over an 8,000-year period. Interpolating these values into the standard Gaussian logistic growth equation for biological population growth suggests that the Indo-European language family has evolved at a rate equivalent to the budding off of a new language from each existing language, on average, about once every 1,600 years—or about as long as it has taken modern English and Scots to evolve out of Anglo-Saxon. (We might, of course, expect geographical dispersion to have been important in addition, as in the case of the modern Latinate languages.)

It seems that the question as to why languages should be so labile in this respect has rarely been asked. From an evolutionary point of view, however, this is a singularly puzzling phenomenon. After all, if language evolved to allow individuals to exchange information, one might expect stability to be an important consideration, especially in a multi-generational community where the 'wisdom of the ancients' might provide an important resource for survival. Yet dialects seem to evolve with a speed that approximates the scale of a generation (Barrett et al. 2002). Since it is not beyond the wit of evolution to have produced language structures that are more resistant (if not totally resistant) to corruption in this way, the implication is that the corruptibility of language is precisely the whole point (and has been deliberately selected for).

One answer, of course, is simply drift: the gradual accumulation of accidental mutations (mispronunciations, unintended slippages of meaning) over long periods of time. Such a process is relatively slow, however, and would not be expected to lead to quite such a degree of differentiation in so short a space of time. If the process is not accidental, then it must be deliberate, and deliberate in this context means 'under the influence of selection'. What selection processes could promote such high rates of language change?

The most plausible selection pressure is likely to be the need to differentiate communities. The key problem faced by all intensely social organisms that depend on cooperation for successful survival and reproduction is the freerider—the individual who takes the benefits of cooperation but does not pay the costs (Enquist and Leimar 1993; Dunbar 1999). Nettle and Dunbar (1997; Nettle 1999) have argued that dialects are particularly well designed to act as badges of group membership that allow everyone to identify members of their exchange group: dialects are difficult to learn well, generally have to be learned young and change sufficiently rapidly that it is possible to identify an individual not just with a locality but also within a generation within that locality. They used a simple spatial model (in which a dialect was a very simple six-digit barcode attached to each individual) to show that a rate of dialect change approaching 50 per cent per generation was required to ensure that individuals who had to exchange resources with each other in order to reproduce were not exploited by freeriders.

Nettle (1999) analysed the world distribution of languages in relation to a number of climatic variables (including latitude, ambient temperature and the length of the growing season) and showed that there were more languages per unit area, each with a smaller number of speakers, in low-latitude habitats (with long growing seasons) than in high-latitude habitats (with short growing seasons). These results may reflect the influence of two (not entirely mutually exclusive) selection pressures. One is that, in more seasonal habitats (such as those at high latitudes), individuals need a wider network of social contacts in order to buffer themselves against the vagaries of the natural environment. Since famine is likely to be both unpredictable and more frequent, the trading network needs to be wide enough to ensure that at least someone you know well enough to trade with is living in an area of surplus during famine years. Reciprocal arrangements across this large set of individuals ensures that food supplies are always available. Being able to communicate well enough to trade is, of course, important; but being able to signal descent from a common ancestor by using the same language may be crucial in facilitating the willingness to engage in long-term reciprocal exchange with strangers.

In low-latitude habitats in the tropics, where the growing season is more or less continuous, famine is a rare event and individuals are more self-sufficient. Because such habitats are capable of sustaining very high population densities, individuals living there incur high costs from the effects of competition and crowding, not to say the perennial risk of individuals using sheer muscle to raid and steal from their more industrious neighbours (the classic producer/scrounger problem). To buffer themselves against these costs, they need to ensure that their social networks work extremely well. Small networks are more effective in this respect because it is possible to bond small groups more effectively than large ones; in addition, it is easier to identify members of the group through badges like dialects. So we might anticipate pressure to diversify languages in the equatorial regions.

Beyond this, however, we have little real understanding of the processes involved in either dialect change or language evolution, or for that matter in the functions that these processes subserve. We assume that these functions are largely social, and we have some understanding of the types of process that can precipitate language change (trade, colonization, emulation of culturally or economically superior groups, etc.), but by and large there is little other than conjecture to explain why these processes exist or why they should work in the way they do. Part of the problem is, of course, the temporal scale on which these changes occur (generations in the case of dialects, perhaps millennia in the case of languages). Inevitably, this makes it all but impossible for us to observe the process first-hand. These are, nonetheless, fruitful areas for future exploration.

Notice that this explanation for language instability is based on the same argument as that for language evolution, namely the importance of bonding small communities or social groups. In terms of parsimony, this provides considerable support for the social bonding argument, since the alternative hypothesis for the evolution of language (technical information exchange) cannot explain why languages should diversify so easily. It requires some other unrelated explanation to account for this feature of language behaviour.

## Conclusions

In this chapter I have focused on two questions about language that have received a great deal less attention than they deserve. One was how language evolved in the first place, and the other was why languages are so prone to diversification. My aim here has been to identify neglected issues that need more detailed study, rather than to provide conclusive answers. In offering some suggestions as to what those answers might be, I have endeavoured to offer what I think are plausible directions in which that research might go.

Although there has been some desultory discussion of the first issue (how language came to evolve at all), much of this has focused on a rather odd form of communication (gesture) and paid very little attention to what is actually involved in communication (e.g. its cognitive requirements). I have tried to suggest some reasons why gestural theories of language origins do not make sense. Whatever the origins of language, it is clear that some additional steps are needed between the basal primate communication system from which language must have evolved and language as we know it today. I have previously suggested that this intermediate step (or steps) is likely to have involved the use of primate-like vocalizations in chorusing. I presented some additional arguments favouring this suggestion, one of which is the way that communal singing seems to trigger endorphin release (thus mimicking social grooming). It is clear that once a vocal channel of this kind is in fluent operation, it requires very little (as Nowak's models have shown) to push it that one crucial step further, to the stage at which meaning is attached to sounds (especially given that primates seem already capable of doing that).

The second issue that I highlighted (the rapidity with which languages diversify, first into dialects and then into true languages) has received almost no attention. This is particularly surprising because this remarkable characteristic seems to run counter to everything that we commonly suppose language evolved to do (namely to facilitate the exchange of information). The implication is that language's corruptibility has been directly selected for. I suggested that this might have been to facilitate group bonding by allowing community members to identify each other more easily (the 'dialects as social badges' argument). This explanation has the distinct merit of allowing both the evolution of language and the evolution of language diversification to be derived from the same basic idea—the suggestion that language evolved to subserve an essentially social function (the bonding of what are, by primate standards, large communities).

## FURTHER READING

The best general summary of my ideas about the evolution of language as a device for social bonding is my book *Grooming, Gossip and the Evolution of Language* (Dunbar 1996). Much of the empirical data on which this hypothesis depends is scattered in the primary research journals; since this is difficult to track down, the book provides an accessible non-technical summary. Deacon's *The Symbolic Species* (1997) provides an additional excellent review of the neuroanatomical and other aspects of language origins, while offering a particular spin on the social language theme. Nettle's *Language Diversity* (1999) is a particularly important volume, summarizing a great deal of his work on how and why languages diversify so rapidly.

13

# Launching Language:
# The Gestural Origin of Discrete Infinity

*Michael Studdert-Kennedy and Louis Goldstein*

## Introduction

'Human language is based on an elementary property that also seems to be biologically isolated: the property of discrete infinity' (Chomsky 2000: 3). 'Discrete infinity' refers to the property by which language constructs from a few dozen discrete elements an infinite variety of expressions of thought, imagination, and feeling. The property 'seems to be biologically isolated', because it is unique among systems of animal communication. From another point of view, however, it is not isolated at all, but rather an instance of a general principle common to all natural systems that 'make infinite use of finite means' (Humboldt 1836/1999: 91), including physics, chemistry, genetics, and language, namely, 'the particulate principle of self-diversifying systems' (Abler 1989).

## The Particulate Principle

According to the particulate principle, the only route to unbounded diversity of form and function is through a combinatorial hierarchy in which discrete elements, drawn from a finite set, are repeatedly permuted and combined to yield larger units higher in the hierarchy and more diverse in structure and function than their constituents. The particulate units in physical chemistry include atoms, ions, and molecules; in biological inheritance, chemical radicals, genes and proteins; and in language, gestures (as will be argued below), segments, syllables, words, and phrases.

Preparation of this chapter was supported in part by Haskins Laboratories and by grants NICHD HD-01994, NIDCD DC-00403 to Haskins Laboratories.

A parallel between languages and genetic systems has repeatedly been re-marked by physicists (e.g. Schrödinger 1944), linguists (e.g. Jakobson 1970), and biologists (e.g. Jacob 1977; Pollack 1994). Jacob, for example, wrote:

> Linguistics has furnished genetics with an excellent model. The image which best describes heredity is that of a chemical message...written...with the combination ...of just four chemical radicals. The four units...are combined and permuted in-finitely, just as are the letters of the alphabet throughout the length of a text. As a phrase corresponds to a segment of text so does a gene correspond to a segment of the nucleic acid fiber.   (Jacob 1977: 187).

Like Jakobson (1970), Jacob emphasized that for such a system to work its basic units must themselves be devoid of meaning or function. In language, only if phonetic units have no meaning can they be commuted across con-texts to form new words with new meanings.

Jacob went on to observe that the principle of combining discrete units to form successive levels of a hierarchy 'is not limited to language and he-redity...[but]...appears to operate in nature each time there is a question of generating a large diversity of structures using a restricted number of building blocks' (1977:188). But he did not try to explain why systems as ap-parently diverse as language, physics, and genetics converge on a common structural principle. That was left to Abler (1989), who first recognized cor-respondences among Fisher's (1930) genetical theory of natural selection, the atomic theory of physical chemistry, and Humboldt's (1836/1999) de-scription of language.

Fisher (1930) reasoned that if parents' characteristics were to blend in their offspring, they would vanish in an average; variation, critical to the process of natural selection, would then decrease from one generation to the next. In fact, of course, variation is conserved, or even increased, across generations, and parental characters lost in one generation may reappear unmodified in the next. From such facts Fisher (like Mendel before him) inferred that biological inheritance was necessarily effected by a particu-late mechanism: unbounded biological diversity can only be maintained by permutation and combination of discrete genetic entities.

Abler (1989) saw that Fisher's logic of particulate combination applied to physics, chemistry, and language no less than to genetics. Moreover, Humboldt's characterization of the language hierarchy could be extended to these other domains: all four achieve unbounded diversity by 'a synthetic [i.e. combinatorial] process...[that]...creates something...not present *per se* in any of the associated constituents' (1836/1999: 67). Novel struc-tures and functions arise at each level of a hierarchy because units do not blend and disappear, but combine as integral units to form new integral units, whose properties are not limited by, and cannot be predicted from, the properties of their constituents.

We cannot derive the fire-extinguishing properties of water from the combination of hydrogen (which burns) and oxygen (which sustains burning), nor the properties of proteins from the genes that control their formation. In language, we cannot derive the meaning of a word from the phonetic elements that compose it, nor the meaning of a phrase from the lexical meanings of its words without regard to their syntax. Indeed, it is precisely because the properties of units at each new level cannot be derived from the properties of their constituents that successive levels in the language hierarchy (phonology, morphology, syntax) are independent and subject to their own characteristic rules of combination.

Thus, the particulate principle rationalizes and generalizes across diverse domains the combinatorial mechanisms and the independence of successive levels in a hierarchy that standard linguistic theory adopts as axioms of linguistic analysis. The principle is a mathematical constraint to which any system that has the property of discrete infinity necessarily conforms. That is why, despite their different modalities, signed and spoken languages arrive at analogous hierarchies of phonology (or sign formation) and syntax (Klima and Bellugi 1979). By assimilating language to other particulate domains, we do not ignore the unique properties of syntax and phonology essential to their functions. We do, however, emphasize the roots of language in biophysics, and the critical importance for both lexicon and syntax of the prior evolution of phonetic capacity.

## Discrete Phonetic Units as Conditions of a Lexicon and Syntax

Discrete phonetic units of some kind must have emerged relatively early in the evolution of language. For, as Bickerton remarks, 'syntax could not have come into existence until there was a sizable vocabulary whose units could be organized into complex structures' (1995: 51). And a sizeable vocabulary could not have come into existence until holistic vocalizations had been differentiated into categories of discrete phonetic units that could be organized into words. A critical early step into language, therefore, was (as it still is) the breakthrough into words, or symbolic verbal reference, by means of a particulate phonetics (Studdert-Kennedy 1998; 2000).

Less often remarked, though no less important, syntax could also not have come into existence until there was a code, a phonetic form, for short-term storage of words, independently of their meaning and syntactic function, during preparation of an utterance by a speaker and comprehension of an utterance by a listener. Independent phonetic segments, devoid of meaning, are indeed taken for granted by virtually every approach to the evolution of syntax. Berwick (1998), for example, in his account of the development of hierarchic syntactic concatenation of words by the operator 'Merge' in the minimalist framework starts his derivation with a 'bag' of unordered words, each marked by independent phonetic, formal, and semantic features. Kirby (2000), for another example, models the emergence of syntax from 'holistic' utterances associated with decomposable meanings. His initial utterances are semantically holistic, but consist 'physically' of discrete symbols, randomly concatenated into strings of 'phonetic gestures'. Thus, a necessary condition of compositional syntax (discrete phonetic units) is included in the initial conditions: compositionality can only emerge, because 'holistic' utterances readily fractionate along the fault lines of their discrete components. Similarly, Wray (2000), deriving words from holistic utterances as a first step into syntax, assumes that 'arbitrary phonetic representation developed not in the service of words, but of complete, [semantically] holistic utterances . . . long before words or grammar appeared' (293). Thus, phonetic break-points between portions of a semantically holistic utterance (portions that, in Kirby's and Wray's models, eventually become words, if they happen to correlate with presupposed break-points in the field of reference) are built into the utterance.

Where, then, do these phonetic break-points come from? What is the physical basis for phonetic segments? The standard units, consonants and vowels, will not do, because they and their descriptive features are purely linguistic and therefore precisely what an evolutionary account must explain. What we require is a prelinguistic unit of motor action that takes on linguistic form and function as it is put to communicative use.

## The Nature of Phonetic Units

Modern humans, speaking English at a comfortable rate, produce 120–180 words/minute, or 10–15 phonetic segments/second. On the face of it, such a rate makes prohibitive demands on both listening and speaking. For the speaker, the rate would seem to exceed by far that of any other motor activity (with the possible exception of skilled musical performance); for the listener, 10–15 units/second are close to the rate at which acoustic pulses merge into a low-pitched buzz. The puzzle begins to resolve itself, however, once we recognize that the perceptually salient units of speech are not phoneme-sized phonetic segments but syllables, and that syllables, carriers of speech melody and rhythm, are intricate patterns of simultaneous or overlapping gestures.

Such rates can be achieved only if separate parts of the articulatory machinery—muscles of the lips, tongue, velum, etc..—can be separately controlled, and if . . . a change of state for any one of these articulatory entities, taken together with the current state of others, is a change to another phoneme . . . it is this kind of parallel processing that makes it possible to get high-speed performance with low-speed machinery.    (Liberman et al. 1967: 446)

Similarly, sign languages and finger spelling depend on rapid sequences of movement distributed across arm, wrist, and fingers.

Evidently, particulate language, whatever its modality, requires an integral neuro-anatomical system of discrete, independently movable parts that can be coordinated to effect rapid sequences of motor action, and it is the actions of these individual parts, the gestures, that constitute the basic units of language. We do not, of course, have to suppose that the particulate structure of either the vocal apparatus or the hand evolved specifically for language. The task, rather, is to understand how referential communication exploited, refined, and, as it were, emancipated into language a particulate machinery that had already evolved in primates for other purposes.

Our initial hypothesis, following MacNeilage (1998), is that speech, *as a motor function*, draws on phylogenetically ancient mammalian oral capacities for sucking, licking, swallowing, and chewing. Sucking (the defining mammalian behaviour), licking, and tongue action in preparing food for swallowing presumably initiated neuroanatomical differentiation of the mammalian tongue; the evolution of speech carried the process further, differentiating tongue tip, tongue body, and tongue root into independent organs of phonetic action. Cyclical lowering and raising of the jaw for mastication laid the neural ground for early homologues of the repetitive lip and tongue smacks, calls, and cries that we observe in modern monkeys and apes. From these, by hypothesis, arose the hominid protosyllable, precursor of the modern child's unit of babble, and of the modern adult's unit of rhythm and melody. Such a protosyllable can be viewed as a gesture, that is, as constriction and release of one of the vocal organs, set in the context of an overall vocal tract posture and combined with phonatory action.

Thus, early hominids chanced on the particulate principle by adopting for communicative use an apparatus already divided neuroanatomically into discrete components. Constrictions of the distinct parts could be associated with distinct meanings. At first, utterances would necessarily have been holistic, employing (perhaps repeated) constrictions of a single vocal organ, superimposed on a characteristic posture of the vocal tract as a whole. Actions of distinct organs could not yet be permuted and combined to form different utterances. Pressure for differentiation would have grown as demands for lexical increase were placed on an apparatus equipped for a limited number of discrete speech actions. The initial impetus for 'reuse' of articulators (and so for the emergence of combinatorial mechanisms) would then have come from the simple fact that articulators were few in number (cf. Lindblom 1998; 2000).

Once gestures of distinct organs (and their neurophysiological support) had evolved as discrete, combinable units, expansion of the phonological system could have occurred by sociocultural processes in at least two ways without any further genetic change. First, expansion of the set of discrete gestures could have resulted from differentiating the actions of a given organ into distinct types through a process of mutual attunement, or accommodation, between individual speakers/hearers. Second, increasingly complex gestural structures could have emerged, somewhat as we now observe the process in the developing child.

In what follows, we begin by justifying the gesture as the basic unit of phonological structure, and then go on to show how the basic set of gestures might have been expanded and elaborated in specific languages by sustained speaker–hearer interactions.

## Articulatory Phonology

Within the approach of articulatory phonology (e.g. Browman and Goldstein 1992; 1995a), the most primitive (atomic) units of phonological structure are hypothesized to be units of articulatory action, or *gestures*. Since units of articulatory action can be defined and in principle observed even in the absence of a mature phonological system, this approach allows the description of phonology and pre-phonological vocal behaviour with the same set of primitives. This stands in sharp contrast to most theories of phonological structure, in which the units (features) are defined relationally, in *opposition* to one another (e.g. Jakobson et al. 1951/1963), and therefore cannot exist, as such, in the absence of a developed system. From the perspective of articulatory phonology, we can observe the development of a phonological system in the child (and, by inference, the broad course of its evolution in the species) as atomic action units come to be combined systematically into the ions and molecules (roughly, segments and syllables) of mature phonological form.

The principal theoretical motivation behind articulatory phonology is to reconcile the apparently incompatible phonological and physical descriptions of speech. The phonological description views speech as composed of a small number of discrete (particulate), permutable, context-independent units, while the physical description finds continuous, context-dependent variation in a large number of articulatory, acoustic, auditory, or neural variables. The discovery of this incompatibility (through the development of techniques for physical analysis of speech in the mid-twentieth century) led theoreticians to remove phonological units from the domain of direct physical observation altogether by hypothesizing that they correspond to mental (cognitive) events but *not* physical ones. This view is well captured by Hockett's (1958) familiar Easter egg analogy. Phonological units, in the mind of the message sender, are represented by a row of brightly coloured, but unboiled, Easter eggs. The act of talking involves smashing the eggs through a wringer. The message receiver then must act like a forensic expert, reconstructing the original egg sequence from the bits of shell and knowledge of possible sequences.

While contemporary views of the relation between the phonological and physical descriptions of speech are less colourful than Hockett's, they are similar theoretically. Phonological units are seen as exclusively mental (e.g. Pierrehumbert 1990), and implementation rules (e.g. Keating 1990) are posited to translate these discrete mental representations to their physical interpretations. Since there are no generally accepted constraints to which implementation rules must adhere (Clements 1992), the relation between the phonological and physical structures of speech can even be construed as arbitrary.

Apart from its inelegance, this bipartite view of phonology and speech has substantive problems. First, as Fowler (1980) has argued, it would be extremely odd for a communication system to evolve in which the essential properties of the system's units are destroyed in the act of conveying them. Second, as Browman and Goldstein (1995*a*) point out, reciprocal constraints prevail between the structure of phonological systems and the physical properties of phonological units, exactly as would be expected if the phonological and physical descriptions of speech were, in fact, low and high dimensional descriptions of a single self-organized complex system (e.g. Kugler and Turvey 1987). For example, languages that contrast voicing in labial and coronal stops often lack a voiced dorsal stop. This tendency can be attributed to the relative difficulty of sustaining phonation under the aerodynamic conditions imposed by the different closures (Ohala 1983). More generally, suspension of contrast (neutralization) can often be understood with reference to physical constraints. At the same time, physical properties associated with some phonological units may vary cross-linguistically as a function of the nature and number of contrasting units in the language's inventory. For example, the amount of context-dependence exhibited by a particular vowel may vary as a function of the number of other vowels in a language's system (Manuel 1990). Thus, certain constraints on the physical properties of speech stem from the structure of phonological inventories, and certain constraints on phonological inventories stem from physical properties of speech.

*The Gestural Hypothesis: Units of Information
as Units of Action*

Articulatory phonology resolves the apparent incompatibility between the phonological and physical (phonetic) descriptions of speech by hypothesizing that speakers deploy discrete, context-independent units of action, permuting and combining them to form the contrasting words of the language. Thus, articulatory phonology equates the most basic units of phonological information with units of vocal tract action. In this view, the apparent continuity of speech stems from the fact that measurements of speech record the *results* of the activities that go on in speakers' vocal tracts when they talk, but not the activities themselves.

As an example of an action unit, consider the word *bad*. If we examine what takes place in the vocal tracts of speakers of English as they produce this word, we expect to see changes in the moment-to-moment positions of all the vocal tract articulators, and differences in that pattern of changes across speakers and across contexts in which the word is spoken. However, we also expect something to be the same across speakers and contexts. At the beginning of the word, the two lips come together to form a seal: the upper lip is displaced downward and the lower lip is displaced upward, due both to their own muscular activations and to the raising of the mandible. This pattern of activities constitutes a distinct action unit, a lip closure *gesture*. By hypothesis, this same action unit is deployed by speakers when they produce words like *mat*, *pint*, and *lob*, in coordination with other gestural actions.

A lip closure gesture is not identical to a traditional phonological segment or feature. It is clearly 'smaller' than a segment, as several different segments involve lip closure gestures (/p/, /b/, /m/). It is also 'larger' than a feature, as it would be specified as [labial], [−continuant], [+consonantal]. However, it is a viable choice as the basic permutable element (atom) of phonological structure, because larger structures (syllables, words) can then be viewed as molecules composed of (coordinated) gestural atoms, and segments can be viewed as coherent substructures of gesture (corresponding to ions in the chemical analogy) that recur in many distinct molecules Not only is such a view of phonological structure possible, but Browman and Goldstein (e.g. 1992; 1995*b*) have argued that certain kinds of phonological alternation (casual speech, allophony due to syllable position) are better described with gestural structures than with traditional feature-based representations. Gafos (2002) has shown the necessity of positing gestural structures in accounting for some cases of more abstract allomorphy.

## Bases for Discrete Action Units: Organs of the Vocal Tract

The nature of the action units posited by articulatory phonology and the basis for their particulation can provide some insight into how such a system could have evolved. The human vocal tract can be viewed as composed of a small number of relatively independent constricting devices: the lips, tongue tip/blade, tongue body, tongue root, larynx, and velum. While these devices, or 'organs' as we shall refer to them, share some common articulatory components (e.g. jaw movement contributes to lip, tongue tip, and tongue body constrictions), each of these organs can constrict without causing a concurrent constriction of one of the other organs. Action units are hypothesized within articulatory phonology to be constriction gestures performed by one of these organs.

Gestures of different organs 'count' as different and provide the basis for phonological contrast. Thus, the word *bad* begins with a gesture of the lips organ. It contrasts with the word *dad*, which begins with a constriction of the tongue tip organ instead of the lips organ, and also with the word *pad*, which adds a gesture of the larynx organ to that of the lips. Organs are then functional synergies or coordinative structures composed of several articulators. A gesture of the lips organ involves a functional synergy comprising the upper lip, lower lip, and jaw articulators.

Of course, not all contrasting speech gestures differ in the organ employed. For example, the words *tick*, *thick*, and *sick* all begin with gestures of the tongue tip organ, and most vowels are composed of tongue body gestures (with or without a lips gesture). Gestures of a given organ may also differ in the metric properties of the constrictions that they create—for example, in the degree of constriction (*tick* vs. *sick*), or in the location within the vocal tract of the constriction (*sick* vs. *thick*). However, evidence from the nature of phonological systems and from phonological development in children suggests that the most basic phonological contrasts are those that involve different organs.

### Discrete Organs in Phonologies

The phonological system of almost every language includes a contrast between labial (lips), coronal (tongue tip), and dorsal (tongue body) stops, and a contrast of nasality (velum) in stops is almost as common. These are all between-organ contrasts. The importance of organs can also be seen in the phonological alternations that languages exhibit. Within the theory of feature geometry, features are arranged in a hierarchy that captures their behaviour in phonological processes, such as spreading (assimilation) and delinking (neutralization of contrast). A higher feature node dominates its dependents in the sense that when a given feature node is involved in an operation (such as spreading), all the dependent nodes necessarily participate in the operation as well. The converse is not true. Feature nodes that refer to the organ employed stand near the top of this feature hierarchy. For example, in McCarthy's (1988) geometry, nasal, laryngeal, and place (in turn branching into labial, coronal, dorsal, and pharyngeal) are branches at the top of the feature tree.

### Discrete Organs in Children's Systematic Errors

Evidence for the somatotopic decomposition of the vocal tract into distinct organs comes from systematic correspondences between children's and adults' utterances. Corpora of such correspondence 'errors' have little to offer on this topic, however, if they simply classify and tabulate errors according to standard phonetic features. Olmsted (1971), for example, collected perhaps the largest such corpus: over 3,000 errors from half-hour interviews with 100 English-speaking children, aged from 1;3 to 4;6. He classified consonant errors by place, friction, voice, and nasality, and found place errors to be far the most common. If we rescore the data (Olmsted 1971: 71, table 13), however, classifying responses by whether the correct organ (lips, tongue tip, tongue body, velum) was employed or not, friction errors are now most frequent, with organ and voice errors roughly equal. Yet we still cannot tell whether any particular error resulted from paradigmatic substitution of a wrong gesture or from syntagmatic shift of the correct gesture to a wrong point in the utterance (Studdert-Kennedy and Goodell 1995; Studdert-Kennedy 2002). For an adequate test of the somatotopic hypothesis we need a different type of corpus. First, we need data from children during that narrow developmental window in which they are attempting their first words, before they have gained enough control over the amplitude and timing of their gestures to make random adult-like speech errors. Second, we need full transcriptions of a child's actual utterances and their presumed targets, so that we can determine the nature of the errors.

Data of this kind have come from several sources, although they have not been analysed in terms of gestures (e.g. Menn 1983; Vihman and Velleman 1989). Ferguson and Farwell (1975), for example, showed that standard phonetic transcriptions of a child's earliest words (first fifty) are quite variable phonetically: the initial consonant in a given word (or set of words) produced by the child is transcribed as different phonetic units ('segments') on different occasions. Ferguson and Farwell (1975) argue that the variability is too extreme for the child to have a coherent phonemic system, with allophones neatly grouped into phonemes, and that the basic unit of the child's production must therefore be the word, not the segment. If we assume further, and not incompatibly, that the child is not producing segments at all, but rather (gross) constriction gestures of the vocal organs that vary in their exact degree and location, it turns out that children are remarkably consistent in the organs they move for a given word (Studdert-Kennedy 2002). Thus, children appear to be acquiring a relation between actions of distinct organs and lexical units very early in the process of developing language. This relation is apparently not mediated by a phonological unit specified for additional features. At this stage, it is simply organ identity itself that infants employ to differentiate lexical items.

### Discrete Organs in Vocal and Facial Imitation

The preceding discussion takes for granted a capacity for vocal imitation, or vocal learning, that is unique among primates to humans (Hauser 1996). The role of imitation in language acquisition is often discounted, because children developing syntax do very much more than repeat what they have heard. Whatever more they do depends, however, on their first building a repertoire of words by copying their companions. For the modern child (as perhaps for the early hominid), vocal imitation is the enabling step into a particulate phonetics without which a particulate (or compositional) syntax could never develop (Studdert-Kennedy 2000). To imitate a spoken utterance, imitators must first find in the acoustic signal information specifying which organs moved, as well as where, when, and how they moved with respect to each other, and must then engage their own articulatory systems in a corresponding pattern of action. Such a skill requires, as suggested above, that the vocal tract be represented somatotopically in the brain (Lindblom 1998; 2000). How might such a representation have evolved?

Here the facts of facial imitation, also unique among primates to humans (Hauser 1996), lend insight. Particulation of the mouth and face into distinct organs can be seen in the facial mimicry of which very young infants, even neonates (one hour old), have been shown to be capable (Meltzoff and Moore 1997). In these experiments, when the experimenter performs an action with some facial organ (e.g. lips, tongue, eyes), the infant produces some movement of the same organ, even though the movement may not be the same as that of the adult (at least not on the first try). For example, if the experimenter protrudes the tongue from the mouth sideways, the infant will also protrude the tongue, but perhaps straight out. Thus, for neonates, the first step into facial mimicry is to match organs: infants evidently individuate their own orofacial organs and identify them with another's. Organ identification is a fundamental part of Meltzoff and Moore's (1997) model of early facial mimicry for which the infant has only optic information about the model and only kinaesthetic information about his/her own face.

Organ identification, as we have just seen, is also the first step in a child's early attempts to imitate words. The close link between facial expression and vocal tract configuration was indeed remarked by Darwin (1872/1998: 96), and is evidenced in modern humans by audiovisual interactions in the perception of speech and by their capacity for lip- or speech-reading. Neural support for facial and vocal imitation may come from 'mirror neurons' of the type discovered for manual action in the macaque monkey by Rizzolatti et al. (1996). Such neurons discharge not only when an animal engages in an action, but also when it observes another animal engage in the same action.

We do not have space to develop the argument here (see Studdert-Kennedy 2002), but we are encouraged by the salience of facial expression in human communication to hypothesize that: (i) facial imitation evolved in the mimetic culture of early hominids (Donald 1991) by duplicating a mirror system already established for manual action in lower primates, and (ii) vocal imitation evolved by coopting and extending the facial mirror system with its characteristic somatotopic organization. Evidence consistent with activation of a mirror neuron system in the perception of speech has indeed already come from transcranial magnetic stimulation studies (Fadiga et al. 2002), and we believe that our hypotheses concerning the mechanisms of facial and vocal imitation should be amenable to further experimental tests by brain imaging and other techniques.

In any event, it is evidently the capacity for vocal imitation that supports mutual attunement of individuals' speech. And it is by attunement, we hypothesize, that the universal set of organ-based gestures is enlarged and elaborated in specific languages.

## Expansion of Phonological Systems by Attunement Among Individuals: Particulating Gestures with a Common Organ

Within articulatory phonology, gestures are modelled as task-dynamical systems that regulate the formation of constrictions by one of the vocal tract organs (e.g. Saltzman and Munhall 1989). The dynamic specification for each gesture includes a spatial goal, which is the equilibrium position for the (point attractor) dynamics. In tongue tip gestures, for example, a goal is specified for two vocal tract variables: tongue tip constriction degree (distance of the tongue tip from the palate) and tongue tip constriction location (location of the constriction along the palate/upper tooth surface). So the oral constriction gestures at the beginning of words like *tick, thick, sick* differ in degree and location, which, of course, define continuous metric spaces. How do these spaces come to be partitioned into discrete regions (or points) that can serve as contrasting gestures?

One possible answer is that partitioning emerges when multiple vocal tracts *attune* themselves to one another by mutual vocal mimicry. While we can describe speech actions as events within an individual (I-language), the actions must also be shared by the members of a speech community (E-language) (Chomsky 1986). Attunement can be observed in phonological development as early as 6 months of age, when infants begin to modify their speech actions (babbling) so as to approximate quantitative properties of the speech to which they are exposed. Of course, particulation of a continuum by attunement to the ambient language depends on the continuum being already partitioned in that language. From an evolutionary perspective, we must ask where this partitioning comes from in the first place.

### Consonant Gestures

We can test the possible evolutionary role of attunement in partitioning a continuum by means of computational simulations in which none of the participants (or 'agents') begins with the continuum broken into categories. Browman and Goldstein (2000) report the results of such simulations in which all agents begin by producing the points along some continuum at random, with uniform density distribution along the continuum. The agents attune by comparing their gestures to the gestures they perceive their partner(s) producing, and if gestures match within some criterion, agents increase the likelihood of reproducing the matching continuum points. The outcome of repeated cycles of interchange depends on the nature of the articulatory–acoustic mapping. If the articulatory continuum is homogeneous but the acoustic 'continuum' includes regions of spectral discontinuity, non-random partitioning can emerge. For example, suppose that the acoustic consequences of a tongue tip constriction-degree continuum have the step-like, quantal character described by Stevens (1989): 'stable' regions of silence, turbulent noise, and low amplitude vocalic phonation, in which there is little acoustic change as constriction degree is varied, bordered by 'unstable' regions of rapid acoustic change. When agents recover their partner's constriction degree from the acoustics of such a system in the presence of noise, their behaviour converges on three 'categories' corresponding to stops, fricatives, and glides. That is, the initial random distribution of constriction degree across the continuum changes to a tri-modal distribution, with peaks in the stable regions. Only these stable regions afford attunement. Thus, the discrete categories are density distribution maxima that emerge when randomly acting vocal tracts attune to one another, under constraints of the articulatory–acoustic mapping. They are attractors of the dynamic system of attunement.

On the other hand, if both articulatory and acoustic continua are homogeneous, the outcome of attunement is that agents simply converge on a narrow range of values along the articulatory continuum. This might be the case for within-organ contrasts such as constriction location among tongue tip consonant gestures (e.g. dental, alveolar, retroflex). Organ differentiation and articulatory–acoustic mappings will lead a language to evolve a stop gesture produced with the tip of the tongue. As speakers attune, however, they converge on a range of locations for tongue tip stops that is narrower than the range of anatomically possible tongue tip stops. Such a narrowed range could then become a point of reference for other gestures that might emerge along the continuum, and so be subject to competition with these new gestures for motor and perceptual efficiency. Notice that, on this account, gestures are in no sense targets aimed at by the emerging system. They arise from random search, driven by pressures on speakers to enlarge their shared lexicons and to 'sound like one another'.

### Vowel Gestures

Vowels are not entirely straightforward from the perspective of constricting organs. On the one hand, all vowels involve gestures of the tongue body, with differences between vowels being analysable as differences of constriction location (palatal, velar, uvular, pharyngeal) or degree. On the other hand, the vowel /u/ involves a lip gesture in addition to a tongue body gesture, and it would be possible to analyse the vowel /a/ as involving a tongue root gesture in addition to a tongue body gesture. Thus, the most basic vowels found in almost all human languages (/i, a, u/) can be viewed as gestures that employ distinct organs. With these vowels as reference points other vowels could emerge by attunement, as described above.

De Boer (2001) has proposed a related process of vowel evolution. He presents simulations of how vowel systems like those of human languages (in the number of vowels and their distribution) can self-organize through the interaction of agents attempting to imitate one another within the constraints of a modern vocal tract. None of the agents possesses a vowel system at the outset, and vowels are added to growing systems both at random and in response to imitative failures. Strikingly, pressure to add vowels to a system does not come from the need for linguistic contrast (which has no role in these simulations), but from the need to optimize imitative success. However, the form of the simulation is such that it presupposes the concept of discrete units, so that the issue of where agents get the idea that vowels are discrete points within their articulatory capabilities is never addressed.

Another approach to the emergence of vowels, viewed as tongue body gestures that differ parametrically, comes from the theoretical work of Carré and Mrayati (1995). These authors show that the vocal tract as an acoustic transmission system can be divided into distinctive regions that have the property of maximal acoustic efficiency: for constrictions within these regions, small changes in vocal tract area result in large perturbations of formant frequencies away from the values for a neutral tube. They explain the evolution of vowel systems on this basis: vowel gestures are characterized by (acoustically) maximally efficient constrictions and thus fall within the distinctive regions. Carré and Mrayati (1995) do not themselves distinguish within- from between-organ differentiation of gestures. For them, the basis for particulation is purely acoustic. However, while the relevant attunement simulations have not been done, it is not hard to imagine that differential acoustic efficiency would cause the attunement process to partition tongue body constrictions into the locations (although not the degrees) that are observed in vowel systems.

## Expansion of Phonological Systems by Development of Complex Gestural Structures

Evidence for another possible way of expanding the phonology of a language comes from the order of acquisition of English consonants. As we have seen, gestures in articulatory phonology are atoms that combine with other gestures to form molecules corresponding to larger phonological units such as segments, syllables, feet, or words. Combining gestures in volves establishing a pattern of (temporal) coordination among them. Words such as *cab* and *back* are composed of the same gestures coordinated in different ways. Segment-sized phonological units (vowels or consonants) can be viewed as ions, combinations of atoms that recur in many different molecules. Crucially, segmental ions differ in their internal complexity. Some consonant segments are composed of single gestures, while others are complex, requiring the coordination of multiple gestures. This difference in complexity has implications for language change (cf. Lindblom and Maddieson 1988), but also, as argued below, for the course of phonological development, and so, by hypothesis, for the evolution of phonological structures (cf. Kohler 1998).

While the order of acquisition of individual consonant segments in English-learning children varies widely across individuals, Dinnsen (1992) finds that all children's consonant inventories (in a sample of forty), at any point in time, fall into one of the five classes of segments shown here in (1a–e). These classes exhibit increasing levels of complexity: Dinnsen defines the classes as elaborations of a feature 'tree', with additional features being recruited at each level. Here we will see that the succession of levels follows from the inter-gestural coordination required to produce the segment types of each level.

(1)   (a)   stops (voiceless unaspirated), nasals, glides and [h];
      (b)   all segments in (a), plus voicing contrast in stops;
      (c)   all segments in (b), plus fricatives and affricates;
      (d)   all segments in (c), plus one liquid ([l] or [r]);
      (e)   any English segment.

The consonants in level a are just what we would expect if children were producing poorly controlled, uncoordinated gestures of vocal tract organs. Note that (i) the almost interchangeable appearance of stops and glides of the same organ in the phone classes of children's first fifty words (Ferguson and Farwell 1975) suggests that degree of constriction is not being systematically controlled; (ii) the fact that stops are voiceless unaspirated indicates that oral and laryngeal gestures are not being coordinated—voiceless unaspirated is the expected phonatory consequence of a stop closure in the absence of any consonant-specific laryngeal adjustment; (iii) a gesture of the larynx (glottal abduction) in the context of regular phonation will be perceived as [h]; (iv) a nasal stop will be perceived if a child happens to lower the velum and effect an oral constriction at roughly the same time: a wide range of overlaps can be perceived as nasal. This interpretation is consistent with the appearance of oral and nasal stops of the same organ within phone classes of Ferguson and Farwell (1975).

The appearance of aspirated stops at level (b) heralds the ability to coordinate gestures in a stable fashion: production of aspirated stops requires coordination of an oral constriction gesture with glottal abduction. Interestingly, level (b) tends to occur exactly when the first CVC words (as opposed to CV) are found in children's productions (Stoel-Gammon 1985). Note that a CV word can be produced by a single organ forming a constriction and release without any precise coordination of consonant and vowel gestures. By contrast, a CVC word requires precise inter-gestural coordination—either consonant gestures to vowel or consonant gestures to each other. Otherwise, the temporal structure of a CVC syllable will not be perceived.

The voiceless fricatives and affricates of level (c) could not be expected any earlier than the aspirated stops of level (b), as they require laryngeal–oral coordination. But why do fricatives come later? Note first that the most common early fricative in children's inventories is [s], for which the positioning of the lower teeth (by action of the mandible) must be coordinated with the tongue tip constriction against the alveolar ridge, a problem in coordination that does not arise in the production of stops or glides. A second possible reason for the delay is that fricatives and affricates require more precise control of constriction degree than do stops or glides, in order to form a constriction just narrow enough to produce turbulence.

At level (d), the child begins to produce a liquid ([r] or [l], but not both). English liquids involve coordination of multiple oral constriction gestures which is presumably more difficult than coordination of oral and laryngeal gestures seen at level (b). Evidence for liquids as a coordination problem comes from children's common 'errors' of omission. For example, [w] for [r] follows from omission of the tongue tip gesture, [j] for [l] from omission of the tongue body retraction gesture. On this account, other kinds of segments (non-liquids) that require coordination of multiple oral constriction gestures should also be acquired relatively late, as is evidently the case for click consonants (Herbert 1990). Conversely, liquid consonants that do *not* require coordination of multiple oral constrictions, could be acquired relatively early, as seems to be the case for the trilled [r] of K'iché (Ingram 1992).

At level (e), a second liquid is acquired. If coordinated gestures of mechanically linked organs is difficult (level (d)), then imposing more than one pattern of coordination on those organs should be even more difficult.

In summary, empirical regularities in the order of acquisition of English consonant segments support the hypothesis that gestures are the basic units of phonological structure, and that phonological development is a cumulative, hierarchical process by which gestures are combined into larger and increasingly complex structures.

## Summary and Conclusions

According to the particulate principle of self-diversifying systems, a necessary condition for the evolution of discrete infinity—a property on which every language, spoken or signed, is based—was the evolution of a finite set of discrete phonetic units. Since production is logically prior to perception, a necessary condition for the evolution of these units was an integral neuroanatomical system of potentially independent movable parts (such as the hand, the face, or the vocal apparatus) that could be differentiated to effect rapid, coordinated strings of discrete motor actions, or gestures.

Under pressure for mutually intelligible imitative exchange, the vocal apparatus differentiated evolutionarily into six neuromotorically independent structures (lips, tongue tip, tongue body, tongue root, velum, and larynx) capable of effecting discrete changes in vocal tract configuration. By hypothesis, these organs provide the universal biologically determined base for discrete units of phonetic action in every language. Different languages then expand and elaborate the basic gestural phonology by sociocultural processes of attunement among speakers through mutual vocal mimicry, resulting in differentiation of gestures produced by a given organ and coordination of gestures into larger complex structures. The imitative processes of learning to speak and of mutual attunement among speakers of a language may have evolved by means of specialized, somatotopically organized systems of mirror neurons of the type observed for manual action in the macaque monkey.

### FURTHER READING

We have here taken it for granted that early evolutionary steps into language entailed differentiation of the primate vocal apparatus and its neural support, under pressure for lexical growth, but we did not consider proximate pressures (e.g. toward ease of articulation and perceptual distinctiveness) that must have shaped the morphology, modes of action, and acoustic outputs of the apparatus.

For an assessment of the role of an enlarged pharynx, relative to that of other primates, in increasing phonetic scope, see Fitch (2000). For an account of the role of acoustic contrast in differentiating the vocal tract, see Carré and Mrayati (1995). For discussion of language-independent neuromuscular and biomechanical constraints on the evolution of a basic, universal set of consonants and vowels, see Kohler (1998).

A crucial step in speech evolution, according to MacNeilage, was adoption of the open/close alternation of the primate mandible (previously used for chewing and for communicative lip/smacks) as the 'frame' for syllables. Subsequent expansion entailed differentiation of the syllable into its segmental 'content', consonants and vowels. For the 'frame/content' theory of speech production, see MacNeilage (1998).

Lindblom, also concerned with derivation of phonetic segments and gestures from non-linguistic precursors, sees phonological evolution as a tug-of-war between speaker and listener, fought through words and their phonetic components, as they compete for survival within a language. Lindblom conceptualizes the strengths of phonetic segments or gestures as ratios of articulatory cost to perceptual benefit, minimized by competition and summed across the phonological system to a least squares criterion. He quantifies articulatory cost, in principle, in terms of the metabolic costs of learning, remembering and executing a phonetic unit, in such a way that *reuse* of a unit reduces its cost. Thus, discrete gestures and combinatorial mechanisms emerge automatically, not only because gestures are executed by a small number of discrete organs, as argued above, but also because gestural cost–benefit ratios are reduced by repeated use of a small number of units. For full development of these ideas, see Lindblom (1992; 1998; 2000).

Finally, for selected readings on articulatory phonology, see the following: basic motivation (Browman and Goldstein 1986), application to sound change (Browman and Goldstein 1991; Gick 1999), fluent speech (Browman and Goldstein 1990; Zsiga 1995), syllable structure and prosody (Browman and Goldstein 2000; Byrd and Saltzman 1998; Gick in press).

# 14

# Motor Control, Speech, and the Evolution of Human Language

*Philip Lieberman*

## Introduction

The mark of evolution is always evident in biology. The anatomy and brains of humans have been shaped in the course of evolution to make human language and thought possible and the comparative method, based on the principles and procedures introduced by Charles Darwin, can decode the evolutionary process. The study of humans and other species, their communicative and cognitive abilities and the biological substrate regulating these behaviours explicate human language and cognitive ability. Working in this biological framework, drawing on evidence from many independent studies, I will attempt to demonstrate that the mark of evolution is evident when one examines the anatomical specializations and neural mechanisms that make human speech and language possible.

I shall note studies that show that the neuroanatomical structures of the human brain that regulate speech production form crucial parts of the networks that confer syntactic and cognitive ability. I shall also present the argument that brain mechanisms adapted for adaptive motor control were the starting point for the evolution of human language—the initial set point being upright, bipedal locomotion, walking. Natural selection directed at enhancing motor control mechanisms for speech production, then, played a central part in the evolution of the human brain and its unique behavioural attributes, language, and higher cognition.

Charles Darwin introduced more than the theory of evolution by means of natural selection. He also provided the research paradigm that has yielded virtually all our knowledge concerning the evolution of human beings and other species. Darwin noted the mark of evolution discernible when the biological bases of behaviour in related living species are compared. For example, we cannot directly observe either the behaviour or brains of the extinct reptiles who were ancestral to early mammals or early mammals. However, although contemporary reptiles are not identical to these extinct reptilian species, they retain skeletal features of the fossil species. Therefore, reasonable inferences concerning the evolution of the morphology and physiology governing the behavioural traits that differentiate mammals from their extinct reptilian ancestors can be derived by studying and comparing the anatomy, physiology, and brains of contemporary reptilian species with mammals.

Evolutionary biologists differentiate 'primitive' and 'derived' characteristics (aspects of morphology, physiology and behaviour). A primitive human feature is a hand that has five digits (frogs, monkeys and apes have five digits). In contrast, humans have "derived" skeletal morphology that confers the ability to walk upright—a feature absent in apes. The evolution and biological bases of the derived characteristics that enable walking have been established by comparing the skeletal morphology and muscles of living apes with the skeletal remains of extinct hominid fossils. These skeletal features in living apes differ from those of early hominids who lived million of years ago; apes lack the derived skeletal morphology that confers the ability to walk upright with minimal muscular effort for long periods of time. The analysis is facilitated by the fact that apes occasionally walk upright for short distances. No absolute distinction exists in this regard—to a degree apes walk, though not in the same manner as humans.

The study of the evolution of human language has been vexed by the claim that human language bears no relationship to the communications of any living species. If this were true, comparative studies would be useless, the biological bases of human language and cognition being disjoint from those of all other species. This claim is not new; it can be seen in various creation myths and is reified in the work of Noam Chomsky and his adherents. The uniqueness of human language has been linked at various times to claims that no other living species can understand or communicate by means of words (Chomsky 1976) or that only human language makes use of syntax (Calvin and Bickerton 2000). However, these claims are refuted by comparative studies of the communicative and linguistic abilities of dogs, apes, birds, and other species. These studies show that the biological bases of human linguistic ability depend both on primitive characteristics shared with many species and on certain derived anatomical and neural properties. It is first necessary to determine the primitive characteristics of human language in order to establish the derived biological features of the human brain and body that make us human.

## The Biological Mosaic of Human Language

Many discussions of the evolution of language posit a sudden transition or transitions from animal communication to 'language', or from 'proto-language' to 'full language'. However, the mosaic of primitive and derived biological mechanisms that presently confer human linguistic ability suggests that language had a complex, graded evolutionary history. The intermediate stages that led to human language are not readily apparent because the species ancestral to modern *Homo sapiens* are extinct. At one time lexical ability, naming, was thought to be one of the hallmarks of language (Chomsky 1976). The ability to name objects, actions and states of being clearly is a primitive feature of human language that must have been present in the earliest hominid species some five million years ago. Since present-day chimpanzees can acquire about 150 words and devise new words as well as modifying the meaning of words that they already have (Gardner and Gardner 1969; 1984), it is clear that this aspect of human language stretches back to the dawn of hominid evolution. Again, although virtually all theoretical linguists have focused on syntax and rule-governed processes as the novel aspects of human language, chimpanzees can understand sentences, deriving meaning from the sentence's syntax (Gardner and Gardner 1994; Savage-Rumbaugh and Rumbaugh 1993). It is most unlikely that the earliest hominids' command of syntax was less than that of language-using chimpanzees. Therefore, lexical ability and the neural bases of these attributes of human language are primitive characteristics.

Although apes can, to a degree, comprehend human speech, they cannot vocally reply. Despite many attempts, spanning several centuries, to teach apes to talk, they cannot speak. Human speech achieves its productivity by altering the sequence in which a limited number of speech sounds occur. The words *see* and *me* contain the same vowel; the initial consonants signify different concepts. Changing vowels also signifies different words—*sue, ma, sit, mat.* Human speakers are able voluntarily to alter the sequence of articulatory gestures that generate the meaningful speech sounds, the 'phonemes' that convey the words of their language. Different languages have particular constraints, but any neurologically intact child raised in a 'normal' environment learns to speak his or her native language or languages. It is clear that apes lack the neural capacity freely to alter the sequence of muscle commands that generate phonemes. Anatomical limitations, which will be discussed below, limit the range of phonetic forms that apes could produce. However, acoustic analysis of the vocal signals that they produce in a state of nature reveals many of the segmental phonetic elements that could be used to form spoken words. Chimpanzee vocalizations, for example, include segments that could convey the consonants [m], [b] and [p] and the vowel of the word *but* (Lieberman 1968; Hauser 1996). Computer modelling studies (Lieberman et al. 1972) show that the chimpanzee speech apparatus could produce the sounds [n], [d], and [t] and most vowels other than [i], [u], and [a] (the vowels of *tea, too,* and *ma*), but chimpanzees are unable voluntarily to speak any English words. Field observations show that chimpanzee vocal signals are bound to particular emotions or situations. They even have great difficulty suppressing their calls in situations where that would appear to be warranted (Goodall 1986).

Thus, if we follow the logic of evolutionary biology, we must conclude that speech production is a derived characteristic of human language. And, although virtually all theoretical linguists have focused on syntax for the past forty years, we must account for the evolution of human speech. Interestingly, converging evidence from the neurophysiological research discussed below suggests that the structures of the human brain implicated in speech production play a crucial part in conferring human syntactic and cognitive ability (Lieberman 2000).

## Speech and Language

In most discussions of human language little attention is paid to the rate of information transfer. At normal speaking rates, twenty to thirty phonemes are transmitted from a speaker to listeners. This rate exceeds the temporal resolving power of the human auditory system; individual non-speech sounds merge into a buzz at rates in excess of fifteen per second (Liberman et al. 1967). Indeed, it is difficult even to count more than seven sounds per second. Studies of the evolution of the physiology of speech production indicate that human anatomy has been modified to facilitate this process.

A short discussion of the anatomy and physiology of speech production may be useful in understanding the contribution of human speech to human linguistic ability. The physiology of speech production has been studied since the time of Johannes Muller (1826). The lungs power speech production. The outward flow of air from the lungs is converted to audible sound by the action of the larynx, which during 'phonation' generates a series of quasi-periodic puffs of air. The larynx operates in much the same manner as the diaphragm of a harmonica, producing acoustic energy which is then shaped into the notes of a musical composition by the air passages above it. The harmonica's air passages each act as a filter, allowing maximum acoustic energy through at a particular frequency or 'note'. The airway above the human larynx filters the acoustic energy produced by the larynx, except that we perceive the frequencies at which maximum acoustic energy as different speech sounds. Acoustic energy can also be generated by air turbulence at a constriction, in much the same manner as the source of acoustic energy in a flute or organ.

The shape of the airway above the larynx, termed the 'supralaryngeal vocal tract' (SVT), continually changes as a person talks. Major changes in its shape are produced by the tongue. The human tongue, which extends down into the pharynx, can be depressed or elevated backwards or forward to change dramatically the shape of the SVT (Nearey 1979). The position and degree of constriction of the lips and vertical position of the larynx, which can move up or down about 25mm, can also change the SVT configuration. The filtering properties of the SVT are determined by its shape and overall length (Chiba and Kajiyama 1941; Fant 1960; Stevens 1972). Peak energy can potentially be transmitted through the SVT at particular 'formant frequencies'. Systematic research since the end of the eighteenth century (Hellwag, 1781) shows that the phonetic properties of many speech sounds are determined by these formant frequencies. The vowel [i] of the word *see*, for example, differs from the vowel of [a] of the word *ma* solely because of its different formant frequencies. The pitch of a person's voice, which is determined by the rate at which the vocal cords of the larynx open and close, is irrelevant.

The rapid transmission rate of human speech follows from the 'encoding' of these formant frequencies. As a person talks, the SVT continually changes its shape, thereby changing the formant frequencies. For example, it is impossible to move the lips open instantaneously to produce the consonant [b] of the word *bat*. Nor can a person's tongue instantly move from the position that produces the formant frequencies that convey the vowel of *bat* to produce the [t] sound. The articulatory gestures that make up the initial and final consonants and vowel of *bat* are necessarily melded together forming a syllable. It is impossible to isolate the formant frequencies of the individual sounds that constitute the syllable. In the case of the word *bat*, all three phonemes are transmitted together, 'encoded', as a unit. Thus, the pattern of formant frequencies generated as a syllable is produced changes at a slower rate than the individual sounds of speech.

The rapid information transfer rate and complex conceptual and syntactic properties of human language derive from the encoded nature of human speech. The transmission of information at the slow non-speech rate precludes long or complex sentences; research at the Haskins Laboratories in the 1960s showed that listeners forgot the beginning of a sentence before reaching its end when arbitrary acoustic signals were used in place of speech. Moreover, the listeners' full attention was occupied with transcribing these non-speech signals. In contrast, human listeners effortlessly derive the sounds that make up the 'encoded' syllables of speech by means of a process that involves an implicit neural representation or knowledge of the constraints of speech production (Liberman et al. 1967). The decoding process must take into account the length of the SVT that produced a speech sound (Nearey 1979), since the absolute values of the formant frequencies produced by a long SVT are lower than those of a shorter SVT.

Studies of this aspect of the speech-decoding process have shed light on the evolution of human language. The perceptual mechanism that allows humans to perform this feat has a long evolutionary history; it is a primitive mechanism used by other species to gauge the size of a conspecific by listening to its vocalizations (Fitch 1997). For example, the length of a monkey's SVT is highly correlated with its height and weight. A larger monkey's vocalizations therefore have lower formant frequencies than a smaller monkey's, and other monkeys can gauge an individual's size by listening to it. Human beings can make similar estimates of body size and height (Fitch 1994).

This primitive perceptual mechanism, which is apparent at 3 months in infants (Lieberman 1984), is used by human listeners to decode the formant frequency patterns that transmit human speech. Talkers having different SVT lengths can produce different formant frequencies when they say the same words; listeners, therefore, may confuse one word for another when faced with the problem of rapidly shifting from one speaker's voice to another's (Peterson and Barney 1952; Hillenbrand et al. 1995). Under most conditions, human listeners rapidly adjust to the speech produced by speakers having different SVT lengths, the most effective vowel sound for this purpose is that of the word *see*—[i] in phonetic notation. Many studies have shown that [i] is the 'supervowel' of human speech. It is less often confused with other sounds (Peterson and Barney 1952; Hillenbrand et al. 1995). Words formed around [i] are correctly pronounced more often by young children as they acquire their native languages (Olmsted 1971). The vowel [i] is less confused because it yields an optimal reference signal from which a human listener can determine the length of the supralaryngeal vocal tract of a speaker's voice (Nearey 1979; Fitch 1994; 1997).

Anatomically modern human beings are the only living species who can produce the vowel [i] (Lieberman and Crelin 1971; Lieberman et al. 1972; Lieberman 1968; 1975; 1984; Carré et al. 1995). The adult human tongue and SVT differ from those of all other living animals (Negus 1949). In other species, including apes, the body of the tongue is long and relatively flat and fills the oral cavity. The non-human larynx is positioned high and can lock into the nasopharynx forming an air pathway sealed from the oral cavity, thereby enabling an animal simultaneously to breathe and drink. In contrast, the posterior portion of human tongue in a mid-sagittal view is round and the larynx is positioned low. The resulting human supralaryngeal airway has an almost right-angle bend at its midpoint. As we talk, extrinsic tongue muscles can move the tongue upwards, downwards, forwards or backwards—yielding abrupt and extreme changes in the cross-sectional area of the human supralaryngeal airway at its midpoint (Chiba and Kajiyama 1941; Fant 1960; Nearey 1979). The vowel sounds that occur most often in the languages of the world (Jakobson 1940; Greenberg 1963), [i], [u], and [a]—the vowels of the words *see*, *do*, and *ma*—can only be formed by the extreme area function discontinuities at the midpoint of the human supralaryngeal vocal tract (Lieberman et al. 1972; Stevens 1972; Carré et al. 1995).

Examination of the skulls of apes, newborn human infants, and the fossil remains of early hominids shows that the base of the skull and hard palate (the roof of the mouth) support SVTs in which the tongue is positioned almost entirely within the oral cavity. Acoustic analyses of the speech of human infants and living apes and computer modelling of the possible range of speech sounds show that they and the reconstructed SVTs of early australopithecine hominids could not have produced the vowel [i] as well as the vowels [u] and [a] (Lieberman 1975). Between birth and age 6, the skeletal structure and soft tissue of the human skull, tongue, and other aspects of human anatomy that define the SVT restructure. The human face moves backward from its birth position; the human face is almost in line with the forehead (D. Lieberman and McCarthy 1999). Present-day apes follow the opposite growth path. As they mature, their faces gradually project forward, yielding long mouths and long, thin tongues positioned almost entirely in the mouth. Early australopithecine hominids followed the non-human growth path and could not have produced the full range of human speech sounds. Neanderthal hominids, who survived until about 35,000 years ago, also appear to have followed the non-human growth trajectory (Vleck 1970; D. Lieberman 1998).

The primary life supporting functions of the mouth, pharynx, throat and anatomical components of the SVT are eating, swallowing and breathing. These functions are, as Darwin (1859) and anatomists such as Negus (1949) noted, impeded by the human SVT. The low position of the adult human larynx and the shape and position of the human tongue results in food being propelled past the opening of the larynx. Food lodged in the larynx can result in death. Chewing is also less efficient in the shorter human mouth; our teeth are crowded and molars can become impacted and infected, resulting in death in the absence of dental intervention. The only selective advantage that the human SVT yields is to enhance the neural process that estimates SVT length by making it possible to speak the vowel [i]. This must yield the conclusion that speech was already the mode of linguistic communication well before the appearance of anatomically modern *Homo sapiens*. In other words, the neural capacity to produce speech must have been in place before the evolution of the modern human SVT. It also is likely that fairly complex syntactic ability existed well before the appearance of modern human beings since the neuraoanatomical structures of the human brain that regulate speech production also are implicated in syntactic ability.

## The Neural Bases of Human Language

Clearly, the human brain must have some derived characteristics that allow us to produce voluntary speech conveying referential information. Human beings can also use alternate systems such as manual sign language, or even tactile stimulation in the case of Helen Keller. The traditional (Broca's-Wernicke's area) model for the neural bases of language was proposed by Lichtheim (1885); Broca's area of the cortex hypothetically regulates speech production, while Wernicke's area, a posterior region located near cortical regions associated with auditory and visual perception, is hypothetically concerned with language comprehension. Many recent versions of the standard theory (e.g. Caplan 1987), which take into account the difficulties that Broca's aphasics have in comprehending distinctions in meaning conveyed by syntax, claim that syntactic and other rule-governed processing occur in Broca's area; Wernicke's area is viewed as part of a system concerned with lexical access. However, it is becoming apparent that these locationist models are wrong. Neurobiological studies show that complex behaviours, such as talking, comprehending a sentence, or even preventing an object from hitting your eye, are regulated by 'functional neural systems' that link activity in many different neuroanatomical structures.

For example, a class of functional neural systems rapidly integrate sensory information with the present state of the organism and past experiences to achieve timely motor responses. Monkey brains have a neural system adapted to keeping foreign objects from hitting their eyes. Visual and tactile signals are channelled to the putamen, a subcortical basal ganglia structure, which interrupts ongoing motor activity; the monkey's arms deflect objects moving towards its eyes (Graziano et al.,1994). However, the monkey's putamen is not dedicated to this task as one might expect in 'modular' theories (Pinker 1986). The putamen is implicated in many other activities; populations of putamenal neurons form part of neural circuits regulating other aspects of motor control, as well as emotion and cognition (Alexander et al. 1986; Parent 1986; Cummings 1993; Marsden and Obeso 1994).

It is important to note that the putamen, in itself, is not the neural basis of the monkey's 'close object intercept' system. A neural 'circuit' linking activity in cortex and the basal ganglia regulates this aspect of monkey behaviour. Although various areas of cortex and particular subcortical structures can be identified that perform particular 'computations', such as storing visual information for a brief period or sequencing motor commands, complex behaviours are generally not regulated in one localized part of the brain. A neural circuit is formed by a segregated neuronal population in one neuroanatomical structure that projects to neuronal populations in other neuroanatomical structures. Complex behaviours generally are regulated by distributed networks consisting of neural circuits linking many neuroanatomical structures throughout the brain.

The model proposed in Lieberman (2000) for the nature and evolution of the neural bases of human speech and language takes into account these neurophysiological principles and studies on speech and language deficits resulting from damage to the human brain, as well as imaging studies of neurologically intact subjects. The data of these studies and neurophysiological and comparative studies of the brains and behaviour of other species suggest that the human brain contains a 'functional language system' (FLS) that evolved to regulate the production and comprehension of spoken language. Like other functional neural systems, the FLS evolved to produce timely responses to environmental challenges and opportunities. The linguistic functions executed by the FLS include rapidly sequencing speech motor commands and sequencing the cognitive sets and conceptual knowledge necessary to generate and comprehend sentences having complex syntax. Vocal language (or alternate forms such as manual sign language) allows humans to act or think rapidly, transcending the limits on information flow imposed by the auditory and visual systems. Articulatory rehearsal, a process of 'silent-speech', also maintains the information and concepts represented by words in verbal working (the short-term mental computational space in which sentence comprehension takes place) (Baddeley 1986; Awh et al. 1996). Neural circuits linking cortex and the basal ganglia also allow humans to form and shift abstract concepts in 'nonlinguistic' domains.

The neural bases of human linguistic ability, in which the subcortical basal ganglia play a key role, cannot be dissociated from other aspects of human behaviour. Converging evidence from comparative studies of many species show that basal ganglia carry out at least three functions:

(1) They are involved in learning particular patterns of motor activity that yield a reward.
(2) They play a part in sequencing the individual elements that constitute a motor programme.
(3) They interrupt an ongoing sequence, contingent on external events signalled by sensory inputs.

## Experimental Studies on the Neural Bases of Language

### Aphasia

Although the Broca's–Wernicke's theory for the neural bases of language is still cited in virtually all texts that discuss the evolution of language, it is wrong. The basis for this theory was the study of the effects of brain damage on language. Broca's and Wernicke's syndromes are real, but they are not caused by damage to the regions of the human neocortex named to memorialize the aphasiologists who first described some of the signs of these syndromes—sets of possible linguistic deficits. As the aphasiologists Stuss and Benson (1986) recognized, permanent language loss never occurs in the absence of major subcortical brain damage. The destruction of Broca's and Wernicke's areas causes transient language loss and other behavioural deficits. However, patients generally recover with six months. In contrast, damage to basal ganglia results in permanent aphasia..

Aphasiology began with Paul Broca's (1861) study of a patient who had suffered a series of strokes that destroyed an anterior area of neocortex, 'Broca's area', as well as massive subcortical damage. The most apparent linguistic deficit of the syndrome is laboured, slurred speech. However, a number of other disruptions to normal behaviour that characterize Broca's aphasia have since been noted, including deficits in fine manual motor control and oral apraxia (Stuss and Benson 1986). Broca's aphasics often have difficulty executing either oral or manual sequential motor sequences (Kimura 1993). 'Higher-level' linguistic and cognitive deficits also occur in this aphasic syndrome. The utterances produced by Broca's aphasics were traditionally described as 'telegraphic'. When telegrams were a means of electrical communication, the sender paid by the word and 'unnecessary' words were eliminated. Hence the utterances of English-speaking aphasics who omitted 'grammatical' function words and tense markers had the appearance of telegrams. Aphasic telegraphic utterances were thought to be the consequence of the patient's minimizing difficulties associated with speech production. The presence in Broca's aphasics of language comprehension deficits that involve syntax was established by studies starting in the 1970's. Broca's aphasics have difficulty comprehending distinctions in meaning conveyed by syntax (cf. Blumstein 1995). Anomia, word-finding difficulties, are common in aphasics who have damage to the frontal areas of the cortex described by Broca, as well as in other forms of aphasia. Patients having Broca's syndrome are unable to name an object or a picture of an object though they appear to be fully aware of its attributes; moreover, they usually have no difficulty classifying words along semantic dimensions. Indeed, Broca's aphasics often rely on semantic knowledge to comprehend the meaning of sentences that have moderately complex syntax. Non-linguistic deficits characterize Broca's syndrome; Kurt Goldstein (1948) stressed the cognitive deficits that often occurred in aphasic patients. Goldstein referred to the loss of the 'abstract capacity', deficits in planning, in deriving abstract criteria, and in 'executive capacity generally associated with frontal lobe activity' (Stuss and Benson 1986; Fuster 1989; Grafman 1989).

Wernicke in 1874 described different language deficits that hypothetically resulted from damage to a posterior area of cortex. The speech of Wernicke's aphasics is fluent, but they again have comprehension deficits. Though their speech is not distorted, it is often 'empty'. The syllabic structure of real words may be altered by sound substitutions, e.g. *poy* for *boy*, and by apparent semantic substitutions, e.g., *girl* for *boy*. In some cases neologisms are produced, nonexistent words that conform to the phonetic constraints of the speaker's language, e.g. *toofbay* (Blumstein 1995).

Although aphasia is generally thought to be the result of cortical damage, aphasiologists have wondered whether subcortical brain damage is responsible for these syndromes almost from the publication of Broca's studies. Marie (1926), for example, claimed that subcortical lesions were implicated in the deficits of aphasia. This was plausible since the middle cerebral artery is the blood vessel most involved in the strokes that result in aphasia and the thin-walled lenticulostriate branch of this artery that supplies the basal ganglia is exceedingly vulnerable to rupture. Brain-imaging techniques have resolved this issue. It has become apparent that subcortical neural structures are *necessary* elements of a functional human language system. As Stuss and Benson note in their review of studies of aphasia (1986: 161), damage to:

> the Broca area alone or to its immediate surroundings . . . is insufficient to produce the full syndrome of Broca's aphasia . . . The full, permanent syndrome (big Broca) invariably indicates larger dominant hemisphere destruction . . . deep into the insula and adjacent white matter and possibly including basal ganglia

Moreover, subcortical damage that leaves Broca's area intact can result in Broca-like speech production deficits (Naeser et al. 1982; Alexander 1987; Dronkers et al. 1992; Mega and Alexander 1994). The brain damage traditionally associated with Wernicke's aphasia includes the posterior region of the left temporal gyrus (Wernicke's area), but often extends to the supra marginal and angular gyrus, again with damage to subcortical white matter below (Damasio 1991). Indeed, premorbid linguistic capability can be recovered after complete destruction of Wernicke's area (Lieberman 2000). As D'Esposito and Alexander (1995: 41), in their study of aphasia deriving from subcortical damage, conclude 'That a *purely* cortical lesion—even a macroscopic one—can produce Broca's or Wernicke's aphasia has never been demonstrated'.

### Neurodegenerative Diseases

Diseases such as Parkinson's disease (PD) and progressive supranuclear palsy (PSP) result in major damage to subcortical basal ganglia, sparing cortex until the late stages of these diseases (Jellinger 1990). Therefore, studies of the behavioural effects of these diseases can illuminate the role of basal ganglia. The primary deficits of these diseases are motoric; tremors, rigidity, and repeated movement patterns occur. However, linguistic and cognitive deficits also occur. In extreme form a dementia different in kind from Alzheimer's occurs (Albert 1974; Cummings and Benson 1984; Xuerob et al. 1990). The afflicted patients retain knowledge but are unable to form or change cognitive sets. Deficits in the comprehension of syntax have been noted in independent studies of PD (e.g. Lieberman 2000; Lieberman et al. 1990; 1993; Grossman et al. 1991; 1992; Natsopoulos et al. 1993). A pattern of speech production, syntax, and cognitive deficits similar in nature to those typical of Broca's aphasia can occur in even mild and moderately impaired PD patients (Harrington and Haaland 1991; Lange et al. 1992; Lieberman et al. 1992; Morris et al. 1993; Taylor et al. 1990).

### Sequencing

In the era before Levadopa treatment was available to offset the dopamine depletion that is the immediate cause of Parkinson's disease, thousands of operations were performed. In a seminal paper Marsden and Obeso (1994) review the effects of these surgical interventions and similar experimental lesions in monkeys. They note that the basal ganglia appear to have two different motor control functions.

First, their normal routine activity may promote automatic execution of routine movement by facilitating the desired cortically driven movements and suppressing unwanted muscular activity. Secondly, they may be called into play to interrupt or alter such ongoing action in novel circumstances .... Most of the time they allow and help cortically determined movements to run smoothly. But on occasions, in special contexts, they respond to unusual circumstances to reorder the cortical control of movement. (p. 889)

But reviewing the results of many studies that show that basal ganglia circuitry implicated in motor control does not radically differ from that implicated in cognition, Marsden and Obbeso conclude:

> ... the role of the basal ganglia in controlling movement must give insight into their other functions, particularly if thought is mental movement without motion. Perhaps the basal ganglia are an elaborate machine, within the overall frontal lobe distributed system, that allow routine thought and action, but which responds to new circumstances to allow a change in direction of ideas and movement. Loss of basal ganglia contribution, such as in Parkinson's disease, thus would lead to inflexibility of mental and motor response ...   (p. 893)

Sequencing deficits are apparent in the speech of Broca's aphasia, PD, and people suffering hypoxia, oxygen deficits most probably affecting the basal ganglia. The primary speech production deficit of Broca's syndrome is deterioration of the sequencing between independent articulators, principally between the larynx and the tongue and lips. This is apparent in the primary acoustic cue, 'voice onset time' (VOT), that differentiates stop consonants such as [b] from [p] in the words *bat* and *pat*. VOT is defined as the interval between the 'burst' of sound that occurs when a speaker's lips open and the onset of periodic phonation produced by the larynx. The sequence between lip opening and phonation must be regulated to within 20 msec. Broca's aphasics are unable to maintain control of the sequencing laryngeal and SVT gestures; their intended [b]s may be heard as [p]s, [t]s as [d]s, and so on (Blumstein et al. 1980; Baum et al. 1990). Control of duration is preserved in Broca's aphasics since the intrinsic duration of vowels is unimpaired (Baum et al. 1990). The production of the formant frequency patterns that specify vowels is virtually normal (Ryalls 1986; Baum et al. 1990). Since formant frequency patterns are determined by the configuration of the supralaryngeal vocal tract (tongue, lips, larynx height) we can conclude that the control of these structures is unimpaired in aphasia.

VOT sequencing deficits similar in nature to those of Broca's syndrome occurred in PD subjects who also had sentence comprehension deficits similar to Broca's syndrome (Lieberman et al. 1992) in hypoxic subjects (Lieberman et.al. 1994; 1995) and when focal bilateral damage to the putamen occurred (Pickett et al. 1998). Moreover, subjects having impaired basal ganglia function from these causes have difficulty forming and shifting cognitive sets on tests such as the odd man out test (Flowers and Robertson 1985), which involves forming a conceptual category—for example, sorting pictures by their shapes and then switching to sorting them by size. Cognitive perseveration occurs when subjects are asked to shift their sorting strategy. Cognitive perseveration also occurs in life-threatening situations for hypoxic subjects climbing Mount Everest (Lieberman et al. 1994; 1995). In a PD study Grossman et al. (1991) found correlated deficits in sequencing manual motor movements and linguistic operations in a sentence comprehension task. Manual motor sequencing deficits have been noted in many studies of PD (e.g. Cunnington et al. 1995).

## Evolution of the Neural Bases of Language

The central role of the subcortical basal ganglia in the functional language system suggests a posssible starting point for the evolution of the FLS. As Darwin (1859) and Mayr (1982) noted, evolution is a miser that often adapts an existing structure to serve a 'new' end. For example, the bones of the mammalian middle ear derive from the hinged mandibles of reptiles (Mayr 1982). Circuits linking cortex and the basal ganglia of the human brain continue to regulate motor control. However, the basal ganglia and cortex also play a part in regulating cognitive activity, integrating sensory information with the brain's store of knowledge. The FLS may derive from natural selection that enhanced biological fitness through rapid motor responses to environmental challenges and opportunities (Lieberman 1984; 1985; 1998; 2000). This premise is not novel. Lashley (1951) suggested that a common neural substrate regulated the 'syntactic organization' of motor behaviour and thought. In short, the FLS may derive from neural systems that produced timely motor responses in response to changing environmental challenges and opportunities.

Comparative studies provide some clues to the probable evolution of the FLS. Aldridge et al. (1993) determined the role of basal ganglia in rodent grooming patterns. The 'syntax' of the grooming sequence is regulated in basal ganglia; damage to cortex or cerebellum does not affect the grooming sequence. In contrast, destruction of basal ganglia disrupts these grooming sequences, but does not affect the gestures that make up a grooming sequence. Aldridge et al. (1993: 393) conclude:

Hierarchal modulation of sequential elements by the neostratum may operate in essentially similar ways for grooming actions and thoughts, but upon very different classes of elements in the two cases . . . Our argument is that very different behavioural or mental operations might be sequenced by essentially similar neural processes.

In monkeys, the basal ganglia are elements of neural circuits that acquire and regulate learned sequential motor responses. Kimura et al. (1993) showed that basal ganglia interneurons in monkeys were implicated in learning conditioned motor tasks. Learned motor activity in response to an external arbitrary stimulus is a cognitive act.

Independent studies of basal ganglia activity in animals and humans (e.g. Alexander et al. 1986; Middleton and Strick 1994; Cunnington et al. 1995) show that cortical to basal ganglia circuits regulate cognition, projecting to prefrontal cortical areas implicated in 'executive control' and dual-task performance. In short, the mark of evolution is evident in the neural substrate of human language. Basal ganglia and cortex work together. Similar circuits regulate motor control and cognitive acts. Basal ganglia mechanisms that date back to anurans confer the ability for adaptive motor control. Organisms are able to learn motor responses to environmental opportunities.

The study of family KE, which putatively showed specific deficits in grammar (Gopnik and Crago 1991) consistent with the 1990s version of Chomsky's innate Universal Granmar (Pinker 1994) demonstrates the role of basal ganglia in both motor control and language. In many of the members of the large extended family KE, the anomalous expression of a gene (Lal et al. 2001) results in bilaterally small Caudate nuclei. The afflicted individuals suffer from extreme orofacial apraxia, corresponding speech production deficits, and general deficits in the comprehension of grammar. They cannot simultaneously protrude their lips and tongue or repeat two words and have general difficulty comprehending distinctions in meaning conveyed by syntax. When the general cognitive abilities of members of family KE are assessed using standard tests many, but not all, are impaired compared to their afflicted siblings (Vargha-Khadem et al. 1995; 1998). However, the results of cognitive tests such as the odd man out (Flowers and Robertson 1985), which are sensitive to the cognitive inflexibility that results from basal ganglia damage, have not yet been reported.

The elaboration of basal ganglia circuits may be one of the factors that confers the motor and cognitive flexibility necessary to produce human speech and regulate complex syntax. One of the first of the derived features that differentiates human beings from apes is walking. Walking is a learned behaviour (Thelen 1984). Deterioration of walking and upright balance is one of the primary deficits of PD (Hoehn and Yahr 1967), deriving from impaired basal ganglia function. Perhaps Natural Selection for basal ganglia that facilitate learning and sequencing the complex movements necessary to walk upright was the factor that started the process which ultimately lets us communicate in this manner (Hochstadt, pers. comm.).

## FURTHER READING

I have noted above that in the course of evolution the primate vocal tract was modified. The exact course of the morphological changes involved in this process is unclear, but the speech producing morphology capable of producing the full range of human speech is a species-specific human attribute that appeared in the last 150,000 years. For further reading on this topic see Lieberman and Crelin (1971) and Lieberman (1984; 2000).

Readers having an interest in evolutionary theory will find the facsimile of the first edition of Charles Darwin's *On the Origin of Species* (1859/1964) readable and informative, as well as Ernst Mayr's (1982) superb synthesis of evolutionary theory. Informative data on lexical and syntactic abilities is presented in the peer-reviewed papers of Gardner and Gardner (1984) and Savage-Rumbaugh et al. (1993).

Current theories concerning the neural bases of complex behaviours are presented in Mesulam (1990), Greybiel (1995; 1997), Greenberg et al. (2000) and Lieberman (2000). A review of recent experimental data and theories concerning the nature and evolution of the neural bases of human language is in Lieberman (in press). The particular role of subcortical structures in the cortical–striatal–cortical regulating complex behaviours is discussed in these papers and books and in the seminal paper of Marsden and Obeso (1994). The relation between motor control and the evolution of cognitive ability is discussed in Lieberman (1975; 1984; 1985; 2000; 2002) and Kimura (1993).

# 15

# From Language Learning to Language Evolution

*Simon Kirby and Morten H. Christiansen*

## Introduction

There are an enormous number of communication systems in the natural world (Hauser 1996). When a male Túngara frog produces 'whines' and 'chucks' to attract a female, when a mantis shrimp strikes the ground to warn off a competitor for territory, even when a bee is attracted to a particular flower, communication is taking place. Humans as prodigious communicators are not unusual in this respect. What makes human language stand out as unique (or at least very rare indeed: Oliphant 1998) is the degree to which it is *learned*.

The frog's response to mating calls is determined by its genes, which have been tuned by natural selection. There is an inevitability to the use of this signal. Barring some kind of disaster in the development of the frog, we can predict its response from birth. If we had some machine for reading and translating its DNA, we could read off its communication system from the frog genome. We cannot say the same of a human infant. The language, or languages, that an adult human will come to speak are not predestined in the same way. The particular sounds that a child will use to form words, the words themselves, the ways in which words will be modified and strung to-gether to form utterances—none of this is written in the human genome.

Whereas frogs store their communication system in their genome, much of the details of human communication are stored in the environment. The information telling us the set of vowels we should use, the inventory of verb stems, the way to form the past tense, how to construct a relative clause, and all the other facts that make up a human language must be acquired by observing the way in which others around us communicate. Of course this does not mean that human genes have no role to play in determining the structure of human communication. If we could read the genome of a human as we did with the frog, we would find that, rather than storing details of a communication system, our genes provide us with mechanisms to retrieve these details from the behaviour of others.

From a design point of view, it is easy to see the advantages of providing instructions for building mechanisms for language acquisition rather than the language itself. Human language cannot be completely innate because it would not fit in the genome. Worden (1995) has derived a speed limit on evolution that allows us to estimate the maximum amount of information in the human genome that codes for the cognitive differences between us and chimpanzees. He gives a paltry figure of approximately 5 kilobytes. This is equivalent to the text of just the introduction to this chapter.

The implications of this aspect of human uniqueness are the subject of this chapter. In the next section we look at the way in which language learning leads naturally to language variation, and what the constraints on this variation tell us about language acquisition. In the third section, 'Sequential learning', we introduce a computational model of sequential learning and show that the natural biases of this model mirror many of the human learner's biases, and help to explain the universal properties of all human languages.

If learning biases such as those arising from sequential learning are to explain the structure of language, we need to explore the mechanism that links properties of learning to properties of what is being learned. In the fourth section, 'Iterated learning and the origins of structure', we look in more detail at this issue, and see how learning biases can lead to language universals by introducing a model of linguistic transmission called the *Iterated Learning Model*. We go on to show how this model can be used to understand some of the fundamental properties of human language syntax.

Finally, we look at the implications of our work for linguistic and evolutionary theory. Ultimately, we argue that linguistic structure arises from the interactions between learning, culture, and evolution. If we are to understand the origins of human language, we must understand what happens when these three complex adaptive systems are brought together.

## From Universals to Universal Bias

When a behaviour is determined to a great extent genetically, we do not expect there to be a great deal of variation in that behaviour across members of a species. The whines and chucks a frog makes should not vary hugely from frog to frog within the same species. If there is variation in a behaviour, then this must either be down to genetic variability or to some interaction between the organism's phenotype and the environment that is itself variable. Clearly the same does not hold for behaviours that are learned. In fact, we can see a learned behaviour as being one in which the environment interacts in a complex way with the phenotype. It is natural, therefore, that human language, involving as it does a large learned component, should exhibit a great deal of variation.

There are currently around 6,000 human languages (although this number is dropping fast). Each is distinct in its phonology, lexicon, morphology, and syntax. In fact, this severely underestimates cross-linguistic variation. It is likely that no two people have exactly the same linguistic system or idiolect. Does this mean that, for every child, the environment is different? The answer is yes: the linguistic environment is made up of the set of all utterances that that child hears, and this is unlikely to be the same for any two learners. This in itself makes human language very unusual even compared with other learned behaviours in nature. The environment that the learner interacts with consists of the 'output' of that learning by other individuals. We will return to examine the significance of this later.

### Constraints on Variation

Linguistics is interested not only in documenting cross-linguistic variation, but also in uncovering the fundamental *limits* on that variation. There are a number of obvious universal facts about languages. For example:

*Digital infinity*. Languages make 'infinite use of finite means' (Humboldt 1836, cited in Chomsky 1965: 8—see also, Studdert-Kennedy and Goldstein, in Chapter 13 above). Despite having a fairly small inventory of sounds, language is open-ended, allowing us to produce an unlimited range of utterances.

*Compositionality*. The meaning of an utterance is some function of the meanings of parts of that utterance and the way they are put together. For example, you can understand this sentence by virtue of understanding the meanings of its component words and phrases, and how they are put together. Although we can use non-compositional expressions (e.g. the meaning of many idiomatic expressions must be stored holistically), all languages are compositional to a great extent.

Whilst these universals are immediately obvious facts about language, there are other constraints on variation that cannot be uncovered by looking at one or two languages. In order to discover more subtle patterns of variation, typologists categorize languages into a taxonomy of types. For example, one type might divide the set of languages into those that use prepositions (such as English) and those that use postpositions (such as Hindi). What is interesting is that typically some combinations of types are common, whereas others are rare or completely absent in a representative sample of the world's languages. For example:

> *Branching direction.* Dryer (1992) has shown that languages with a consistently branching word order are more common than those with inconsistent branching. This means that languages whose direct object follows the verb are almost certain to have prepositions, for example.

There is a third sort of universal properties that are not as obvious as compositionality, for example, but are typically discovered without looking at a large sample of languages. By very detailed analysis of only a handful of languages, certain regularities become clear that can be used to predict the grammaticality of a variety of constructions. Whereas typology relies heavily on cross-linguistic comparison, evidence for these universals comes mainly from comparing patterns of grammaticality *within* a language. If a large range of seemingly unrelated grammaticality facts can be captured by a simple generalization, and furthermore, if that generalization can be applied to other unrelated languages, then this is taken as evidence that the generalization captures a universal property of language. A classic example of this type of universal is:

> *Subjacency.* The subjacency condition limits the distance that an element in a sentence can be moved from its canonical position (though see the next section for a different perspective on this universal). 'Movement' is used here to refer to the relationship between the position of *who* in *Mary loves who* and in *Who does Mary love*, for example. The ungrammaticality of sentences such as *Who did Mary tell you when she had met* is predicted by the subjacency condition. Interestingly, subjacency applies universally, although the precise details of the measurement of 'distance' varies from language to language (see Kirby 1999 for a more detailed discussion).

## Acquisition as Explanation

It is helpful to think of the various aspects of a theory of human language in terms of the sorts of question that they answer. A major goal of the sorts of cross-linguistic comparison and in-depth syntactic analysis summarized in the previous section is finding an answer to a *what* question:

*What* constitutes a possible human language?

Having uncovered some of the features of language—the constraints on possible variation—a second question naturally follows:

*Why* are human languages the way they are, and not some other way?

This is the fundamental explanatory challenge for the study of language.

As we argued earlier, it is the unique status of language as a learned communication system that leads to cross-linguistic variation. Given this, it seems sensible to look to learning to explain *why* that variation is constrained in the way that it is. From the viewpoint of Chomskyan generative linguistics (see e.g. Chomsky 1986), language universals are determined by the structure of our language-learning mechanism, which is in turn determined by our genetic endowment.

In the Chomskyan approach, the characterization of language universals (i.e. the description of linguistic variation) is replaced by the characterization of set of constraints on language, termed 'Universal Grammar' (UG). Jackendoff (2002) notes that the exact meaning of UG is somewhat ambiguous in the literature. We use it here to refer to the prior biases that the language learner brings to the task of learning language—in some sense, the knowledge of the hypothesis space of the language learner before hearing a single utterance.

Generative linguistics suggests that a correct description of language universals and a correct description of UG are one and the same thing. In other words, the 'what' and the 'why' questions can be rolled together. So, for example, if the Subjacency Principle captures linguistic variation correctly, then the Subjacency Principle forms part of UG. Thus, on this account, by careful examination of languages, one may uncover a property of the language-learning part of our brain—the human Language Acquisition Device.

In itself, this is not quite enough to constitute a theory of language learning. UG gives us a model of a child's initial biases relative to language acquisition, but does not tell us how these interact with the child's experience of language to give her knowledge of Hungarian, for example. There are a number of different approaches to language learning in the generative programme, but a shared characteristic of these approaches is that they downplay the complexity of the learning task. In fact, the term 'learning' is usually avoided in favour of 'acquisition'. On this view, if all the child is doing is setting a few switches on their already provided UG rules, then perhaps 'learning' is too grand a term, suggesting that the process of picking up knowledge of one's native language is more open-ended than it is.

This may seem as though we are merely tinkering with terminology, but in fact opting for 'acquisition' over 'learning' hides another key feature of the generative approach to explanation. If the constraints on cross-linguistic variation are due to constraints on learning, which in turn are due to our genetic endowment, we have not made any claims that these learning constraints are specific to language. After all, a genetic predisposition to learn in a particular way may well affect how we develop competency in a range of different domains. But by removing *learning* from the process of language development, one implicitly allows linguistic constraints to become domain-specific. That is, if language is considered to be acquired rather than learned, the principles of UG can be kept pristine from the influence of other learning biases. From this perspective, it therefore makes sense to assume that linguists can uncover properties of our biology by examining languages alone.

In short, the generative approach contends that what makes a human language is determined by innately coded domain-specific knowledge. In this chapter, we present a different perspective on language acquisition. First, we will argue that some—perhaps many—of the learning biases we bring to language learning may not be domain-specific. From this perspective, it may be unhelpful to think of the prior knowledge of the child as Universal *Grammar*, with all the language-specific connotations of the term. Perhaps Universal Bias is a better term. Either way, it should be clear that what we (and many other language researchers) are studying is the set of constraints and preferences that children have and which are brought to bear on the task of language learning.

After showing that we need to be extremely cautious in assuming that constraints on language learning are necessarily language-specific, we go on to show that innateness alone cannot be the determinant of linguistic structure. Our findings suggest that an explanatory theory of language must be evolutionary in the widest possible sense.

## Sequential Learning

The appeal to language-specific learning biases seems a reasonable response to the challenge of explaining universals such as branching direction and subjacency. How else can we explain universals that appear to be unique to language? A surprising alternative arises when we look at a computational modelling work involving simple recurrent networks, SRN (Elman 1990).

### SRNs and Language Learning

An SRN is a type of neural network that can be used to learn sequences of items. Like many types of network, it consists of three layers of 'nodes' wired together with weighted connections. One of these layers, the input layer, represents an element in a sequence that the network is currently attending to. Another layer, the output layer, represents the network's expectation of what the next element in the sequence will be. The weights on the connections between nodes encode the network's knowledge of sequences. The networks can be *trained* on a set of sequences, which amounts to tuning the connection weights in such a way that the network is good at next-element prediction. Importantly, the SRN has a way of building up a representation of the structure of a sequence as it goes along from element to element. This representation acts like a memory, and enables the prediction of the next element in a sequence to depend not only on the current element but also on those that have gone before.

SRNs are interesting to us because they can be used to model many aspects of language learning (see Christiansen and Chater 2001*b*, for a review) as well as sequential learning (Cleeremans 1993). While language learning certainly involves more than simply predicting the next word in a sentence, infants do appear to acquire aspects of their language through such statistical learning (e.g. Saffran et al. 1996). Importantly, SRN behaviour after training closely resembles that of humans. In particular, the types of sequence that cause difficulty for the networks seem also to be ones that humans find hard. Christiansen and Chater (1999), for example, show that sequences generated by a grammar with multiple centre-embeddings are hard for networks to learn correctly. The same sorts of structure are difficult for us too: consider the difficulty in understanding *balls girls cats bite see fall* which contains multiple clauses embedded inside each other. Crucially, the

SRNs fit the human data without having to invoke external performance constraints as is the case with more traditional non-statistical models.

But what if the linguistic universals that have been said to require a language-specific UG fall naturally out of the prior biases of an SRN in a similar way?

### SRNs and Learning-Based Constraints

Christiansen and Devlin (1997) set out to explore this possibility. They look at the earlier-mentioned branching-direction universal, which captures the fact that some basic word orders are far more commonly attested in the world's languages than others. There have been a number of suggestions in the literature for how this might be accounted for by a domain-specific account of UG (see e.g. Giorgi and Longobardi 1991). Another influential approach explains word-order universals such as this one by appealing to a particular model of parsing (Hawkins 1994; Kirby 1999).

If, however, the pattern of cross-linguistic variation matches the prior bias of an SRN, then we have a strong argument for rejecting a domain-specific or parsing-based explanation. To test this, Christiansen and Devlin (1997) constructed thirty-two simple grammars; each exhibited a different word order, but they were in all other aspects identical. The grammars, though simple, posed non-trivial tasks for a neural network that is learning to predict sequences. For example, they were recursive, allowing sentences where adpositional phrases were embedded within other adpositional phrases. They included grammatical number marked on nouns, verbs, and genitive affixes, as well as a requirement that verbs agree with their subjects, and genitives agree with the possessing noun.

Each of the thirty-two grammars can be used to generate a corpus of sentences. In the Christiansen and Devlin (1997) model, these sentences are actually sequences of grammatical categories and an end-of-sentence marker. So, for example, a possible sequence in one of the grammars that mirrors English word order might be 'singular noun + singular verb + preposition + singular noun + singular genitive + singular noun + end-of-sentence marker'.[1] In order that these corpora can be used to train SRNs, each grammatical category is arbitrarily assigned one input node and one output node

---

[1] To see that this is like English word order, consider a sentence like *The man sat on the cat's mat*. It should be noted that one of the simplifications in the grammars used is that determiners are ignored.

in the networks. So, if node 1 corresponds to a singular noun, and node 2 to a singular verb, then we should expect a network successfully trained on an English-like language to activate node 2 of the output layer after being shown the first word of the sequence above. Actually, things are a little more complicated than this, because a singular verb is not the only thing that can follow a sentence-initial singular noun. For example, a singular genitive affix is also possible. In this case, then, the network should show a *pattern of activation* that corresponds to all the grammatical continuations of a sentence that started with a singular noun in the language that it is trained on.

The ability of a network correctly to predict continuation probabilities after being trained on a corpus can be used to calculate a *learnability score* for a grammar. We can think of this score as reflecting how well each of the thirty-two grammars fits the prior bias of an SRN sequential learner. The surprising result is that these scores are significantly correlated with the degree of branching consistency in the grammars used. Furthermore, a direct comparison with the cross-linguistic data (using Dryer's 1992 statistically controlled database) shows that the prior bias of an SRN is a good predictor of the number of languages that exhibit each word-order type. Simulations by Van Everbroeck (1999) and Lupyan and Christiansen (2002), involving SRNs learning grammatical role assignments from simple sentences incorporating noun/verb inflection and noun case-markings, respectively, provide further support for explaining language-type frequencies using learning-based constraints.

This is clearly a striking result. Why postulate a domain-specific constraint if the data that constraint should account for is predicted by a general model of sequential learning? One potential criticism might be that the SRN is so different from the human brain, that we can draw no parallels between its learning bias and humans. Aside from the fact that the same argument could be applied to any model of UG, there are reasons to suspect that the human prior is not so far from the network's.

Christiansen (in preparation; described in Christiansen and Ellefson 2002) employs an experimental paradigm known as artificial language learning (ALL) to demonstrate this. Two grammars were chosen from the set on which the neural networks were trained: one which exhibited a word order that was rare cross-linguistically (and which the SRNs found hard to learn), and one which exhibited a more common pattern. The grammars generated 'sentences' made up of strings of arbitrary symbols.

In the first phase of the experiment, subjects are exposed to strings generated from one of the grammars. They are asked to read and reproduce the strings, but were not told that the strings conformed to any kind of rule set or grammar. During the test phase, however, they were told that the strings were part of a language. They were presented with novel strings and asked to say whether they were grammatical or not.

Christiansen found that subjects trained on the grammar with rare word order were significantly worse at determining grammaticality than those trained on the more common type of word order. This was despite the fact that the word order of the latter grammar did not have anything in common with English word order (in fact, the difficult grammar had more features in common with English).

The ALL experiment suggests that the SRNs' sequential learning biases mirror those of humans, and that this non-language-specific bias underpins at least some of the constraints on cross-linguistic variation. Further evidence for this comes from experiments by Christiansen et al. (in preparation) that look at the ability of subjects with agrammatic aphasia to learn sequences in ALL experiments. Agrammatic aphasics are characterized by a reduced ability to deal with grammatical constructions in language. There have also been suggestions (Grossman 1980) that they have a more general deficit in their ability to deal with sequential learning tasks (see also Lieberman, Chapter 14 above). In the ALL experiments, normal subjects were able to determine the grammaticality of strings generated by a simple finite-state grammar, whereas the aphasic subjects could not perform above chance. This provides further evidence for a strong connection between language learning and sequential learning.

What these experiments with ALL and SRNs show us is that we should be careful about ascribing universal properties of language to a domain-specific innate bias. We argue that an explanation that appeals to non-linguistic biases should be preferred where possible. Simplistic, all-or-nothing explanations should be avoided, however. Whereas the above results suggests that some language universals may derive from non-linguistic sequential learning biases, others may require language-specific biases for their explanation. We submit, however, that this is an empirical question that cannot be settled a priori.

We end this section with a brief look at another language universal that has been claimed to reflect a domain-specific UG principle: subjacency. Christiansen and Ellefson (2002) used a similar paradigm of SRN and ALL experiments to see if the patterns of grammaticality predicted by the subjacency principle actually reflect biases from sequential learning. In particular, they wanted to see if a language with the grammaticality pattern of

English would be easier for an SRN to learn than a language with an unnatural grammaticality pattern (i.e. one in which subjacency-violating constructions were permitted).

The following types of string make up the English-like language (English glosses are given for clarity—the networks were trained on category names, and human subjects were presented with arbitrary symbols, as with the word-order experiment):

(1)  Everybody likes cats.
(2)  Who (did) Sara like?
(3)  Sara heard (the) news that everybody likes cats.
(4)  Sara asked why everybody likes cats.
(5)  What (did) Sara hear that everybody likes?
(6)  Who (did) Sara ask why everybody likes cats?

The unnatural language looks like this (note the last two examples, which are ungrammatical in English since they violate the subjacency principle):

(7)   Everybody likes cats.
(8)   Who (did) Sara like?
(9)   Sara heard (the) news that everybody likes cats.
(10)  Sara asked why everybody likes cats.
(11)  *What (did) Sara hear (the) news that everybody likes?
(12)  *What (did) Sara ask why everybody likes?

As with the SRN simulations, and ALL experiments for word order, human subjects and neural networks behaved the same way. The unnatural language was significantly harder for subjects to learn, and resulted in significantly greater network error, than the English-like language.

These two string sets obviously do not capture all the possible distinctions between a subjacency-obeying language and a hypothetical subjacency-violating language. There are many more predictions about grammaticality that follow from the principle. Nevertheless, the learning biases reflected in this language universal are not likely to be completely domain-specific.

## Iterated Learning and the Origins of Structure

In the previous section we examined one approach to answering the question: why are languages the way they are? We argued that the range of cross linguistic variation is ultimately determined by our innate learning biases, many of which may not be specific to language (see Fig. 15.1).

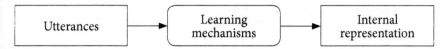

FIG. 15.1 Language learning. The child's learning mechanisms take linguistic data (utterances) and generate some kind of internal representation of a language. The range of possible languages is determined by the structure of the learning mechanisms (i.e. its prior biases).

There is something missing from this picture, however. We have looked at the way in which particular types of learning respond to particular sets of data. We have not said anything about where these data come from. In most models of learning, the data are considered to be given by some problem domain. In Christiansen's simulations, for example, the training data are provided by the experimenter.[2] It is the ability of the networks or experimental subjects to learn the data, and the limitations they exhibit, that is of interest.

### The ILM

Kirby (2000) has suggested that we need to look more carefully at where the training data come from if we want a truly explanatory account of the structure of language. What makes language unique in this regard is that the data that make up the input to learning are themselves the output of that same process (see Fig. 15.2). This observation has led to the development of a model of language evolution—the Iterated Learning Model (ILM)—that builds this in directly (Kirby and Hurford 2002).

The ILM is a multi-agent model that falls within the general framework of situated cognition (Brighton et al. in press). It treats populations as consisting of sets of individuals (agents), each of which learns its behaviour by observing the behaviour of others (and consequently contributes to the experience of other agents' learning). We can contrast this approach with the

---

[2] Though see Christiansen and Dale (in press) for simulations in which the learning biases of the SRN force languages to change over generations of learners to become more easily learnable.

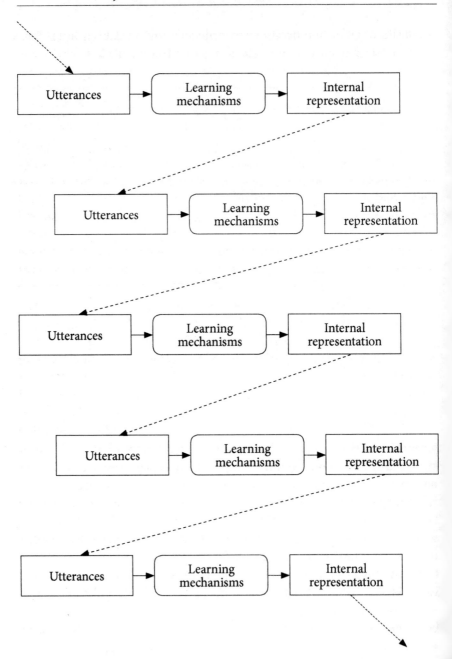

FIG. 15.2 Iterated learning. The input to learning is the product of the acquired lan guage of others. There are now two dynamical systems that contribute to the range of possible languages: individual learning, and social/cultural transmission.

idealizations of the homogeneous speech community of Chomsky (1965).

The ILM is evolutionary in the sense that the linguistic behaviour of agents is *dynamic*. It is not predetermined, but emerges from the process of repeated use and acquisition from generation to generation. This is not biological evolution taking place over the lifetime of a species, rather the process of linguistic transmission operating on a historical/cultural timescale.

Iterated learning is discussed in depth in Kirby (1999), where it is used to demonstrate how biases (such as the bias for consistent branching direction) actually lead to language universals, and why some classes of universal (specifically, implicational or hierarchical ones) can only be explained from the viewpoint of iterated learning (for possible psychological and connectionist underpinnings of this view, see Christiansen 1994). In the following section we will summarize recent research suggesting that some of the more fundamental properties of language arise from the process of repeated acquisition and use.

### The Origins of Compositionality

A striking property of human language that sets it apart from most other communication systems is the way in which the meaning of an utterance is composed of the meanings of parts of that utterance.[3] This *compositional* nature of language is so fundamental to the syntax of human language, it is rarely considered a target of explanation. Yet it is compositionality that gives language its open-ended expressivity (consider how difficult it would be if you had to give every communicatively relevant situation a unique name).

Whilst compositionality is rife in language, it is not everywhere equally. For example, idiomatic expressions are not as clearly compositional. The sentence *John bought the farm* can be read compositionally as being information about someone called John buying a farm. In some dialects of English, however, it can be taken to mean that John has died. This second meaning of *bought the farm* is holistic, rather than compositional.

Similarly, if we look at the morphology of a lot of languages, we can see both holism and compositionality. In the English past tense there are regu-

---

[3] We note, however, that during normal language comprehension each word in a sentence may not contribute to its overall meaning – sentence interpretation appears to be at least partially underspecified. For a review of the relevant literature from psycholinguistics and computational linguistics, see Sanford and Sturt (2002).

lar verbs, such as *walk/walked*, and irregulars, such as *go/went*. The former express their past tense compositionally through the addition of the affix *-ed*, whereas the latter are holistic.

Where did compositionality come from and what maintains it in language? If language was once holistic (as Wray 1998 suggests), by what mechanism did this change? Kirby (2000) suggests that the answer to these questions lies in understanding the process of iterated learning.

To test this, Kirby and others (Batali 1998; 2002; Brighton and Kirby 2001; Brighton 2002; Kirby and Hurford 2002; Kirby et al. 2002; Tonkes 2002) have implemented the ILM as a computational simulation. Modelling situated cognition, and dynamic systems in nature more generally, using computational techniques is an increasingly popular methodology (see e.g. Kirby 2002 for a review of the ways in which it has been used in evolutionary linguistics). Computer modelling gives us an easy way to uncover the relationship between the components of a complex system—in this case individual learners—and the emergent outcomes of their interactions.

The simulation models are typically made up of these components:

(1) A population of agents.
(2) A space of possible signals (usually strings of symbols).
(3) A space of possible meanings (usually some kind of structured representation, such that some meanings are more similar to each other than others).
(4) A production model. This determines how, when prompted with a meaning, an agent uses its knowledge of language to produce a signal.
(5) A learning model. The learning model defines how an individual agent acquires its knowledge of language from observing meaning–signal pairs produced by other agents.[4]

Within this broad framework, there is a lot of variation in the literature. In particular, a range of different learning models have been employed: from symbolic approaches such as grammar induction (Kirby and Hurford 2002) to connectionist approaches such as SRNs (Batali 1998). A key point in common between the various ILM simulations is that the language of the

---

[4] The fact that the learners are given meanings as well as signals seems unrealistic. Ultimately, simulations of the process of iterated learning will need to enrich the model with *contexts*. Whereas meanings are private and inaccessible to learners, contexts are public and may allow the inference of meanings. See e.g. Steels et al. (2002) and Smith (2001) for discussion of these fascinating extensions to the model.

population persists only by virtue of its constant use and acquisition by individual agents.

These simulations are usually initialized with a random language. That is, the agents initially produce purely random strings of symbols for every meaning that they are prompted with. Excepting some highly improbable set of random choices, this language will be purely holistic. That is, there will be nothing in the structure of the strings of symbols that corresponds to the structure of the meanings being conveyed.

In the simulations, these intial random languages are typically highly unstable. A snapshot of the language of the population at one point in time will tend to tell you very little about what the language will be like at a later point. It is easy to see why. Unless a learning agent is exposed to the output of a particular speaker for every possible meaning, there is no way in which that learner will be able to reproduce accurately the language of that speaker.

In some simulations, holistic languages can be made to be stable, but only if the learners hear so much data that they are guaranteed to observe a speaker's utterance for every possible meaning. However, this is a highly unrealistic assumption. Much is made in the language acquisition literature of the 'poverty of the stimulus' (see Pullum and Scholz 2002 for a review), suggesting that learners are exposed to an impoverished subset of the total language. Aside from this, in reality the range of possible meanings is open-ended, so complete coverage cannot be assumed.

In the initial phases of the simulations, then, the language of the population tends to vary widely and change rapidly. The striking thing is that eventually some part of the language will stabilise, being passed on faithfully from generation to generation. How long the simulations run before this happens depends on a variety of factors relating to the particular simulation's population dynamics (we return to this briefly in the next section). Nevertheless, for a broad range of different simulation models with different learning models and so on, these pockets of stability always seem to occur.

As the simulations continue, more and more of the language increases in stability, until eventually signals corresponding to the entire meaning space are passed on reliably from generation to generation without the learner being exposed to the whole language. These final languages invariably use a compositional system to map meanings to strings (and vice versa).

How are these languages stable, and why are they compositional? The initial holistic languages are unstable by virtue of the poverty of the stimulus, which acts as a bottleneck on the transmission of language. In the early stages of the simulation, the population is essentially randomly searching around the space of possible meaning–string pairs, driven by the learners' failure to generalize (non-randomly) to unheard meanings. At some point a learner will infer some form of non-random behaviour in a speaker and use this to generalize to other meanings. In the first instance, this inference of 'rule-like' behaviour will actually be ill-founded (since the speaker will have been behaving randomly). Nevertheless, this learner will now produce utterances that are, to a small extent, non-random.

Languages generated by a learner who has generalized are themselves generalizable by other learners. The key point is that the aspects of the language that are generalizable in this way are more stable. This is because, by definition, generalizations can be recreated each generation without the need for complete coverage in the training sample. In other words, because a generalization can be used for a range of meanings, and can be learned by exposure to a subset of those meanings, generalizations can pass through the learning bottleneck more easily. As Hurford (2000) puts it, 'social transmission favours linguistic generalization'.

As a result of this differential in stability between random and non-random parts of the linguistic system, the movement towards a language made up of generalizations is inevitable. The meanings in the model have internal structure, as do the signals. What is learned is the mapping between these two spaces. Generalizations about this mapping have to utilize the fact that the spaces are structured. It is unsurprising, therefore, that a compositional system of mapping is inevitable. In a compositional language, the internal structure of a meaning to be conveyed is non-randomly related to the structure of the string that conveys that meaning.[5]

These results link compositional structure in language to the impoverished stimulus that a learner faces. Poverty-of-stimulus arguments are usually used to suggest the need for a strongly constraining prior provided 'in-advance' by our biology. For us, however, the restricted richness of input is actually the engine that drives the evolution of language itself. Language adapts to aid its own survival through the differential stability of generalizations that can be transmitted through the learning bottleneck.

Further evidence that this type of evolution through iterated learning explains some aspects of language structure is given by Kirby (2001).

---

[5] The stability of particular representations in an iterated-learning scenario is related to their compressibility (see Brighton 2002; Teal and Taylor 1999).

Returning to the fact that language is not 100 per cent compositional, Kirby wonders whether the ILM might explain not only why compositionality emerges but also why some kinds of holism persist. In this simulation, the frequency with which particular meanings are used by the agents is not uniform. In other words, some meanings turn up more frequently than others (in contrast to earlier work, where every meaning has an equal chance of making it through the bottleneck). This is clearly a more realistic assumption, mirroring the fact that the real world is 'clumpy'—that some things are common, others rare. With this modification in place, the result is a language that utilizes both compositional structure and holistic expressions.

The simple language in the simulation can be compared to a morphological paradigm. The infrequently used parts of the paradigm tend to be compositionally structured, whereas the more frequent parts are holistic.[6] The simulation appears to be an appropriate model for language since there is a clear correlation between frequency and irregularity in morphology (the top ten verbs in English by frequency are all irregular in the past tense).

Once again, we can understand these results in terms of the pressure on language to be transmitted through a learning bottleneck. Frequently used expressions may be faithfully transmitted even if they are idiosyncratic simply by virtue of the preponderance of evidence for them in the data provided to the child (or to the learning agents in the simulation). Infrequent expressions, on the other hand, cannot be transmitted in this way, and must instead form part of a larger paradigm that ensures their survival even if they do not form part of the linguistic data observed by the learner. This also explains why languages vary in the degree to which they admit irregulars. A completely regular paradigm is perfectly stable, so there is no reason to expect that language must have irregulars (e.g. Chinese is noted for the rarity of irregularity). On the other hand, if, for reasons such as language contact or processes of phonological erosion, irregulars make their way into a language, the pressure to regularize them will be strongest in the low-frequency parts of the system.

---

[6] An interesting by-product of the introduction of frequency biases to the meaning space is the removal of a fixed endpoint to the simulations. The language in this model is always changing - but not so much that speaker-to-speaker intelligibility is degraded. This is another way in which these simulation results seem to mirror what we know about language more accurately. Further research on the dynamics of language change in these models is needed. For example, see Niyogi and Berwick (1997) and Briscoe (2000) for discussion of iterated learning and the oft-noted logistic time-course of change.

## Implications and Conclusion

We have argued that human languages vary because they are learned. The key to understanding the constraints on this variation (and hence the universal structural properties of language) lies in a characterization of the properties of the language learner. The generativist approach to language equates language universals with domain-specific, innate constraints on the learner—essentially drawing a direct relationship between universals and UG.

We agree that it is important to give an account of the initial state of the language learner: the Universal Biases. However, it is premature to assume that such an account must be domain-specific. Results from artificial language-learning experiments and neural-network models show that some universals can be explained in terms of biases relating to sequential learning in general.

The results from the Iterated Learning Model have further implications for an explanatory account of linguistic structure. Typically, the relationship between prior (i.e. innate) bias and linguistic variation is assumed to be transparent and one-to-one. That is, the 'problem of linkage' (Kirby 1999) is left unsolved. Iterated learning is an *evolutionary* approach to the link between bias and structure, in that it looks at the dynamical system that arises from the transmission of information over time. It no longer makes sense to talk about language structure as being purely innately coded. Instead, learning bias is only one of the forces that influence the evolutionary trajectory of language as it is passed from generation to generation. Other factors include the number of utterances the learner hears, the structure of the environment, social networks, and population dynamics.[7] If we are right, we cannot infer the nature of the learning biases purely through linguistic analysis. In some sense, the learning biases act as the environment within which languages themseleves adapt (Christiansen 1994); and without understanding the process of adaptation, we cannot be sure that any particular theory of language acquisition makes the correct predictions.

---

[7] We do not describe them here for lack of space, but a comparison of different ILM simulations (e.g. Batali 1998 and review in Hurford 2002) reveals the importance of vertical versus horizontal transmission in the population. When learners mainly learn from adults, the language changes relatively slowly and may take many generations to stabilize on a structured system. If, on the other hand, there is a lot of contact between learners, structure can emerge very rapidly. It has been suggested that modelling work may provide insights into the very rapid emergence of languages like Nicaraguan Sign Language (Ragir 2002).

So far in this chapter we have said little about the biological evolution of the language learner, instead focusing on the evolution of language itself as it passes from learner to learner. Pinker and Bloom (1990), in their classic paper on language evolution, take a complementary stance by arguing that evolution by natural selection is the key explanatory mechanism (see also Pinker, Chapter 2 above). The logic of their argument runs as follows:

(1) The principal features of language appear to be adapted to the task of communication
(2) These features arise from a domain-specific, innate language faculty.
(3) The only known mechanism that can deliver a biological faculty with the appearance of design is evolution by natural selection.

Obviously, if our conclusions are right, we cannot use this logic for at least some of the properties of language since they are neither innate nor domain-specific. The question of whether the language faculty is well adapted to communication would be the subject of another chapter; but Kirby (1999) argues that in fact *dis*functionality may be the hallmark of innate aspects of language.

Does biological evolution have any role to play in an explanatory account of language structure? Despite our reservations about the adaptationist approach of Pinker and Bloom, we agree that it must (how else can we understand human uniqueness?). To our 'what' and 'why' questions, we add a third:

*How* did language come to exhibit the structure that it does?

It makes little sense to mount an explanation in terms of natural selection for the aspects of learning that are not language-specific. We could, however, try to uncover evolutionary pressures that shaped more general biases such as those that underpin sequential learning (see e.g. Conway and Christiansen 2001). Much structure in biology is the result of exaptation rather than adaptation (although a combination of both forces is likely to be common). Perhaps the mechanisms for sequential learning evolved earlier in our species history, and more recently were employed to assist language learning (see Lieberman, Chapter 14 above). Perhaps the relevant biases are spandrels arising as a bi-product of other aspects of our biology. Currently, there is no way of telling.

Another possible line of enquiry might be to look at what the evolutionary prerequisites for iterated learning itself might be. The simulation models mentioned in the previous section take a lot of things for granted.

There are three principal assumptions:

(1) Agents have structured representations of the world.
(2) Learners have some way of inferring the meaning of a particular signal. At least some of the time, they can mind read.
(3) Speakers are inclined to communicate about an open-ended range of topics.

By systematically varying the representations of meanings to which the agents in the ILM have access, we are able to see under which circumstances structured mappings between meanings and signals will emerge. Brighton (2002) shows, using mathematical models, that compositional languages have a stability advantage over holistic ones when agents have the capacity to form representations of the environment that are multi-dimensional—in other words, when they are highly tuned to structural similarities between different environmental states. The question of how learners have access to these intended meanings (at least some of the time) is tackled by researchers working on negotiation of communication that is grounded in the environment (see e.g. Cangelosi 1999). For example, a growing body of research exploring learned communication between robots (Steels and Kaplan 2002) may provide answers to these questions. Finally, we need to understand why humans communicate to such a degree. Lessons may be learned here from the ethology literature, research on other closely related primates, and work looking more carefully at the selective benefits of human language (see e.g. Dunbar, Chapter 12 above).

It is also possible that learning mechanisms may undergo biological adaptation after they have been employed for a particular task. Briscoe (Chapter 16 below) looks at this possibility in some detail (though from the viewpoint of a language-specific UG). We can consider an extension to the iterated-learning model that also allows for genetic specification of different learning biases which may lead to coevolution of language and language learners (see Fig. 15.3). This provides a potential framework for understanding the evolution of language-learning mechanisms that may include both domain-general and domain-specific components.

A complete theory of language evolution will necessarily be multi-faceted. We should not expect a single mechanism to do all the work. Ontogenetic development, cultural transmission, and evolution of prior biases may all conspire to give rise to human language. As we begin to understand the central role of learning in these three interacting complex dynamic systems,

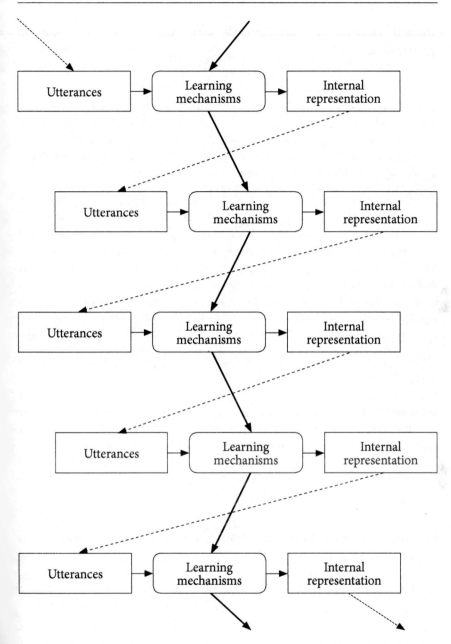

FIG. 15.3 Evolutionary iterated learning. The cognitive mechanisms (and therefore prior biases) of an individual language learner are provided through genetic transmission, which necessarily involves selection. The structure of language arises from the interaction of three systems of information transmission.

we will get closer to a fundamentally evolutionary understanding of linguistic structure.

### FURTHER READING

In this chapter we have shown how a number of different computational techniques can be used to explore the concept of innate learning biases. This approach to language evolution is rapidly gaining in popularity as complex simulations become more practical. A review of this literature, looking in particular at so-called 'artificial life' models, is Kirby (2002). Cangelosi and Parisi (2002) and Briscoe (2002) are two collections of recent research into this area.

For some of the work we describe, simple recurrent neural networks act as models of learning. See McLeod et al. (1998) for a hands-on introduction to using neural networks—including the SRN—to model cognitive behaviour. Cleeremans (1993) presents SRN simulations of sequential learning. Christiansen and Chater (2001a) provides a comprehensive introduction to connectionist modelling of many different kinds of psycholinguistic phenomenon.

We use the term 'bias' throughout this chapter to describe the preferences a learner has before the linguistic experience of that learner is taken into account. A technical discussion about bias from the perspective of statistics and the relationship between bias and learning mechanisms can be found in Batali (1998). This book is also an excellent textbook on various approaches to modelling learning more generally.

We talk about both 'language universals' and 'learning biases' when discussing what evolutionary linguistics needs to explain. The former term is typically used in the field of linguistic typology (see e.g. Croft 1990), whilst the latter within generative approaches is associated with the notion of a Universal Grammar (see Jackendoff 2002 for an accessible introduction). Newmeyer (1998) reviews a fundamental divide between 'functional' and 'formal' approaches in linguistics that is commonly associated with the typological and generative perspectives.

# 16

# Grammatical Assimilation

*Ted Briscoe*

## Introduction

In this chapter I review arguments for and against the emergence and maintenance of an innate language acquisition device (LAD) via genetic assimilation. By an LAD, I mean nothing more or less than a learning mechanism which incorporates some language-specific inductive learning bias in favour of some proper subset of the space of possible languages. Genetic assimilation is a neo-Darwininan mechanism by which organisms can appear to inherit acquired characteristics. Genetic assimilation of grammatical generalizations exemplified in the environment of adaptation of the LAD facilitates more rapid and robust grammatical acquisition by first-language learners. I will develop a co-evolutionary account of this process in which natural languages are treated as complex adaptive systems undergoing often conflicting selection pressures, some of which emanate from the LAD, which itself evolved in response to (proto)languages in the environment of adaptation.

The existence of an innate LAD has not gone unquestioned, and it is probably the case that some arguments proposed in its favour are either questionable or wrong (e.g. Pullum and Scholz 2002; Sampson 1989; 1999). I will argue that all remotely adequate extant models of grammatical acquisition do presuppose an LAD (in the above weak sense), and that genetic assimilation is the most plausible account of its emergence and maintenance. These arguments do not constitute a proof either of the existence of an innate LAD or of its evolution by genetic assimilation. However, they do suggest that the onus is on non-nativists to demonstrate an adequate, detailed

I am grateful to the editors for providing helpful feedback, both from them and from their students on the first draft, which helped me improve this one. Remaining errors or infelicities are entirely my responsibility.

account of grammatical acquisition which does not rely on an LAD, and on non-assimilationists to propose a detailed, plausible alternative mechanism for the evolution of the LAD.

I use the more succinct phrase 'grammatical assimilation' as short-hand for the more ponderous 'genetic assimilation of grammatical information into the LAD'. The general concept of genetic assimilation is described and discussed in more detail in the fourth section, 'Grammatical assimilation', where some arguments for and against grammatical assimilation are also presented. The next section below reviews work on grammatical acquisition and presents the case for the existence of the LAD. The section on 'Linguistic evolution' outlines an account of languages as complex adaptive systems, and spells out several consequences for models of grammatical assimilation. The final section describes and evaluates extant simulations of genetic and grammatical assimilation.

## Grammatical Acquisition

Adequate accounts of grammatical acquisition during first-language learning must satisfy at least the following desiderata. First, there is the desideratum of *coverage*: models should support acquisition of any attested grammatical system and adequately characterize the range of possible mappings from meaning to form in attested systems. A reasonable requirement, given current knowledge, is that the model be capable of learning the mappings for a proper subset of indexed languages, including those exhibiting cross-serial dependencies (e.g. Joshi et al. 1991). This rules out much work which purports to address the issue of grammatical acquisition (e.g. extant work based on (recurrent) neural networks), as these have only been shown to make (graded) grammaticality judgements for small language fragments (e.g. Lawrence et al. 1996), and not to recover mappings from meaning to form. Second, models must work with *realistic input*: grammatical acquisition is based on finite positive but noisy input; that is, learners are exposed to a finite sequence of utterances drawn from mixed and non-stationary sources, as speech communities are never totally homogeneous or static (e.g. Milroy 1992). Many models instead assume a single non-noisy stationary source, or equivalently a finite sequence of 'triggers' drawn from the target grammar to be acquired (e.g. Gibson and Wexler 1994). Third, models should work with *realistic input enrichment*: many assume that each 'trigger' is paired reliably with its correct meaning (logical form) and that the learner never hypothesizes an incorrect pairing. Such assumptions may facilitate formal learnability results for inadequate algorithms, but they presuppose, implausibly, that the context of utterance during learning is always highly determinate and redundant—or, equivalently, that the learner knows when to ignore input (e.g. Osherson et al. 1986: 100). Fourth, models should account for *selectivity* in acquisition: learners do not acquire 'covering' grammars of the input, but instead reject noise and other random or very infrequent data in favour of a single consistent grammar (e.g. Lightfoot 1999). Fifth, models should display *accuracy*: learners do not 'hallucinate' or invent grammatical properties regardless of the input, though they do (over)generalize and, in this sense, 'go beyond the data'.

If accuracy is defined in terms of formal learnability (e.g. Bertolo 2001; Niyogi 1999) from realistic, finite, positive but noisy sentence-meaning pairs over a hypothesis space with adequate coverage, even when drawn from a single stationary target grammar, then some form of inductive bias in the acquisition model is essential (see also Nowak et al. 2001).[1] In much current work on grammatical acquisition within the Principles and Parameters (P&P) framework (e.g. Chomsky 1981; Gibson and Wexler 1994), inductive bias takes the form of a restricted finite hypothesis space of grammars within which individual grammars are selected by setting (finite-valued) parameters. There may also be additional bias in terms of default initial settings for a subset of parameters, creating a preference ordering on grammars in the hypothesis space (e.g. Chomsky 1981: 8 f.). P&P models, which do not incorporate a statistical or quantitive component, are not able to deal adequately with noisy input (e.g. Briscoe 1999; 2002). There is a well-known formulation of inductive bias in terms of Bayesian probabilistic learning theory (see e.g. Mitchell 1997: 154 f. for an introduction). Bayes theorem provides a general formula and justification for the integration of prior bias with experience, and it has been demonstrated that an accurate prior supports learnability from finite noisy data over infinite (though restricted) hypothesis spaces (e.g. Horning 1969; Muggleton 1996).

Bayesian learning theory is a general domain-independent formulation of learning. The most general formulation of learning in this framework (Kolmogorov Complexity) posits a learner able to learn any generaliza-

---

[1] The term inductive bias is utilized in the field of machine learning to characterize both hard constraints on the hypothesis space considered by a learner, usually imposed by a restricted representation language for hypotheses, and soft constraints which create preferences within the hypothesis space, usually encoded in terms of cost metric or prior probability distribution on hypotheses (e.g. Mitchell 1997: 39 f.).

tion with a domain-independent bias (the so-called 'universal prior') in favour of the smallest, most compressed hypothesis (e.g. Li and Vitanyi 1997). However, nobody has demonstrated that this general formulation could, even in principle, result in a learning algorithm capable of accurately acquiring a specific grammar of a human language from realistic input. Horning's (1969) work is restricted to the (infinite) class of stochastic context-free grammars, which violates the coverage desideratum introduced above, as cross-serial dependencies are not covered. However, Muggleton's (1996) proof is defined over a restricted form of stochastic logic programme which does meet the coverage desideratum. Furthermore, both Horning and Muggleton require that the prior distribution over grammars in the hypothesis space is *accurate*, in the sense that it defines a preference metric over hypotheses that leads the learner to the correct target grammar given realistic input (i.e. generalization due to inductive bias from the input is correct).

A prior distribution or cost metric encoding a preference for smaller, more compressed grammars will, in general, select ones that predict the grammaticality of supersets of the learning input. The exact form of the representation language in which candidate grammars are couched and/or the addition of factors other than just size to the prior distribution or cost metric will determine which of the grammars generating a superset of the input is acquired by the learner. This is where domain-specific inductive bias appears to be unavoidable if the desideratum of learning accuracy is to be met. And thus, this is the basis on which an LAD, in the sense outlined in the introduction above, is unavoidable in any adequate account of grammatical acquisition (see also Briscoe, forthcoming).

## Linguistic Evolution

First-language learners are not typically exposed to homogeneous data from a static speech community. Though major and rapid grammatical change is relatively rare, learners typically hear utterances produced by members of other speech communities, and the learning period is sufficiently extended that they may be exposed to ongoing linguistic change within a single community. A major tenet of generative diachronic linguistics is that first language acquisition is the main engine of grammatical change because faced with such mixed data, learners can acquire grammars that are distinct from those of the previous generation (e.g. Lightfoot 1999: 77 f.).

We can model the development of the 'external' (E-)language of a speech community as a dynamical system in which states encode the distribution of 'internalized' individual grammars (and lexicons) (i.e. I-languages or idiolects) within the community. In such dynamical systems, transitions between states are defined in terms of changes in the distribution of internalized grammars or I-languages (Briscoe 1997; 2000*b*; Niyogi and Berwick 1997). If there is inductive bias in first-language acquisition (regardless of its provenance), then languages are best characterized as adaptive systems, because learners will preferentially select linguistic variants which are easier to learn and thus more adaptive with respect to the acquisition procedure (Briscoe 1997; 2000*b*; Kirby 1998). However, linguistic selection of this kind does not come exclusively from language acquisition. Other, often conflicting selection pressures created by the exigencies of production and comprehension mean that the fitness (or adaptive) landscape for language is complex and dynamic, with no fixed points or stable attractors (Briscoe 2000*b*). For example, a functional pressure for more parsable linguistic variants (Briscoe 2000*a*; Kirby 1999) may be counterbalanced by a social pressure to produce innovative variants (Nettle 1999) or a functional pressure to produce shorter utterances (Lindblom 1998). Thus individual languages are complex adaptive systems on rugged, dynamic, and multipeaked fitness landscapes, in the sense of, for example, Kauffmann (1993).

Linguistic evolution proceeds via cultural transmission (i.e. first-language acquisition) at a faster rate than biological evolution. The populations involved are generally smaller (speech communities, rather than entire species), and language acquisition is a more flexible and efficient method of information transfer than genetic mutation. Clearly, vocabulary learning and, at least, peripheral grammatical development are ongoing processes that last beyond childhood, so that linguistic inheritance is less clearly delineated or constrained than the biological mechanisms of genetic evolution.

Several consequences emerge from the evolutionary account of languages as adaptive systems which must be taken into consideration by any plausible account of grammatical assimilation. First, several researchers have considered what type of language acquisition procedure could not only underlie accurate learning of modern human languages but also predict the emergence of protolanguage(s) with undecomposable form–meaning correspondences and the (subsequent) emergence of protolanguage(s) with decomposable (minimally grammatical) sentence–meaning correspondences (e.g. Oliphant 2002; Kirby 2002; Brighton 2002). They conclude that the language acquisition procedure must incorporate inductive bias resulting in generalization, and consequent regularization of the input, in order that repeated rounds of cultural transmission of language regularize random variations into consistent and coherent communication systems.[2]

Moreover, the account of languages as adaptive systems entails that linguistic universals no longer constitute strong evidence for an LAD. Deacon (1997), Kirby and Hurford (1997), and others make the point that universals may equally be the result of convergent evolution in different languages as a consequence of similar evolutionary pathways and linguistic selection pressures. For example, the fact that in attested languages irregularity is associated with high-frequency forms is unlikely to be a consequence of a nativized constraint, and much more likely to be a universal consequence of the fact that low frequency irregular forms are less likely to be reliably learned by successive generations of first-language learners (Kirby 2001).

## Grammatical Assimilation

If there is an LAD, then it is legitimate to ask how this unique biological trait emerged. There are only two clearly distinct possibilities compatible with modern evolutionary theory: some degree of exaptation of pre-existing traits combined with saltation and/or genetic assimilation (e.g. Bickerton 2000).

### Genetic Assimilation

Genetic assimilation is a neo-Darwinian (and not Lamarckian) mechanism supporting apparent 'inheritance of acquired characteristics' (e.g. Waddington 1942; 1975). The fundamental insights are that: (1) plasticity in the relationship between phenotype and genotype is under genetic control; (2) novel environments create selection pressures which favour organisms with the plasticity to allow within-lifetime developmental adaptations to the new environment; and (3) natural selection will function to 'canalize' these developmental adaptations by favouring genotypic variants in which the relevant trait develops reliably on the basis of minimal environmental

---

[2] Newport (1999) reports the results of experiments on sign language acquisition from poor and inconsistent signers which clearly exhibits exactly this bias to *impose* regularity where there is variation unconditioned by social context or other factors.

stimulus, providing that the environment, and consequent selection pressure, remains constant over enough generations.[3]

A simple putative example of genetic assimilation is the propensity to develop hard skin on certain regions of the body on the basis of quite limited environmental stimulation. Selection for individuals who developed hard skin more rapidly, and subsequent canalization of this trait prevented infection, aided mobility, and so forth. A more complex putative case is Durham's (1991) example of gene–culture interaction resulting in extended lactose tolerance in human populations in which animal husbandry is well established. The ability to consume milk in maturity was selected for in an environment in which it was one of the most reliable and beneficial sources of nutrition.

One form of plasticity in primates is the ability to learn from the environment. The Bayesian learning framework provides a general and natural way to understand and model how more and more accurate prior distributions over hypothesis spaces with better and better 'fit' with the environment can evolve. Staddon (1988) and Cosmides and Tooby (1996) both argue at length that Bayesian learning theory is an appropriate framework for modelling learning in animals and humans and that evolution can be understood within this framework as a mechanism for optimizing priors to 'fit' the environment and thus increase fitness.

### Genetic Assimilation of Grammatical Information

Pinker and Bloom (1990) develop a gradual assimilationist account of the evolution of the LAD. However, they rely heavily on linguistic universals as their evidence. Waddington himself suggested earlier that genetic assimilation provided a possible mechanism for the evolution of an LAD:

If there were selection for the ability to use language, then there would be selection for the capacity to acquire the use of language, in an interaction with a language-

---

[3] Waddington's work on genetic assimilation is a neo-Darwinian refinement of an idea independently proposed by Baldwin, Lloyd Morgan, and Osborne in 1896, and often referred to as the Baldwin Effect (see Richards 1987 for a detailed history). Waddington refined the idea by emphasizing the role of canalization and the importance of genetic control of ontogenetic development—his 'epigenetic theory of evolution'. He also undertook experiments with *Drosphila subobscura* which directly demonstrated modification of genomes via artificial environmental changes (see Jablonka and Lamb 1995: 31f. for a detailed and accessible description of these experiments).

using environment; and the result of selection for epigenetic responses can be, as we have seen, a gradual accumulation of so many genes with effects tending in this direction that the character gradually becomes genetically assimilated.   (1975: 305 f.)

Briscoe (1999; 2000*a*) speculates that an initial acquisition procedure emerged via recruitment (exaptation) of pre-existing (pre-adapted) general-purpose (Bayesian-like) learning mechanisms to a specifically linguistic cognitive representation capable of expressing mappings from decomposable meaning representations to realizable, essentially linearized, encodings of such representations (see also Bickerton 1998; 2000; Worden 1998). The selective pressure favouring such a development, and its subsequent maintenance and refinement, is only possible if some protolanguage(s), supporting successful communication and capable of cultural transmission (i.e. learnable without an LAD) within a hominid population, had already emerged (e.g. Deacon 1997; Kirby and Hurford 1997). Protolanguage(s) may have been initially similar to those advocated by Wray (2000), in which complete propositional messages are conveyed by undecomposable signals. However, to create selection pressure for the emergence of grammar, and thus for an LAD incorporating language-specific grammatical inductive bias, protolanguage(s) must have evolved at some point into decomposable utterances, broadly of the kind envisaged by Bickerton (1998). Several models of the emergence of syntax have been developed (e.g. Kirby 2001; 2002; Nowak et al. 2000). At the point when the environment contains language(s) with minimal syntax, grammatical assimilation becomes adaptive, under the assumption that language confers a fitness advantage on its users, since assimilation will make grammatical acquisition more rapid and reliable.

Saltations or macro-mutations are compatible with evolutionary theory if a single change in genotype creates a large, highly adaptive change in phenotype, though general considerations predict that such genetic macro mutations are extremely unlikely to be adaptive (e.g. Dennett (1995: 282 f.) Saltationist accounts have been proposed by Chomsky (1988), Gould (1991), Bickerton (1998), Berwick (1998), Lightfoot (2000), and others who variously speculate that the LAD emerged rapidly, in essentially its modern form, as a side effect of the development of large general-purpose brains (possibly in small heads) and/or sophisticated conceptual representations These accounts not only entail that the LAD emerged in a single and extremely unlikely evolutionary step (see e.g. Pinker and Bloom (1990) for detailed counter-arguments) but also neglect the fact that selection pressure is required to *maintain* a biological trait (e.g. Ridley 1990). Without such selection pressure, we would expect a trait to be whittled away by accumulated random mutations in the population (i.e. genetic drift, e.g. Maynard-Smith 1998: 24 f.). However, with such selection pressure, a newly emerged trait will continue to adapt, especially if the environmental factors creating the selection pressure are themselves changing—as languages do. A saltationist account, then, requires the assumption that language, and consequently the ability to learn one fast and reliably with an LAD, confers an adaptive advantage just as much as a gradualist account requires the same assumption. Therefore, even if the first LAD emerged by macro-mutation, evolutionary theory predicts that it may have been further refined by genetic assimilation.

### Counterarguments to Assimilation

Newmeyer (2000) goes one stage further than other saltationists, arguing that, given the assumptions (1) that the LAD incorporates a Universal Grammar based on Government-Binding (GB) theory (Chomsky 1981), (2) that the language(s) extant in the environment of adaptation were exclusively SOV rigid-order languages with grammatical properties similar to their attested counterparts, and (3) that such attested languages do not manifest most of the universal linguistic constraints posited in GB theory, then the LAD, if it exists, could not have emerged as a result of grammatical assimilation and must be the result of saltation. Newmeyer (2002) develops related arguments, for instance arguing that grammatical subordination would have rarely been manifest in preliterate speech communities and therefore that universal constraints relating to such constructions could not have been assimilated. Deacon (1997: 307 f.) argues that, since attested grammatical systems display a trade-off between syntactic and morphological encoding of predicate–argument structure, and since these distinct linguistic devices are also neurally distinct, the changing linguistic environment could not have created consistent selection pressure on either neural mechanism. These arguments all rest on specific assumptions about what precisely is assimilated. However, the account of the LAD in terms of inductive bias, developed in the section above on grammatical acquisition is in no way dependent on any specific linguistic constraints, and does not rest on (speculative) assumptions about linguistic phenomena manifest in the prehistoric environment of adaptation. Assimilation of linguistic constraints or preferences into the LAD only requires that *some* neurally encodable generalizations were manifest in the environment of adaptation for the LAD.

Lightfoot (1999; 2000) argues that the LAD is not fully adaptive and therefore could not have evolved by assimilation, since by definition this is an adaptive process. He uses the example of the putative universal constraint against some forms of subject extraction from tensed embedded subordinate clauses, which prevents the asking of questions like (1).

(1)   *Who do you wonder whether/how solved the problem?

Lightfoot argues that such phenomena show that aspects of the LAD are dysfunctional, since the constraint reduces the expressiveness of human languages, and he provides evidence that the constraint is circumvented by various ad hoc strategies in different languages—in English, such questions become grammatical if the normally optional complementizer *that* is obligatorily dropped, as in (2).

(2)   Who do you think (*that) solved the problem?

He argues that the presence of such a maladaptive constraint entails that the LAD could not have evolved gradually, even though this constraint is a by-product of an adaptive, more general condition on extraction. However, evolutionary theory does not predict that traits will be or will remain optimal. It may be that *any* genetically encodable extraction constraint aiding parsability and/or learnability also has unwanted side effects for expressiveness. Complex fitness landscapes typically contain many local optima which are far more likely to be discovered than any global optimum, should it exist (e.g. Kauffman 1993). Furthermore, a dynamic fitness landscape entails that a once optimal solution can become suboptimal.

Given that grammatical assimilation only makes sense in a scenario in which evolving (proto)languages create selection pressure, Waddington's notion of genetic assimilation should be embedded in the more general one of co-evolution (e.g. Kauffman 1993: 242 f.). Waddington himself (1975: 307) noted that if there is an adaptive advantage to simplifying grammatical acquisition, then we might expect assimilation to continue to the point where no learning would be needed because a fully specified grammar had been encoded. In this case acquisition would be instantaneous and fitness would be maximized in a language-using population. Given a co-evolutionary scenario, in which languages themselves are complex adaptive systems, a plausible explanation for continuing grammatical diversity is that social factors favouring innovation and diversity create conflicting linguistic selection pressures (e.g. Nettle 1999). Genetic transmission, and thus assimilation, will be much slower than cultural transmission, and therefore continued plasticity in grammatical acquisition is probable, because assimilation will not be able to 'keep up with' all grammatical change. Furthermore, too much assimilation will reduce individuals' fitness, if linguistic change subsequently makes it hard or impossible for them to acquire an innovative grammatical (sub)system.

Deacon (1997) and Worden (2002) also assume a co-evolutionary scenario, but argue that genetic assimilation of specifically linguistic, grammatical information is unlikely precisely because languages evolve far faster than brains. Attested languages have shifted major grammatical systems within 1,000 years (or a mere fifty or so generations), so they argue it is far more likely that grammatical systems have evolved to be learnable by a pre-existing general-purpose learning mechanism than that this mechanism adapted to language. The main weakness of this argument is that it fails to take any account of the potential size of the hypothesis space of grammatical systems. The standard classes of languages familiar from formal language theory are infinite, so the hypothesis space for even the regular languages contains an infinite class of regular grammars. Even if we assume, as does the P&P framework, that the *evolved* LAD restricts the class of possible grammars of human languages to be finite, most linguists (implicitly) agree that the hypothesis space remains vast (on the order of thirty million grammars) as they typically posit around thirty binary-valued independent parameters (e.g. Roberts 2001).[4] No amount of rapid change between attested grammatical systems can count as evidence against grammatical assimilation of linguistic constraints which ruled out the many unattested grammars that could *not* have been sampled in the period of evolutionary adaptation (e.g. Briscoe 2000*a*). If, for example, arbitrarily intersecting dependencies of the kind exhibited by the MIX family of context-sensitive languages (e.g. Joshi et al. 1991) were unattested, but a proper subset of grammars in the hypothesis space generated constructions with such dependencies, then assimilation of a hard constraint or preference against

---

[4] Even this degree of finiteness remains controversial (e.g. Pullum 1983). For instance, it would be falsified if a language with a parametrically specified maximum of four syntactically realized arguments developed a predicate, analogous to English *bet* requiring five such arguments: *(np Kim) bet (np Sandy) (np £ 10) (scomp that she would win)* ↝ *(np Kim) bet (np Sandy) (np £ 10) (pp for Red Rum) (vpinf to win)*.

such grammars might be possible. Rapid change within the proper subset of grammars *not* generating MIX languages would not alter the adaptiveness (i.e. learnability advantages) of ruling out or dispreferring the unattested constructions.

There have probably been about 400,000 generations since hominids diverged from chimps. There is an upper bound to the rate at which evolution can alter the phenotype of a given species. The rate of evolution of any trait is dependent on the strength of the selection pressure for that trait, but too much selection pressure causes a species to die out. Estimates of the the upper bound vary between one and 400 bits of new information per generation (Worden 1995; Mackay 1999), creating an upper bound of between 4Kbits and 160Mbits of new genetic information expressed in the species phenotype. If the correct answer is close to the lower estimate, this places severe demands on any account of the emergence of a species-specific LAD, and means that exaptation of pre-existing neural mechanisms will play a critical part of any plausible gradualist scenario. On the other hand, if the higher estimate is closer to the truth, then it appears that there has been time for the *de novo* evolution of quite complex traits. The logic of the speed-limit argument collapses, given a saltationist account based on macro-mutation—a single genetic change brings about a complex of extremely unlikely but broadly adaptive phenotypic changes which spreads rapidly through the population.

A second and related argument is based on the observation that the relationship between genes and traits is rarely one to one, and that epistasis (or 'linkage') and pleiotropy are the norm. In general, the effect of epistasis and pleiotropy will be to make the pathways more indirect from selection pressure acting on phenotypic traits to genetic modifications increasing the adaptiveness of those traits. In general terms, therefore, we would expect a more indirect and less correlated genetic encoding of a trait to impede or perhaps even prevent genetic assimilation. Mayley (1996) presents a general exploration of the effects of manipulating the correlation between genotype (operations) and phenotype (operations) on genetic assimilation. In his model, individuals are able to acquire better phenotypes through 'learning' (or another form of within-lifetime plasticity), thus increasing their fitness. However, the degree to which the acquired phenotype can be assimilated into the genotype of future generations, thus simplifying learning and/or increasing its success, and further increasing fitness, depends critically on this correlation..

## Computational Simulations of Assimilation

One way to explore the arguments and counter-arguments outlined in the previous section is to build a simulation and/or a mathematical model. The latter is, in principle, preferable as analytic models of dynamical systems yield more reliable conclusions (given the assumptions underlying the model), whilst those generated by stochastic computational simulation are statistical (e.g. Renshaw 1991). To date, however, no detailed analytic model of grammatical assimilation has been developed.[5]

Each model consists of an evolving population of individuals. Individuals are endowed with the ability to acquire a trait by learning. However, the starting point for learning, and thus individuals' consequent success, is determined to an extent by an inherited genotype. Furthermore, the fitness of an individual, i.e. the likelihood that individuals will produce offspring, is determined by their successful acquisition of the trait. Offspring inherit starting points for learning (genotypes) which are based on those of their parents. Inheritance of *starting* points for learning prevents any form of Lamarckian inheritance of acquired characteristics, but allows for genetic assimilation, in principle. Inheritance either takes the form of crossover of the genotypes of the parents, resulting in a shared, mixed inheritance from each parent, and overall loss of variation in genotypes over generations, and/or random mutation of the inherited genotype, introducing new variation.

### Genetic Assimilation

Hinton and Nowlan (1987) describe the first computational simulation of genetic assimilation. In their (very abstract) simulation of a population of 1,000 neural networks with twenty potential connections, which can be unset ?, on 1, or off 0, were evolved using a genetic algorithm. The

---

[5] Nowak et al. (2002) briefly describe the general form of a model capable, in principle, of incorporating grammatical assimilation/coevolution. However, the simplifying assumptions required to yield deterministic dynamical update equations make it very difficult to address many of the counter-arguments to assimilation discussed in the previous section. For instance, no counteracting (socio)linguistic selection for diversity/variation is modelled, so the equilibrium point for many instantiations of their model may be a LAD encoding a single grammar/language.

target was a network with all twenty connections set to $\boxed{1}$, but networks were initialized randomly with connection ('gene') frequencies of 0.5 for $\boxed{?}$ and 0.25 for $\boxed{1}$ or $\boxed{0}$ at each position. Each network was able to set $\boxed{?}$ connections through learning (modelled as random search of connection settings) on the basis of 1,000 trials during its lifetime. The fitness of a network was defined as $1 + 19n/1,000$, where $n$ is the number of trials after it has acquired the correct settings, making a network with all $\boxed{1}$ connections initially twenty times fitter than a network which never learned to set them correctly. Reproduction of offspring was by crossover of *initial* connections from two parents whose selection was proportional to their fitness. In the early generations most networks had the same minimum fitness through being born with one or more $\boxed{0}$ settings; however this soon gave way to exponential increases in networks with more $\boxed{1}$ settings, fewer $\boxed{?}$ settings, and no $\boxed{0}$ settings. In the later stages, the increase of $\boxed{1}$ settings and decrease of $\boxed{?}$ settings asymptotes, once the population had evolved to genotypes enabling successful learning.

Hinton and Nowlan point out that the fitness landscape for this model is like a needle in a haystack: only one final setting of all twenty connections confers any fitness advantage whatsoever. Therefore, evolution unguided by learning would be expected to take on the order of $2^{20}$ trials (i.e. genotypes) to find a solution. If increased fitness required evolution of two such networks in the same generation, as would be the case for coordinated communicative behaviour, evolution would be expected to take around $2^{400}$ trials to find a solution. However, with learning, the simulation always converges within 10–15 generations on a viable genotype (i.e. after generating 100–150K networks). Once successful networks appear, their superior performance rapidly leads to the spread of genotypes which support successful learning. However, networks with $\boxed{?}$ settings persist, despite the pressure exerted by the fitness function to minimize the number of learning trials required to find the solution. Hinton and Nowlan suggest that this is a result of weak selection pressure once every network is capable of successful learning. Harvey (1993) analyses the model using the tools of population genetics and argues that, since many settings in genotypes of successful networks derive from the genotype of the first such successful network to emerge, there is a significant chance factor in the distribution of initial settings. When a single successful genotype evolves and dominates subsequent generations, it is possible for a $\boxed{?}$ setting to become 'prematurely' fixated, despite the selective pressure exerted by the fitness function in favour of shorter learning periods. The use of a mutation operator would presumably allow populations to converge to the optimum genotype, provided that selection pressure was strong enough to curtail the effects of subsequent random mutation and genetic drift.

This initial result has been extended by Ackley and Littman (1991), Cecconi et al. (1995), and French and Messenger (1994), variously demonstrating that genetic assimilation can occur without a predefined fitness criterion, can result in complete assimilation of a trait where learning has a significant cost and the environment remains constant, and—when this occurs—can result in loss of the now redundant learning component through (deleterious) genetic drift. An important caveat regarding these positive results is that Mayley (1996) demonstrates that assimilation can be slowed and even stopped if the degree of neighbourhood correlation between genotype and phenotype is reduced. In Mayley's model, individuals have separate encodings of genotype and corresponding phenotype. Learning alters the latter, whilst the directness of the encoding of phenotypes in genotypes and the relationship between learning rules and genetic operators determines the degree of genetic assimilation possible, in interaction with the shape of the fitness landscape and the cost of learning.

## Grammatical Assimilation

The first computational simulation of grammatical assimilation is that of Batali (1994), who demonstrates that the initial weight settings in a recurrent neural network (RNN), able to learn by back-propagation to make grammaticality judgements for sentences generated by a restricted class of unambiguous context-free grammars (CFGs), can be improved by genetic assimilation. An evolving population of RNNs with randomly initialized weights was exposed to languages from this class, and the networks best able to judge sentences from these languages were kept and also used to create offspring with minor variations in their initial settings. RNNs evolved able to learn final weights which yielded much lower error rates for sentences from any of this class of languages. This work is chiefly relevant for its demonstration of the potential for genetic assimilation in a precise computational setting on a non-trivial learning task. The RNN model of grammatical acquisition fails to meet the desiderata identified in the section above on grammatical acquisition, because the RNNs do not model the mapping between form and meaning.

In a related simulation, Livingstone and Fyfe (2000) start with a population of networks able to represent the mapping between undecomposable finite signal–meaning correspondences, and demonstrate that spatially organized networks will genetically assimilate an increased production capacity by switching on further hidden nodes in their networks, given selection for interpretive ability and exposure to a larger vocabulary. They argue that in a spatially organized setting this amounts to a form of kin selection, since networks receive no direct benefit from an increased production ability. They suggest that their approach might be extended to grammatical competence. However, it is difficult to see how, as the network architecture is only able to represent *finite* signal–meaning correspondences.

Turkel (2002) adapts Hinton and Nowlan's (1987) simulation more directly by adopting a P&P model of grammatical acquisition. Individuals in the evolving population are represented by a genotype of twenty binary-valued principles/parameters which can be set to on ($\boxed{1}$), off ($\boxed{0}$), or unset ($\boxed{?}$). $\boxed{?}$ settings represent parameters which are set during lifetime learning, $\boxed{0}/\boxed{1}$ settings represent nativized principles of the LAD. Learned settings of parameters define variant phenotypes of a given genotype, interpreted as different grammars learnable from the inherited variant of the LAD. The fitness of a genotype is determined by the speed with which individuals acquire compatible settings for unset parameters. A population of randomly initialized individuals each with ten parameters attempts to set them in order to communicate with another random individual via the same grammar. Individuals able to communicate are more likely to produce offspring with new genotypes derived from their own by crossover with those of another individual. Populations evolved genotypes which increased the speed and robustness of learning. However, despite the cost of learning, they did not converge on genotypes with no remaining parameters, probably for similar reasons to those identified by Harvey in his analysis of Hinton and Nowlan's original work.

Turkel's approach does not suffer from the weaknesses of neural network-based models, because he does not specify how genotypes encode grammars capable of generating form–meaning correspondences. Turkel, like Hinton and Nowlan, sees the simulation more as an abstract demonstration of how genetic assimilation provides a mechanism for canalizing a trait, and thus as a demonstration of how an LAD might have arisen on the basis of natural selection for communicative success. However, because of the unspecified relationship between genotypes and actual grammars, the only substantive difference from Hinton and Nowlan's model is the use of a frequency-dependent rather than fixed fitness function, which creates an overall lower degree of selection pressure.

Kirby and Hurford (1997) extend Turkel's model by encoding a set of sentences in terms of the principle/parameter settings required accurately to parse them, and by utilizing a modified version of Gibson and Wexler's (1994) Trigger Learning Algorithm. Appropriate parameter settings are learned by individuals as a function (1) of the parsability of individual sentences, where more parsable sentences are generated by grammars defined by $\boxed{1}$ settings at the first 4 loci, and (2) of their distance from the individual's current parameter settings. This introduces linguistic selection into the model, as grammars which generate more parsable sentences can be learned more easily. The initial population consists of individuals with only parameters who are exposed to enough sentences to be able to learn some grammar. As the population evolves, fitness increases through grammatical assimilation of $\boxed{1}/\boxed{0}$ settings which shorten the learning period and therefore increase communicative success.

Kirby and Hurford demonstrate that grammatical assimilation without linguistic selection results in shortening of the acquisition period, but also often results in assimilation of linguistically non-optimal settings in the genotype (i.e. ones yielding grammars generating less parsable sentences). However, in conjunction with linguistic selection, the population converges on a genotype that is compatible with the optimal grammars, because linguistic selection guarantees that the population converges on optimally parsable languages, via the inductive bias built into the learning algorithm, before genetic assimilation has time to fixate individual loci in the genotype. They conclude that functional constraints on variation will only evolve in the LAD if prior linguistic selection means that the constraints are assimilated from an optimal linguistic environment, and thus that natural selection for communicative success is not in itself enough to explain why *functional* constraints could become nativized. This work is important because it develops a co-evolutionary model of the interaction between linguistic selection for variant grammars via cultural transmission and natural selection for variant LADs via genetic assimilation.

Yamauchi (2000; 2001) replicates Turkel's simulation but manipulates the degree of correlation in the encoding of genotype and phenotype. He continues to represent a grammar as a sequence of $N$ principles or parameters but determines the initial setting at each locus from a look-up table which uses $K$ 0/1s (where $K$ can range from 1 to $N-1$) to encode each on/off/unset $\boxed{1}/\boxed{0}/\boxed{?}$ setting (and presumably ensure that all possible genotypes can be encoded). A genotype is represented as a sequence of $N$ 0/1s. A translator reads the first $K$ genes from the genotype and uses the look-up table to compute the setting of the first locus of the phenotype. To compute the setting of the second locus of the phenotype, the $K$ genes starting at the second locus of the genotype are read and looked up in the table, and so on. The translator 'wraps around' the genotype and continues with the first locus when $K$ exceeds the remaining bits of the genotype sequence. Yamauchi claims, following Kauffman (1993), that increases in $K$ model increases in pleiotropy and epistasis. Increased $K$ means that a change to one locus in the genotype will have potentially more widespread and less predictable effects on the resulting phenotype. It also means that there is less correspondence between a learning operation, altering the value of single phenotypic locus, and a genetic operation, potentially altering many in differing ways, or none, depending on the look-up table. For low values of $K$, genetic assimilation occurs, as in Turkel's model; for values of $K$ around $N/2$ genetic assimilation is considerably slowed; and for very high values ($K=N-1$) it is stopped.

Yamauchi does not consider how the progressive decorrelation of phenotype from genotype affects the degree of communicative success achieved or how linguistic systems might be affected. In part, the problem here is that the abstract nature of Turkel's simulation model does not support any inference from configurations of the phenotype to concrete linguistic systems. Yamauchi, however, simply does not report whether decorrelation affects the ability of the evolving population to match phenotypes via learning. The implication, though, is that, for high values of $K$, unless the population starts in a state where genotypes are sufficiently converged to make learning effective, then they cannot evolve to a state better able to match phenotypes and thus support communication. Kauffman's original work with the $NK$ model was undertaken to find optimal values of $K$ for given $N$ to quantify the degree of epistasis and pleiotropy likely to be found in systems able to evolve most effectively. Both theoretical predictions and experiments which allow $K$ itself to evolve suggest intermediate values of $K$ are optimal (where the exact value can depend on $N$ and other experimental factors). But despite these caveats, Yamauchi's simulation suggests that (lack of) correlation of genotype and phenotype with respect to the LAD is just as important an issue for accounts of grammatical assimilation as it is for accounts of genetic assimilation generally, as Mayley (1996) argued.

I have developed a co-evolutionary model and associated simulation (Briscoe 1997; 1998; 1999; 2000*a*; 2002; forthcoming) which supports linguistic selection for grammatical variants, based on learnability, parsability, and/or expressiveness, and natural selection for variant LADs based on communicative success. It incorporates a detailed account of grammatical acquisition, meeting the desiderata of the section above on grammatical acquisition, which in turn supports much more detailed modelling of the grammars acquired. Language agents (LAgts) learn and deploy Generalized Categorial Grammars (GCGs) using a Bayesian learning procedure which acquires the most probable grammar capable of representing the form–meaning mapping manifested by a noisy, finite, unordered sequence of form–meaning pairs generated by other random members of the current population of LAgts.

The starting point for learning is represented by a prior probability distribution over twenty binary-valued principles/parameters defining around 300 viable distinct GCGs. An unset parameter is represented by an unbiased prior (i.e. a uniform distribution over the two possible values), a parameter with a default initial setting by a biased prior capable of being reversed during the learning period, and a principle by a strongly biased prior that cannot be reversed given the amount of data that can be observed during the learning period. Mutation and one-point crossover operators can alter this prior probability distribution, converting unset parameters to default parameters, parameters to principles, and so forth randomly so as not to bias evolution towards any LAD within the space available. The acquired grammar utilizes just those parameters which are consistently expressed in the data, so LAgts can acquire grammars of subset languages. LAgts who communicate successfully with others because they have acquired (partly) compatible grammars reproduce in proportion to their overall relative success. LAgts who have acquired subset grammars or grammars incompatible with that dominant in the population will tend to have lower communicative success. Linguistic variation can be introduced by seeding initial populations with different grammars or by introducing successive migrations of new adult LAgts deploying a grammar different from that currently dominant in order to simulate language contact.

A number of results relevant to grammatical assimilation emerge from this model. First, assimilation occurs when and only when LAgts reproduce according to communicative success. This creates selection pressure for shortening the learning period and making it more robust to noise, so the population assimilates default parameter settings and principles at the expense of unset parameters (Briscoe 1999). Second, populations converge on LADs that further restrict the class of learnable grammars to ones generating subset languages, unless there is an additional conflicting selection pressure on LAgts to acquire more expressive grammars which counteracts the pressure for learnability (Briscoe 2000a). Third, as in the work of Kirby and Hurford (1997), natural selection for communicative success does not guarantee assimilation of functional constraints. However, if parsability inhibits learning or biases the distribution of form–meaning pairs manifest during learning, then assimilated LADs become biased towards more parsable languages (Briscoe 2000a). Fourth, if language change is as rapid as is consistent with maintenance of a speech community (defined as a mean 90% or better communicative success), assimilation still occurs but asymptotes well before the LAD defines a single grammar. In addition, default initial parameter settings (i.e. preferences) are selected over principles (i.e. hard constraints), as subsequent changes can render principles acquired by a proper subset of the population highly maladaptive (Briscoe 1999; 2000a).

Fifth, in a population with a fixed LAD exposed to homogeneous linguistic input manifesting dispreferred parameter settings, successive generations of learners reliably acquire the correct grammar. However, if their input is heterogeneous and manifests conflicting values, the prior distribution assimilated into the evolved LAD will tend to predominate. Briscoe (2002) suggests this can provide the basis for an account of creolization and perhaps other attested major historical change resulting from contact. Finally, Briscoe (forthcoming) progressively decorrelates the effects of the mutation operator from the updating of parameter settings during the learning process. Major decorrelation prevents assimilation, and most mutations which spread are pre-emptive 'side effects' rather than assimilative, causing rapid concomitant linguistic change. Consequently, populations eventually evolve LADs which predefine simple subset languages in which learning is redundant despite natural selection for expressiveness. Intermediate levels of decorrelation slow assimilation and increase the proportion of pre-emptive mutations which spread, but populations are not forced towards subset languages. Low levels of decorrelation have no significant effects, as pre-emptive mutations fail to spread through communities with consistently stable and accurate cultural transmission of language.

## Conclusions

In summary, extant models predict that grammatical assimilation would have occurred given three crucial assumptions. First, communicative success via expressive languages with compositional syntax conferred a fitness benefit on their users. Second, the linguistic environment for adaptation of the LAD consistently manifested grammatical generalizations to be assimilated—rapid linguistic change does not preclude generalizations ruling out or dispreferring areas of the hypothesis space generating unattested constructions. Third, some of these generalizations were neurally and genetically encodable with sufficient correlation to support assimilation. None of the counter-arguments reviewed in 'Counter-arguments to assimilation' above or simulations discussed in the previous section undermines these assumptions. The case for grammatical assimilation as the primary mechanism of the evolution of the LAD remains, in my opinion, strong.

Nevertheless, the co-evolutionary perspective on grammatical assimilation raises two important caveats. First, as languages themselves are adapted to be learnable (as well as parsable and expressive) and as they change on a historical timescale, some of the grammatical properties of human languages were probably shaped by the process of cultural transmission of (proto)language via more general-purpose learning (e.g. Kirby 1998). Secondly, whether the subsequent evolution of the LAD was assimilative, encoding generalizations manifest in the linguistic environment, or pre-emptive, with mutations creating side effects causing linguistic selection for new features, the fit between the inductive bias of the LAD and extant languages is predicted to be very close.

Finally, it is important to emphasize that modelling and simulation, however careful and sophisticated, is not enough to establish the truth of what remains a partly speculative inference about prehistoric events. The value of the simulations, and related mathematical modelling and analysis, lies in uncovering the precise set of assumptions required to predict that grammatical assimilation will or will not occur. Some of these assumptions relate to cognitive abilities or biases which remain manifest today, and these predictions are testable. For example, we have seen that inductive bias is at the heart not only of (grammatical) assimilation but also of any satisfactory model of grammatical acquisition and of the linguistic evolution of modern languages from protolanguage(s). Other assumptions, such as the correlation between genetic and neural encoding, are theoretically plausible but empirically untestable using extant techniques.

## FURTHER READING

Nowak et al. (2002) provide a brief synopsis of formal language theory and learnability theory, and develop evolutionary models of language change and of the emergence of the LAD ('Universal Grammar' in their terms, though they make the point that it is neither universal nor a grammar). Mitchell (1997) provides a more detailed and introductory treatment of learning theory. Joshi et al. (1991) summarize extant knowledge concerning the expressive power of human languages in terms of formal language theory. Jablonka and Lamb (1995) describe Waddington's work and the concept of genetic assimilation. Durham's (1991) theory of gene–culture interactions provides the basis for a co-evolutionary account of grammatical assimilation. Bertolo (2001) is a good collection of recent work in the P&P framework. Cosmides and Tooby (1996) makes the case for integrating the Bayesian learning framework with evolutionary theory as a general model of human learning.

# 17

# Language, Learning and Evolution

*Natalia L. Komarova and Martin A. Nowak*

## Introduction

Throughout the history of life on earth, evolution has come up with several great innovations, such as nucleic acids, proteins, cells, chromosomes, multicellular organisms, the nervous system, and finally—language. Among these innovations, language is the only one which is (presently) confined to one species. What exactly human language is, how we learn it, and how it evolved in its enormous complexity are some of the most fascinating questions of evolutionary biology and cognitive science.

The study of language and grammar dates back to classical India and Greece (Robins 1979). In the eighteenth century, the 'discovery' of Indo-European led to the surprising realization that very different languages can be related to each other, which initiated the field of historical linguistics. Formal language theory emerged only in the twentieth century (Chomsky 1956; 1957; Harrison 1978): the main goals are to describe the rules that a speaker uses to generate linguistic forms (descriptive adequacy) and to explain how language competence emerges in the human brain (explanatory adequacy). These efforts were supported by advances in the mathematical and computational analysis of the process of language acquisition, a field that became known as 'learning theory' (Gold 1967; Vapnik and Chervonenkis 1971; 1981; Valiant 1984; Osherson et al. 1986; Vapnik 1998; Jain et al. 1998). Currently there are increasing attempts to bring linguistic inquiry into contact with various disciplines of biology, including neurobiology (Deacon 1997; Vargha-Khadem et al. 1998), animal behaviour (Smith 1977; Dunbar 1996; Hauser 1996; Fitch 2000), evolution (Lieberman 1984;

Support from the David and Lucille Packard Foundation, the Leon Levy and Shelby White Initiatives Fund, the Florence Gould Foundation, the Ambrose Monell Foundation, the National Science Foundation, and Jeffrey E. Epstein is gratefully acknowledged.

Brandon and Hornstein 1986; Pinker and Bloom 1990; Bickerton 1990; Lieberman 1991; Newmayer 1991; Hawkins and Gell-Mann 1992; Batali 1994; Maynard Smith and Szathmary 1995; Aitchinson 1996; Hurford et al. 1998; Jackendoff 1999; Lightfoot 1999; Knight et al. 2000) and genetics (Gopnik and Crago 1991; Lai et al. 2001). The new aim is to study language as a product of evolution and as the extended phenotype of a species of primates.

There are two common misconceptions of language evolution. The first represents the human language capacity as an undecomposable unit, and states that its gradual evolution is impossible, because no part of it would have any function in the absence of other parts. For example, syntax could not have evolved without phonology or semantics, and vice versa. The other misconception is that language evolution started from scratch some five million years ago, when humans and chimps diverged, and there are virtually no data about it.

Both views are fundamentally flawed. First, all complex biological systems consist of specific components, so that it is often hard to imagine the usefulness of individual parts in the absence of other parts. The usual task of evolutionary biology is to understand how complex systems can arise from simpler ones gradually, by mutation and natural selection. In this sense, human language is no different from other complex traits. Second, it is clear that evolution did not build the human language faculty *de novo* in the last few million years, but used material that had evolved in other animals over a much longer time. Many animal species have sophisticated cognitive abilities in terms of understanding the world and interacting with one another. Furthermore, it is a well-known trick of evolution to use existing structures for new and sometimes surprising purposes. Monkeys, for example, appear to have brain areas similar to our language centres, but use them for controlling facial muscles and for analysing auditory input. Evolution may have had an easy task here to reconnect these centres for human language. Hence the human language instinct is most likely not the result of a sudden moment of inspiration of evolution's blind watchmaker, but rather the consequence of several hundred million years of 'experimenting' with animal cognition.

We can obtain data for language evolution in two ways. We can study the evolution of cognitive abilities and communication in animals (Hauser 1996; Fitch 2000), and we can analyse the enormous evidence provided by the existing human language instinct and its manifestation in 6,000 different languages. The data are in us, similar to genetic evolutionary history being written in our genome.

The perspective of this chapter is to show how methods of formal language theory, learning theory, and evolutionary biology can be combined to improve our understanding of the origins and the properties of human language. In the following section we discuss the key notions of 'language', 'grammar', and 'learning' and present rigorous definitions. We then formulate the 'paradox of language acquisition', and we show that learning theory can demonstrate in what sense Universal Grammar is a logical necessity. In the section on evolution in languages, we present a quantitative approach to questions of communication and evolution. We develop a general framework for the evolution of grammar acquisition and discuss how natural selection acts on Universal Grammar. We define grammatical coherence and find a coherence threshold which gives conditions for a population of speakers to evolve and maintain a stable language. We explore the conditions under which natural selection favours the emergence of a recursive, rule-based grammatical system. We show that our approach can be used to study problems of historical linguistics. In the concluding section we discuss cultural evolution of language as opposed to biological evolution of universal grammar, and come up with a unified description that contains both.

## Formal Language Theory and Applications

Language is a mode of communication, a crucial part of human behaviour, and a cultural object defining our social identity. There is also a fundamental aspect of human language that makes it amenable to formal analysis: linguistic structures consist of smaller units that are grouped together according to certain rules.

The combinatorial sequencing of small units into bigger structures occurs at several different levels. Phonemes are concatenated into syllables and words. Sequences of words form phrases and sentences. Most crucially, the rules for such groupings are not arbitrary and are language specific. Certain word orders are admissible in one language but not in another. In some languages, word order is relatively free but case marking is pronounced. There are always specific rules that *generate* valid or meaningful linguistic structures. Much of modern linguistic theory proceeds from this insight. The area of mathematics and computer science called formal language theory provides a mathematical machinery for dealing with such phenomena.

## Some Important Definitions

We define an *alphabet* as a set containing a finite number of symbols. Possible alphabets for natural languages would be the set of all phonemes or the set of all words of a language. For these two choices one obtains formal languages on different levels, but the mathematical principles are the same. We can also simply consider the binary alphabet {0,1} (see Fig. 17.1).

A *sentence* is defined as a string of symbols. The set of all sentences that can be generated by the binary alphabet is given by {0, 1, 00, 01, 10, 11, 000, ...}. Note that there are infinitely many sentences. More precisely, there are as many sentences as there are integers. Hence, the set of all sentences is 'countable'.

A *language* is a set of sentences. Among all possible sentences, some are part of the language and some are not. A finite language contains a finite number of sentences. An infinite language contains an infinite number of sentences. There are infinitely many finite languages, as many as integers. There are infinitely many infinite languages, as many as real numbers; they are not countable. Hence, the set of all languages is not countable.

A *grammar* is a 'theory' of a language: it is a finite list of rules that specify the language. A grammar is normally expressed in terms of 'rewrite rules', which are of the form: a certain string can be rewritten as another string.

An *alphabet* is a set of symbols: {0,1}
*Sentences* are strings of symbols: 0,1,00,01,10,11,000,001,010,100,101,...
A *language* is a set of sentences: $L=\{000,01100,01010,001110,...\}$
A *grammar* is a finite list of rules that define a language:

$$S \longrightarrow 0A \qquad B \longrightarrow 1B$$
$$A \longrightarrow 1A \qquad B \longrightarrow 0F$$
$$A \longrightarrow 0B \qquad F \longrightarrow \varepsilon$$

FIG. 17.1 The basic objects of formal language theory are alphabets, sentences, languages, and grammars. Grammars consist of rewrite rules: a particular string can be rewritten as another string. Such rules contain symbols of the alphabet (here: 0 and 1), and so called 'non-terminals' (here: S,A,B,F), and a null-element, $\varepsilon$. The grammar in this figure works as follows: Each sentence begins with the symbol S. S is rewritten as 0A. Now there are two choices: A can be rewritten as 1A or 0B. B can be rewritten as 1B or 0F. F always goes to $\varepsilon$. This grammar generates sentences of the form $01^m01^n0$, which means every sentence begins with 0 followed by a sequence of $m$ 1s followed by a 0 followed by a sequence of $n$ 1s followed by 0.

Strings contain elements of the alphabet together with so-called 'non-ter-minals', which are place holders. After iterated application of the rewrite rules the final string will only contain symbols of the alphabet. Figs. 17.1 and 17.2 give examples of grammars.

There are infinitely many grammars, but only as many as integers: any finite list of rewrite rules can be encoded by an integer. Since there are un-countably many languages, only a small subset of them can be described by a grammar. These languages are called 'computable'.

|  | *Grammars* | *Languages* |
|---|---|---|
| Finite-state (regular) | $S \longrightarrow A$ | $L = 0^m 1^n$ |
|  | $A \longrightarrow 0A$ |  |
|  | $A \longrightarrow B$ |  |
|  | $B \longrightarrow 1B$ |  |
|  | $B \longrightarrow \varepsilon$ |  |
| Context-free | $S \longrightarrow 0S1$ | $L = 0^n 1^n$ |
|  | $S \longrightarrow \varepsilon$ |  |
| Context-sensitive | $S \longrightarrow 0AS2$ | $L = 0^n 1^n 2^n$ |
|  | $S \longrightarrow 012$ |  |
|  | $A0 \longrightarrow 0A$ |  |
|  | $A1 \longrightarrow 11$ |  |

FIG. 17.2 Three grammars and their corresponding languages. Finite-state grammars have rewrite rules of the form: a single non-terminal (on the left) is rewritten as a sin-gle terminal possible followed by a non-terminal (on the right). The finite-state gram-mar, in this figure, generates the regular language $0^m 1^n$; a valid sentence is any sequence of 0s followed by any sequence of 1s. A context-free grammar admits rewrite rules of the form: a single non-terminal is rewritten as an arbitrary string of terminals and non-terminals. The context-free grammar in this figure generates the language $0^n 1^n$; a valid sentence is a sequence of 0s followed by the same number of 1s. There is no finite-state grammar that could generate this language. A context-sensitive grammar admits re-write rules of the form $\alpha A \beta \rightarrow \alpha \gamma \beta$. Here $\alpha$, $\beta$, and $\gamma$ are strings of terminals and non-terminals. While $\alpha$ and $\beta$ may be empty, $\gamma$ must be non-empty. The important restriction on rewrite rules of context-sensitive grammars is that the complete string on the right must be at least as long as the complete string on the left. The context-sensitive gram-mar, in this figure, generates the language $0^n 1^n 2^n$. There is no context-free grammar that could generate this language.

### The Chomsky Hierarchy of Languages

There is a correspondence between languages, grammars and machines. The set of all computable languages is described by 'phrase structure' grammars which are equivalent to Turing machines. A Turing machine embodies the theoretical concept of a digital computer with infinite memory (Turing 1936; 1950). For each computable language, there exists a Turing machine which can list as output all sentences of this language.

A subset of computable languages are 'context-sensitive' languages, which are generated by context-sensitive grammars (Fig. 17.2). For each of these languages there exists a machine that can decide if a given sentence is part of the language or not. This can be done by a Turing machine with a finite memory, a so-called 'linear bounded automaton'.

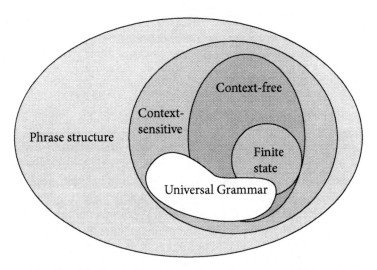

FIG. 17.3 The Chomsky hierarchy and the logical necessity of Universal Grammar. Finite-state grammars are a subset of context-free grammars, which are a subset of context-sensitive grammars, which are a subset of phrase structure grammars, which represent all possible grammars. Similarly, regular languages are a subset of context-free languages, which are a subset of context-sensitive languages, which are a subset of computable languages. Natural languages are considered to be more powerful than regular languages. The crucial result of learning theory is that there exists no procedure that could learn an unrestricted set of languages; in most approaches even the class of regular languages is not learnable. The human brain has a procedure for learning language. This procedure can only learn a restricted set of languages. Universal Grammar is the theory of this restricted set.

A subset of context-sensitive languages are 'context-free' languages, which are generated by context-free grammars (Fig. 17.2), which are equivalent to push-down automata. These are computers with a single memory stack; at any given time they have access only to the top register of their memory.

An even simpler machine can be constructed. A finite-state automaton has a start, a finite number of intermediate states, and a finish. Whenever the machine jumps from one state to the next, it emits an element of the alphabet. A particular run from start to finish produces a sentence. There are many different runs from start to finish, and hence there are many different sentences. If a finite-state automaton contains at least one loop, then it will be able to generate infinitely many sentences. Hence, finite-state automata can generate all finite languages and some infinite languages. Finite-state automata and the corresponding finite-state grammars define the set of regular languages.

Fig. 17.3 shows the Chomsky hierarchy: regular languages are a subset of context-free languages, which are a subset of context-sensitive languages, which are a subset of computable languages, which are disjunct from non-computable languages.

### The Structure of Natural Languages

Natural languages are infinite: it is not possible to imagine a finite list that contains all English sentences. Furthermore, most linguists agree that finite-state grammars are inadequate for natural language. Such grammars are unable to represent long-range dependencies of the form: *If* [*she has an idea for how to solve the problem she has been working on for a long time*] *then* [*she will not go for lunch*]. The string of words between *If* and *then* could be arbitrarily long, and could itself contain more paired if–then constructions. Such pairings ultimately relate to rules that generate strings of the form $0^n1^n$, which require context-free grammars.

The fundamental structures of natural languages are trees. The nodes represent phrases that can be composed of other phrases in a recursive manner. Finite-state automata can only generate a very limited class of trees, which is again an argument that natural grammars are more powerful than finite-state grammars. There is a continuing debate whether context-free grammars are adequate for natural languages (Pullum and Gazdar 1982; Shieber 1985), or whether context-sensitive grammars need to be evoked (Bar-Hillel 1953; Joshi et al. 1975).

One can also define grammars that directly specify which trees are acceptable for a given language. A tree is a 'derivation' of a sentence within the rule system of a particular grammar. The interpretation of a sentence depends on the underlying tree structure. Ambiguity arises if more than one tree can be associated with a given sentence. Much of modern syntactic theory deals with grammars that directly act on tree structures (Chomsky 1984; Sadock 1991; Bresnan 2001; Pollard and Sag 1994). We cannot go into more detail here, but emphasize that all such grammars are ultimately placed somewhere on the Chomsky hierarchy, and that the results of learning theory (to be discussed now) apply to them.

## Learning Theory and a Logical Necessity of Universal Grammar

Learning is inductive inference. The learner is presented with data and has to infer the rules that generate these data. The difference between 'learning' and 'memorization' is the ability to *generalize* beyond one's own experience to novel circumstances. In the context of language, the child learner will generalize to novel sentences never heard before. Any child can produce and understand sentences that are not part of his previous linguistic experience.

### The Paradox of Language Acquisition

Children develop grammatical competence spontaneously, without formal training. All they need is interaction with people and exposure to normal language use. In other words, the child hears a certain number of grammatical sentences and then constructs an internal representation of the rules that generate grammatical sentences. Chomsky pointed out that the evidence available to the child does not uniquely determine the underlying grammatical rules (Chomsky 1965; 1972). This phenomenon is called the 'poverty of stimulus' (Wexler and Culicover 1980). The 'paradox of language acquisition' is that children nevertheless reliably achieve correct grammatical competence (Jackendoff 1997; 2001). How is this possible?

The proposed solution of the paradox is that children learn the correct grammar by choosing from a restricted set of candidate grammars. The 'theory' of this restricted set is Universal Grammar (UG). The concept of an innate, genetically determined UG was controversial when introduced some forty years ago and has remained so. The mathematical approach of learning theory, however, can explain in what sense UG is a logical necessity.

### What is Learnable?

Imagine an ideal speaker–hearer pair. The speaker uses grammar $G$ to construct sentences of language $L$. The hearer receives sentences and should, after some time, be able to use grammar $G$ to construct other sentences of $L$. Mathematically speaking, the hearer is described by an algorithm (or more generally, a function), $A$, which takes a list of sentences as input and generates a language as output.

Let us introduce the notion of a 'text' as a list of sentences. Specifically, text $T$ of language $L$ is an infinite list of sentences of $L$ with each sentence of $L$ occurring at least once. Text $T_N$ contains the first $N$ sentences of $T$. We say that language $L$ is learnable by the algorithm, $A$, if for each $T$ of $L$ there exists a number $M$ such that for all $N > M$ we have $A(T_N) = L$. This means that, given enough sentences as input, the algorithm will provide the correct language as output.

Furthermore, a set of languages is learnable by an algorithm if each language of this set is learnable. We can imagine an algorithm that gives 'English' as output for every input. This algorithm can 'learn' English, but no other language. Hence, we are interested in the question of what set of languages, $L = \{L_1, L_2, ..\}$, can be learned by a given algorithm.

We can now present a key result of learning theory: according to Gold's theorem (Gold 1967), there exists no algorithm that can learn the set of regular languages. As a consequence, no algorithm can learn a set of languages that contains the set of all regular languages. Hence, no algorithm can learn the set of context-free languages, the set of context-sensitive languages, or the set of computable languages. Needless to say, no algorithm can learn the set of all languages.

A common criticism of Gold's framework is that the learner has to identify exactly the right language. For practical purposes, however, it might be sufficient that the learner acquires a grammar which is almost correct. There are various extensions of the Gold framework. For example, the approach of statistical learning theory contains the crucial requirement that the learner converges with a certain probability to a language that is almost the correct language. In the framework of statistical learning theory, it turns out that there exists no procedure that can learn the set of all regular languages. Hence, we obtain the same necessity of an innate UG as before.

Some statistical learning models are motivated by 'informational complexity'. The question is: given a specific amount of information, is there any procedure that can in principle infer the correct language? Other learning theories include the concept of 'computational complexity'. Here the question is: given a specific amount of information, is there any computer programme that can come up with the correct language in reasonable (polynomial) time? Considerations of computational complexity lead to further restrictions on the set of languages that can be learned.

### The Necessity of Innate Expectations

We can now state in what sense there has to be an innate UG. The human child is equipped with a learning algorithm, $A_H$, which enables the child to learn certain languages. This algorithm can learn each of the existing 6,000 human languages and presumably many more, but it is impossible that $A_H$ could learn *any* language. Hence, there is a set of languages that can be learned by $A_H$, and this set must be heavily restricted compared to the set of all possible languages. UG is the rule system that describes the restricted set of languages that is learnable by $A_H$.

Learning theory shows there must be an innate UG, which is a consequence of the particular learning algorithm, $A_H$, used by humans. Discovering properties of $A_H$ requires the empirical study of neurobiological and cognitive functions of the human brain involved in language acquisition. Some aspects of UG, however, might be unveiled by studying common features of existing human languages. This has been a major goal of linguistic research during the last decades. A particular approach is the 'principle and parameter theory', which assumes that the child comes equipped with innate principles and has to set parameters that are specific for individual languages (Chomsky 1981; 1984; Gibson and Wexler 1994; Manzini and Wexler 1987). Another approach is 'optimality theory', where the child has innate constraints, and learning a specific language is ordering these constraints (Prince and Smolensky 1997).

There is some discourse as to whether the learning mechanism, $A_H$, is language-specific or general-purpose (Elman et al. 1996). Ultimately this is a question about the particular architecture of the brain and which neurons participate in which computations; but one cannot deny that there is a learning mechanism, $A_H$, that operates on linguistic input and enables the child to learn the rules of human language. This mechanism can learn a restricted set of languages; the theory of this set is an innate UG.

Hence, the continuing debate around an innate UG should not be whether there is one, but what form it takes (Elman et al. 1996; Tomasello 1999; Sampson 1999). One can dispute individual linguistic universals (Greenberg et al. 1978; Comrie 1981; Baker 2001), but one cannot generally deny their existence.

There is also some discussion about the role neural networks can play in language acquisition. Learning theory clearly shows that there exists no neural network that can learn an unrestricted set of languages. It is well understood that any particular neural network can only learn a very specific set of rules (Geman et al. 1992).

Sometimes it is claimed that the logical arguments for an innate UG rest on particular mathematical assumptions of generative grammars which deal only with syntax and not with semantics. Cognitive (Lakoff 1987; Langacker 1987) and functional linguistics (Bates and MacWhinney 1982) instead are not based on formal language theory, but use psychological objects such as symbols, categories, schemas, and images. This does not, however, remove the necessity of innate restrictions. The results of learning theory apply to any learning process, where a 'rule' has to be learned from some examples. This generalization is an inherent feature of any model of language acquisition and applies to semantics, syntax, and phonetics. Any procedure for successful generalization has to pick from a restricted range of hypotheses.

## Evolution of Language

Let us now formulate a mathematical description of language acquisition. The material presented here is part of a larger effort to use quantitative and computational approaches to study the evolution of language (Cavalli-Sforza and Feldman 1981; Aoki and Feldman 1987; Hurford 1989; Hashimoto and Ikegami 1996; Steels 1996; Kirby and Hurford 1997; Hazlehurst and Hutchins 1998; Nowak and Krakauer 1999; Kirby 2000; Nowak et al. 2000; Kirby 2001; Komarova and Nowak 2001; Cangelosi and Parisi 2002; Christiansen et al. 2002; Christiansen and Dale in press). The sentences of all languages can be enumerated. We can say that a grammar, $G$, is a rule system that specifies which sentences are allowed and which sentences are not.

Universal Grammar, in turn, contains a rule system that generates a set (or a search space) of grammars, $\{G_1, G_2, \ldots, G_n\}$. These grammars can be constructed by the language learner as potential candidates for the grammar that needs to be learned. The learner has a mechanism to evaluate input sentences and to choose one of the candidate grammars that are contained in his search space. The learner cannot end up with a grammar that is not part of this search space. In this sense, UG contains the possibility to learn all human languages (and many more).

More generally, it is also possible to imagine that UG generates infinitely many candidate grammars, $\{G_1, G_2, \ldots\}$. In this case, the learning task can be solved if UG also contains a prior probability distribution on the set of all grammars. This prior distribution biases the learner toward grammars that are expected to be more likely than others. A special case of a prior distribution is one where a finite number of grammars is expected with equal probability and all other grammars are expected with zero probability, which is equivalent to a finite search space.

A fundamental question of linguistics and cognitive science is what restrictions are imposed by UG on human language? In other words, how much is innate and how much is learned in human language. In learning theory, this question is studied in the context of an ideal speaker–hearer pair. The speaker uses a certain 'target grammar'. The hearer has to learn this grammar. The question is: what is the maximum size of the search space such that a specific learning mechanism will converge (after a number of input sentences, with a certain probability) to the target grammar?

In terms of language evolution, the crucial question is what makes a *population* of speakers converge to a coherent grammatical system. In other words, what are the conditions that UG has to fulfil for a population of individuals to evolve coherent communication? In the following, we will discuss how to address this question.

### Language Learning: From Individuals to Populations

Evolution takes place in heterogeneous populations. We have to envisage a population of speakers using slightly different languages. They communicate with each other, which affects their performance, survival, and reproduction. Whether they use language for exchanging information, making plans, cooperative hunting, social bonding, deception, or cooperation, language affects the fitness of individuals. Those that are better at the language game leave more offspring. Otherwise natural selection, which we use here inclusive of sexual selection, could not operate, and the alternative hypothesis would be that language is the product of neutral evolution (or the by-product of some other process), which is unlikely given the extreme complexity of this trait (Pinker 1994).

Imagine a group of individuals that all have the same UG, given by a finite search space of candidate grammars, $G_1,...,G_n$, and a learning mechanism for evaluating input sentences. Let us specify the similarity between grammars by introducing the numbers $s_{ij}$ which denote the probability that a speaker who uses $G_i$ will say a sentence that is compatible with $G_j$.

We assume there is a reward for mutual understanding. The pay-off for someone who uses $G_i$ and communicates with someone who uses $G_j$ is given by

$$F(G_i,G_j)=(s_{ij}+s_{ji})/2.$$

This is simply the average taken over the two situations when $G_i$ talks to $G_j$ and when $G_j$ talks to $G_i$.

Denote by $x_i$ the relative abundance of individuals who use grammar $G_i$. Assume that everybody talks to everybody else with equal probability. Therefore, the average pay-off for all those individuals who use grammar $G_i$ is given by

$$f_i=\sum_{j=1}^{n}x_jF(G_i,G_j).$$

We assume that the pay-off derived from communication contributes to biological fitness: individuals leave offspring proportional to their pay-off. These offspring inherit the UG of their parents. They receive language input (sample sentences) from their parents and develop their own grammar. At first, we will not specify a particular learning mechanism but introduce the stochastic matrix, $Q$, whose elements, $q_{ij}$, denote the probability that a child born to an individual using $G_i$ will develop $G_j$. (In this model, we assume that each child receives input from one parent. It is possible to extend this approach to allow input from several individuals.) The probabilities that a child will develop $G_i$ if the parent uses $G_i$ is given by $q_{ii}$. The quantities, $q_{ii}$, measure the accuracy of grammar acquisition. If $q_{ii}=1$ for all $i$, then grammar acquisition is perfect for all candidate grammars.

The population dynamics of grammar evolution are then given by the following system of ordinary differential equations, which we call the 'language dynamical equations' (Nowak et al. 2001):

(1) $\qquad \dfrac{dx_j}{dt} = \sum_{i=1}^{n} f_i q_{ij} x_i - \phi x_j, \quad j=1,..,n.$

The term $-\phi x_j$ ensures that the total population size remains constant: the sum over the relative abundances, $\Sigma_i x_i$, is 1 at all times. The variable

$$\phi = \sum_{i=1}^{n} f_i x_i$$

denotes the average fitness or *grammatical coherence* of the population. The grammatical coherence is given by the probability that a randomly chosen sentence of one person is understood by another person. It is a measure for successful communication in a population. If $\phi=1$, all sentences are understood and communication is perfect. In general, $\phi$ is a number between 0 and 1.

The language dynamical equation is reminiscent of the *quasi-species equation* of molecular evolution, but has frequency dependent fitness values: the quantities $f_i$ depend on the relative abundances $x_1,.$ .., $x_n$. In the limit of perfectly accurate language acquisition, $q_{ii}=1$, we recover the *replicator equation* of evolutionary game theory. Thus, our model provides a connection between two of the most fundamental equations of evolutionary biology.

### Evolution of Grammatical Coherence

In general, equation (1) above admits multiple (stable and unstable) equilibria (Fig. 17.4). For low accuracy of grammar acquisition (low values $q_{ii}$), all grammars, $G_i$, occur with roughly equal abundance. There is no predominating grammar in the population. Grammatical coherence is low. As the accuracy of grammar acquisition increases, however, equilibrium solutions arise where a particular grammar is more abundant than all other grammars. A coherent communication system emerges. This means that, if the accuracy of learning is sufficiently high, the population will converge to a stable equilibrium with one dominant grammar. Which one of the stable equilibria is chosen depends on the initial condition.

The accuracy of language acquisition depends on UG. The less restricted the search space of candidate grammars is, the harder it is to learn a particular grammar. Depending on the specific values of $s_{ij}$ some grammars may be much harder to learn than others. For example, if a speaker using G has high probabilities formulating sentences that are compatible with many

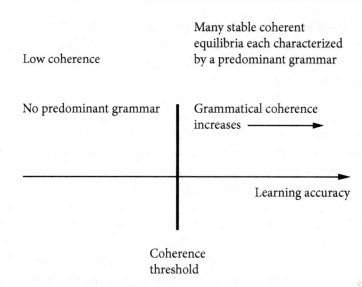

FIG. 17.4 Coherence threshold. When the accuracy of individual learning is low, equation (1) admits a single low-coherence solution with no predominating grammar in the population. As the accuracy of grammar acquisition increases, however, equilibrium solutions arise where a particular grammar is more abundant than all other grammars. A coherent communication system emerges.

other grammars ($s_{ij}$ close to 1 for many different $j$), then $G_i$ will be hard to learn. In the limit $s_{ij}=1$, $G_i$ is considered unlearnable, because no sentence can refute the hypothesis that the speaker uses $G_j$.

The accuracy of language acquisition also depends on the learning mechanism specified by UG. An inefficient learning mechanism or one that evaluates only a small number of input sentences will lead to a low accuracy and hence prevent the emergence of grammatical coherence.

We can therefore ask the crucial question:

*Which properties must UG have such that a predominating grammar will evolve in a population of speakers? In other words, which UG can induce grammatical coherence in a population?*

As outlined above, the answer will depend on the learning mechanism and the search space. We can derive results for two learning mechanisms that represent reasonable boundaries for the actual, unknown learning mechanism employed by humans:

- The memoryless learning algorithm, a favourite with learning theorists, makes few demands on the cognitive abilities of the learner. It describes the interaction between a teacher and a learner. (The 'teacher' can be one or several individuals or the whole population.) The learner starts with a randomly chosen hypothesis (say $G_i$) and stays with this hypothesis as long as the teacher's sentences are compatible with this hypothesis. If a sentence arrives that is not compatible, the learner will at random pick another candidate grammar from his search space. The process stops after a certain number of sentences. The algorithm is called 'memoryless' because the learner does not remember any of the previous sentences nor which hypotheses have already been rejected. The algorithm works, primarily because once it has the correct hypothesis it will not change any more (this is, incidentally, the definition of so called 'consistent learners').
- The other extreme is a batch learner. The batch learner memorizes all sentences and at the end chooses the candidate grammar that is most compatible with the input.

Let us first make a simplifying assumption that all the $n$ grammars are in some sense equally distant from each other. Mathematically speaking this amounts to setting $s_{ij} = s$, some constant, for $i \neq j$, and $s_{ii} = 1$. For the memoryless learner, we can show that grammatical coherence is possible if the number of input sentences, $N$, exceeds a constant times the number of candidate grammars, $N > C_1 n$. For the batch learner, the number of input sentences has to exceed a constant times the logarithm of the number of candidate grammars, $N > C_2 \log n$.

In the more general case, when the similarity coefficients are taken from a uniform distribution, we can prove that for the memoryless learner, the number of sample sentences has to exceed $c_1 n \log n$ in order for the population to maintain grammatical coherence. For the batch learner, this result is $N > c_2 n$. These inequalities define a *coherence threshold*, which limits the size of the search space relative to the amount of input available to the child. A UG that does not fulfil the coherence threshold does not lead to a stable predominating grammar in a population.

The learning mechanism used by humans will perform better than the memoryless learner and worse than the batch learner; hence it will have a coherence threshold somewhere between $N > c_1 n \log n$ and $N > c_2 n$. The coherence threshold relates a life history parameter of humans, $N$, to the maximum size of the search space, $n$, of Universal Grammar.

## List Makers and Rule Finders

Next, let us explore the conditions under which natural selection favours the emergence of a rule-based, recursive grammatical system with infinite expressibility. In contrast to such rule-based grammars, one might consider list-based grammars that consist only of a finite number of sentences. Such list-based grammars can be seen as very primitive evolutionary precursors (or alternatives) to rule-based grammars. Individuals would acquire their mental grammar not by searching for underlying rules, but simply by memorizing sentences and their meaning (similar to memorizing the arbitrary meaning of words). List-based grammars do not allow for creativity on the level of syntax. Nevertheless, whether or not natural selection favours the more complicated rule-based grammars depends on circumstances that we need to explore.

Current human grammars can generate infinitely many sentences, but for the purpose of transmitting information only a finite number of them can be relevant. Natural selection cannot directly reward the theoretical ability to construct infinitely long sentences. Let us therefore consider a group of individuals that use $M$ different sentences (or syntactic structures). Note that $M$ specifies the number of sentences that are relevant from the perspective of biological fitness.

Now imagine individuals that learn their mental grammar by memorizing lists of sentences. We can ask how many sample sentences, $N$, a child must hear for the whole population to maintain $M$ sentences. If all sentences occur equally often, we simply obtain $N > M$.

We can compare the performance of individuals using list-based versus rule-based grammars. Let us use the result for batch learners, which have comparable memory requirements to the list learners, and assume that grammar similarity coefficients, $s_{ij}$, are distributed uniformly between zero and one. Then we obtain that the number of relevant sentences, $M$, has to exceed a constant times the number of the candidate grammars, $n$. We have

$$M > c_3 n.$$

If this condition did *not* hold, it would be more efficient to *memorize* sentences associated with arbitrary meaning. In this case, language would have remained a rather dull communication system without any creative ability on the level of syntax. If, on the other hand, the condition above is satisfied, then rule-based grammars are more efficient than list-based grammars and

will have a fitness advantage. Furthermore, if rule-based grammars are selected, then the potential for 'making infinite use of finite means' comes as a
by-product.

### Applications for Historical Linguistics

The language dynamic equation can be used to study language change in
the context of historical linguistics (Lightfoot 1991; Kroch 1989; Wang
1998; Niyogi and Berwick 1997; de Graff 1999). Here, a good assumption
is that minor language changes are selectively neutral. Hence we can use a
neutral version of our approach possibly in conjunction with stochastic and
spatial population dynamics. It is possible to show that coherence threshold
phenomena calculated for deterministic dynamics of infinite populations
carry over to stochastic dynamics of finite populations.

Of special interest is that, for neutral language dynamics, we find linguistic coherence for

$$u < 1/M,$$

where $u$ is the error rate of language acquisition and $M$ is the effective population size. (This condition was derived under the symmetry assumption
$s_{ij} = s$ so that $q_{ii} = 1 - u$ for all i.) This condition relates the accuracy of language learning by individuals and the size of the linguistic community. If
the linguistic coherence threshold is satisfied, then the language is passed
down the generations with a high degree of accuracy; if the condition is violated, then a language change is likely to occur.

Neutral coherence is an important finding because it explains how linguistic features that do not contribute to communicative fitness (efficacy)
can be fairly homogeneous in a population. Neutral language dynamics
provide an appropriate description for many language changes studied in
historical linguistics, where fitness effects can probably be neglected.

### Cultural Evolution of Language vs. Biological Evolution of Universal Grammar

Evolution of UG requires variation of UG. Thus UG is neither a gramma
nor universal. Imagine a population of individuals using universal gram
mars $U_1$ to $U_M$. Each $U_I$ admits a subset of $n$ grammars and determines a
particular learning matrix $Q^{(I)}$. $U_I$ mutates genetically to $U_J$ with probabilit

$W_{IJ}$. Deterministic population dynamics are given by

$$(2) \quad \frac{dx_{Jj}}{dt} = \sum_{I=1}^{m} W_{IJ} \sum_{i=1}^{n} f_{Ii} Q_{ij}^{(J)} x_{Ii} - \phi x_{Jj} \quad j = 1, .., n \quad J = 1, .., M$$

This equation describes mutation and selection among $M$ different universal grammars. The relative abundance of individuals with Universal Grammar $U_J$ speaking language $L_j$ is given by $x_{Jj}$. At present, little is known about the behaviour of this system. In the limit of no mutation among universal grammars, $W_{II} = 1$, we find that the selective dynamics often lead to the elimination of all but one universal grammar, but sometimes coexistence of different UGs can be observed. Equation (2) describes two processes on different time scales: the biological evolution of UG and the cultural evolution of spoken language (Fig. 17.5).

Only a UG that is sufficiently specific can lead to coherent communication in a population. The ability to induce a coherent language is a major selective criterion for UG. There is also a trade-off between learnability and adaptability: a small search space (small $n$) is more likely to lead to linguistic coherence, but might exclude languages with high communicative pay-off.

Since the necessity of a restricted search space applies to any learning task, we can use an extended concept of UG for animal communication. Therefore, during primate evolution, there was a succession of UGs that finally led to the UG of currently living humans. At some point a UG emerged that

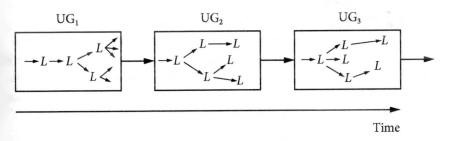

FIG. 17.5 The evolution of language takes place on two distinct time-scales. The larger scale corresponds to the biological evolution of Universal Grammar. Cultural evolution of language takes place on a much shorter time-scale. When modelling cultural evolution of language, we can assume that biological changes in the innate 'hardware' are negligibly small, and treat UG as constant in time and throughout the population. This brings us back to language dynamic equation (1). In order to describe changes in UG, we need to consider biological mutations, equation (2).

allowed languages of unlimited expressibility. Such evolutionary dynamics are described by equation (2) above.

## Conclusions

We have outlined a connection between language, learning and evolution. Ultimately, these three fields of investigation need to be combined: ideas of language theory should be discussed in the context of acquisition, and ideas of acquisition in the context of evolution. The aim is to use evolutionary theory to understand basic design features of human language and the way it is learned. Evolutionary and game-theoretic approaches should outline the gradual emergence of various parts of human language, such as arbitrary signs, words, lexicons, and grammatical rules. Some of these problems constitute individual research projects that can be addressed separately without taking into account the whole unmeasurable complexity of human language. Fascinating questions are: How does a population settle for an arbitrary sign? How does natural selection equip a population with the cognitive machinery for grasping a (linguistic) rule? What is the interplay between the biological evolution of Universal Grammar and the cultural evolution of particular languages? Each of these questions seems to be as rich, for example, as the problem of evolution of cooperation, which is a vast field of evolutionary biology and incidentally tightly linked to language: we speak because we cooperate, we cooperate because we speak.

Cooperation is also required for advancing this line of research. Linguists need to establish more contact with biology, while biologists need to show more concern for human language, which is after all evolution's most interesting invention ever since multicellularity—that is, within the last 500 million years.

### FURTHER READING

A common criticism of Gold's framework is that the learner has to identify exactly the right language. For practical purposes, it might be sufficient that the learner acquires a grammar which is almost correct. There are various extensions of the Gold framework and in particular the approach of statistical learning theory. Here the crucial requirement is that the learner converges with high probability to a language that is almost correct. Statistical learning theory also shows there is no procedure that can learn the set of all regular languages, thereby confirming the necessity of an innate UG.

Classical learning theory was formulated by Gold (1967). Perhaps the most significant extension of the classical framework is statistical learning theory. Here, the learner is required to converge approximately to the right language with high probability. For statistical learning theory, see Vapnik (1998). A deep result, originally due to Vapnik and Chervonenkis (1971) and elaborated since, states that a set of languages is learnable if and only if it has finite VC dimension. The VC dimension is a combinatorial measure of the complexity of a set of languages. Thus if the set of possible languages is completely arbitrary (and therefore has infinite VC dimension), learning is not possible.

Considerations of computational complexity can also be added, where the learner is required to approximate the target grammar with high confidence using an efficient algorithm. Consequently, there are sets of languages that are learnable in principle (have finite VC dimension), but no algorithm can do this in polynomial time: see Valiant (1984).

For more information on the present authors' approach to modelling learning and evolution of language, see the following papers: Nowak and Krakauer (1999) and Nowak, Plotkin, and Krakauer (1999) for language in a game-theoretic framework, Nowak, Krakauer, and Dress (1999) for an error limit in the evolution of language, Nowak, Plotkin, and Jansen (2000) for the evolution of syntactic communication, Komarova and Nowak (2001) for the evolution of a lexical matrix, Nowak et al. (2001) and Komarova, Niyogi and Nowak (2001) for modelling of the evolution of UG, Komarova and Rivin (2002) for the convergence speed of the memoryless learner algorithm, and Komarova and Nowak (2002) for the evolution of language in finite populations.

# References

Abler, W. (1989), On the particulate principle of self-diversifying systems. *Journal of Social and Biological Structures* 12: 1–13.

Ackley, D., and M. Littman (1991), Interactions between learning and evolution. In C. Langton and C. Taylor (eds.), *Artificial Life II*. Menlo Park, Calif.: Addison-Wesley, 487–509.

Adcock, G. J., E. S. Dennis, S. Easteal, G. A. Huttley, L. S. Jermiin, W. J. Peacock et al. (2001), Mitochondrial DNA sequences in ancient Australians: implications for modern human origins, *Proceedings of the National Academy of Sciences*, 98: 537–42.

Aiello, L.C. and R. I. M. Dunbar (1993), Neocortex size, group size and the evolution of language. *Current Anthropology* 34: 184–93.

Aitchison, J. (1996), *The Seeds of Speech*. Cambridge: Cambridge University Press.

Akhtar, N. (1999), Acquiring basic word order: evidence for data-driven learning of syntactic structure. *Journal of Child Language* 26: 339–56.

——and Tomasello, M. (1997), Young children's productivity with word order and verb morphology. *Developmental Psychology* 33: 952–65.

Albert, M. A., R. G. Feldman, and A. L. Willis (1974), The 'subcortical dementia' of progressive supranuclear palsy. *Journal of Neurology, Neurosurgery, and Psychiatry* 37: 121–30.

Albert, R.M., S. Wiener, O. Bar-Yosef, and L. Meignen (2000), Phytoliths in the Middle Palaeolithic deposits of Kebara Cave, Mt Carmel, Israel. *Journal of Archaeological Science* 27(10): 931–47.

Aldhouse-Green, S., and P. B. Pettitt (1998), Paviland Cave: contextualizing the 'Red Lady'. *Antiquity* 72(278): 756.

Aldridge, J. W., K. C. Berridge, M. Herman, and L. Zimmer (1993), Neuronal coding of serial order: syntax of grooming in the neostratum. *Psychological Science* 4: 391–3.

Alexander, G. E., M. R. Delong, and P. L. Strick (1986), Parallel organization of segregated circuits linking basal ganglia and cortex. *Annual Review of Neuroscience* 9: 357–81.

Alexander, M. P., M. A. Naeser, and C. L. Palumbo (1987), Correlations of subcortical CT lesion sites and aphasia profiles. *Brain* 110: 961–91.

Allen, C., and M. D. Hauser (1991), Concept attribution in nonhuman animals: theoretical and methodological problems in ascribing complex mental processes *Philosophy of Science* 58: 221–40.

Alvarez, H. P. (2000), Grandmother hypothesis and primate life histories. *American Journal of Physical Anthropology* 113(3): 435–50.

Ambrose, S. H. (1998), Chronology of the Later Stone Age and food production in East Africa. *Journal of Archaeological Science* 25(4): 377–92.

——(2001), Paleolithic technology and human evolution. *Science* 291: 1748–53.

Andersen, H. (1973), Abductive and deductive change. *Language* 40: 765–93.

Andrew, R. J. (1976), Use of formants in the grunts of baboons and other nonhuman primates. *Annals of the New York Academy of Sciences* 280: 673–93.

Aoki, K., and M. W. Feldman (1987), Toward a theory for the evolution of cultural communication: coevolution of signal transmission and reception. *Proceedings of the National Academy of Sciences USA*, 84: 7164–68.

Appenzeller, T. (1998), Art: evolution or revolution? *Science* 282: 1451.

Aquadro, C. (1999), The problem of inferring selection and evolutionary history from molecular data. In M. T. Clegg (ed.), *Limits to Knowledge in Evolutionary Biology*. New York: Plenum.

Arbib, M. A. (1981), Perceptual structures and distributed motor control. In V. B. Brooks (ed.), *Handbook of Physiology – The Nervous System II: Motor Control*. Bethesda, Md.: American Physiological Society, 1449–80.

——(2002), The mirror system hypothesis for the language-ready brain. In Cangelosi and Parisi, 229–54.

——(2002), The mirror system, imitation, and the evolution of language. In C. Nehaniv and K. Dautenhahn (eds.), *Imitation in Animals and Artifacts*. Cambridge, Mass.: MIT Press, 229–80.

——(ed.) (2003a), *The Handbook of Brain Theory and Neural Networks*, 2nd edn. Cambridge, Mass.: MIT Press.

——(2003b), Schema theory. In Arbib (ed.), 993–8.

——A. Billard, M. Iacoboni, and E. Oztop (2000), Synthetic brain imaging: grasping, mirror neurones and imitation. *Neural Networks* 13: 975–97.

——D. Caplan, and J. C. Marshall (eds.) (1982), *Neural Models of Language Processes*. New York: Academic Press.

——P. Érdi, and J. Szentágothai (1997), *Neural Organisation: Structure, Function, and Dynamics*. Cambridge, Mass.: MIT Press.

Arcadi, A. C. (1996), Phrase structure of wild chimpanzee pant hoots: patterns of production and interpopulation variability. *American Journal of Primatology* 39: 159–78.

——(2000), Vocal responsiveness in male wild chimpanzees: implications for the evolution of language. *Journal of Human Evolution* 39: 205–23.

Arensburg, B., A. M. Tillier, B. Vandermeersch, H. Duday, L. A. Schepartz, and Y. Rak (1989), A Middle Palaeolithic human hyoid bone. *Nature* 388: 758–60.

Armstrong, D. F. (1999), *Original Signs: Gesture, Sign, and the Source of Language*. Washington, DC: Gallaudet University Press.

——W. C. Stokoe, and S. E. Wilcox (1995), *Gesture and the Nature of Language*. Cambridge: Cambridge University Press.

Asfaw, B., Y. Beyene, G. Suwa, R. C. Walter, T. D. White, G. WoldeGabriel, and T. Yemane (1992), The earliest Acheulean from Konso-Gardula. *Nature* 360: 732–5.

Ashton, N., J. McNabb, B. Irving, S. Lewis, and S. Parfitt (1994), Contemporaneity of Clactonian and Acheulian flint industries at Barnham, Suffolk. *Antiquity* 68: 585–9.

Ashton, N. M., J. Cook, S. G. Lewis, and J. Rose (eds.) (1992), *High Lodge: Excavations by G. de G. Sieveking, 1962–8, and J. Cook, 1988*. London: British Museum Press.

Austin, J. L. (1962), *How To Do Things with Words*. Oxford: Clarendon Press.

Awh, E., J. Jonides, R. E. Smith, E. H. Schumacher, R. A. Koeppe, and S. Katz (1996), Dissociation of storage and rehearsal in working memory: evidence from Positron Emission Tomography. *Psychological Science* 7: 25–31.

Baddeley, A. D. (1986), *Working Memory*. Oxford: Clarendon Press.

Baker, M. (2001), *The Atoms of Language*. New York: Basic Books.

Bard, K. A. (1992), Intentional behavior and intentional communication in young free-ranging orangutans. *Child Development* 63(5): 1186–97.

Bar-Hillel, Y. (1953), A quasi-arithmetical notation for syntactic description. *Language* 29: 47–58.

Baron-Cohen, S. (1993), From attention-goal psychology to belief-desire psychology: the development of a theory of mind and its dysfunction. In S. Baron-Cohen, H. Tager-Flusberg, and D. J. Cohen (eds.), *Understanding Other Minds: Perspectives from Autism*. New York: Oxford University Press.

Barrett, L., R. I. M. Dunbar, and J. E. Lycett (2002), *Human Evolutionary Psychology*. Basingstoke: Palgrave.

Bar-Yosef, O., and S. L. Kuhn (1999), The big deal about blades. *American Anthropologist* 101(2): 322–38.

Batali, J. (1994), Innate biases and critical periods: combining evolution and learning in the acquisition of syntax. In R. Brooks and P. Maes (eds.), *Artificial Life 4: Proceedings of the Fourth International Workshop on the Synthesis and Simulation of Living Systems*. Redwood City, Calif.: Addison-Wesley, 160–71.

——(1998), Computational simulations of the emergence of grammar. In Hurford et al., 405–26.

——(2002), The negotiation and acquisition of recursive grammars as a result of competition among exemplars. In Briscoe, 111–72.

Bates, E., D. Thal, and V. Marchman (1991), Symbols and syntax: a Darwinian approach to language development. In N. Krasnegor, D. Rumbaugh, M. Studdert-Kennedy, and R. Schiefelbusch (eds.), *The Biological Foundations of Language Development*. Oxford: Oxford University Press, 29–65.

——and B. MacWhinney (1982), Functionalist approaches to grammar. In L. Gleitman and E. Wanner (eds.), *Language Acquisition: The State of the Art*. Cambridge: Cambridge University Press, 173–218.

Bauer, R. H. (1993), Lateralization of neural control for vocalization by the frog (*Rana pipiens*). *Psychobiology* 21: 243–8.

Baum, S. R., S. E. Blumstein, M. A. Naeser, and C. L. Palumbo (1990), Temporal dimensions of consonant and vowel production: an acoustic and CT scan analysis of aphasic speech. *Brain and Language* 39: 33–56.

Beck, B. B. (1974), Baboons, chimpanzees, and tools. *Journal of Human Evolution* 3: 509–16.

Beitchman, J. H., E. B. Brownlie, A. Inglis, J. Wild, et al. (1994), Seven-year follow-up of speech/language-impaired and control children: speech/language stability and outcome. *Journal of the American Academy of Child and Adolescent Psychiatry* 31: 1322–30.

Benson, D. F., and A. Ardila (1996), *Aphasia: A Clinical Perspective*. Oxford: Oxford University Press.

Bertolo, S. (2001), A brief overview of learnability. In S. Bertolo (ed.), *Language Acquisition and Learnability*. Cambridge: Cambridge University Press, 1–14.

Berwick, R. C. (1998), Language evolution and the Minimalist Program: the origins of syntax. In Hurford et al., 320–40.

Bickerton, D. (1975), *Dynamics of a Creole System*. Cambridge: Cambridge University Press.

——(1981), *Roots of Language*. Ann Arbor, Mich.: Karoma.

——(1990), *Language and Species*. Chicago: University of Chicago Press.

——(1995), *Language and Human Behavior*. Seattle: University of Washington Press.

——(1998), Catastrophic evolution: the case for a single step from protolanguage to full human language. In Hurford et al., 341–58.

——(2000), How protolanguage became language. In Knight et al., 264–84.

——(2001), Linguists play catch-up with evolution. *Journal of Linguistics* 37: 581–91.

——(2002a), Foraging versus social intelligence in the evolution of language. In Wray, 207–25.

——(2003), C-command versus surface minimalism. MS. University of Hawaii.

——(in preparation), *How Syntax Works: Structure, Evolution and the Brain*.

Bischoff, J. L., J. F. Garcia, and L. G. Straus (1992), Uranium-series isochron dating at El Castillo cave (Cantabria, Spain): the 'Acheulean'/'Mousterian' question. *Journal of Archaeological Science* 19: 49–62.

Bishop, D. V. M. (2002), Putting language genes in perspective. *Trends in Genetics* 18: 57–9.

——et al. (1999), Auditory temporal processing impairment: neither necessary nor sufficient for causing language impairment in children. *Journal of Speech, Language, and Hearing Research* 42: 1295–1310.

——T. North, and C. Donlan (1995), Genetic basis of Specific Language Impairment. *Developmental Medicine and Child Neurology* 37: 56–71.

Bishop, J., D. V. M. Bishop, and C. F. Norbury (2001), Phonological processing, language, and literacy: a comparison of children with mild to moderate sensorineural hearing loss and those with specific language impairment. *Journal of Child Psychology and Psychiatry* 42: 329–40.

Blakemore, C. (1991), Computational principles of the cerebral cortex, *Psychologist* 14: 73.

Bloom, L. (1992), *Language Development from Two to Three*. Cambridge: Cambridge University Press.

——(ed.) (2002), *The Transition to Language*. Oxford: Oxford University Press.

Blumstein, S. E. (1995), The neurobiology of language. In J. L. Miller and P. D. Eimas (eds.), *Speech, Language and Communication*. San Diego, Calif.: Academic Press, 339–70.

——W. Cooper, H. Goodglass, H. Statlender, and J. Gottleib (1980), Production deficits in aphasia: a voice-onset time analysis. *Brain and Language* 9: 153–70.

Boëda, E., J. M. Geneste, C. Griggo, N. Mercier, S. Muhesen, J. L. Reyss, A. Taha, and H. Valladas (1999), A Levallois point embedded in the vertebra of a wild ass (*Equus africanus*): hafting, projectiles and Mousterian hunting weapons. *Antiquity* 73(280): 394–402.

de Boer, B. (2001), *The Origins of Vowel Systems*. Oxford: Oxford University Press.

Boesch, C. (1991), Teaching among wild chimpanzees. *Animal Behaviour* 41: 530–2.

——and H. Boesch (1984a), Possible causes of sex differences in the use of natural hammers by wild chimpanzees. *Journal of Human Evolution* 13: 415–40

————(1984b), Mental maps in wild chimpanzees: an analysis of hammer transports for nut cracking. *Primates* 25: 160–70

————(1990), Tool use and tool making in wild chimpanzees. *Folia Primatologica* 54: 86–99

——and M. Tomasello (1998), Chimpanzee and human cultures. *Current Anthropology*, 39(5): 591–614.

Bolus, M., and N. J. Conard (2001), The late Middle Paleolithic and earliest Upper Paleolithic in Central Europe and their relevance for the Out of Africa hypothesis. *Quaternary International* 75: 29–40.

Bonda, E., M. Petrides, S. Frey, and A. C. Evans (1994), Frontal cortex involvement in organised sequences of hand movements: evidence from a positron emission tomography study. *Society for Neuroscience Abstracts* 152: 6.

Bordes, F. (1961), *Typologie du Paléolithique Ancien et Moyen*. Bordeaux: Delmas.

Botha, R. P. (1997), Neo-Darwinian accounts of the origins of language: 1. Questions about their explanatory focus, *Language and Communication* 17: 249–67.

——(2002), Did language evolve like the vertebrate eye? *Language and Communication* 22: 131–58.

Bowler, J. M., R. Jones, H. Allen, and A. G. Thorne (1970), Pleistocene human remains from Australia. *World Archaeology* 2: 39–60.

——and A. Thorne (1976), Human remains from Lake Mungo. In R. L. Kirk and A. G. Thorne (eds.), *The Origin of the Australians*. Canberra: Australian Institute of Aboriginal Studies, 127–38.

Bowles, R. L. (1889), Observations upon the mammalian pharynx, with especial reference to the epiglottis. *Journal of Anatomy and Physiology* 23: 606–15.

Bradbury, J. W., and S. L. Vehrencamp (1997), *Principles of Animal Communication*. Sunderland: Sinauer Associates.

Bradshaw, J. L., and L. J. Rogers (1993), *The Evolution of Lateral Asymmetries, Language, Tool Use, and Intellect*. Sydney: Academic Press.

Brandon, R. N., and N. Hornstein (1986), From icons to symbols: some speculations on the origins of language. *Biology and Philosophy* 1: 169–89.

Bresnan, J. W. (2001), *Lexical-Functional Syntax*, Oxford: Blackwell.

Brighton, H. (2002), Compositional syntax from cultural transmission. *Artificial Life* 8(1): 25–54.

——and S. Kirby (2001), The survival of the smallest: stability conditions for the cultural evolution of compositional language. In J. Kelemen and P. Sosik (eds.), *Advances in Artificial Life*. Heidelberg: Springer.

————and K. Smith (in press), Situated cognition and the role of multi-agent models in explaining language structure. In D. Kudenko, E. Alonso, and D. Kazakov (eds.), *Adaptive agents*. Heidelberg: Springer.

Briscoe, E. (1997), Co-evolution of language and of the language acquisition device. In *Proceedings of the 35th Meeting of the Association for Computational Linguistics*. San Mateo, Calif.: Morgan Kaufmann, 418–27.

——(1998), Language as a complex adaptive system: co-evolution of language and of the language acquisition device. In P. Coppen, H. van Halteren, and L. Teunissen (eds.), *Proceedings of the 8th Meeting of Computational Linguistics in the Netherlands*. Amsterdam: Rodopi, 3–40.

——(1999), The acquisition of grammar in an evolving population of language agents. Electronic transcript of *Artificial Intelligence*, Special Issue 16(3B): *Machine Intelligence*, ed. S. Muggleton, 44–77. www.etaij.org

——(2000a), Grammatical acquisition: inductive bias and coevolution of language and the Language Acquisition Device. *Language* 76: 245–96.

——(2000b), Evolutionary perspectives on diachronic syntax. In S. Pintzuk, G. Tsoula, and A. Warner (eds.), *Diachronic Syntax: Models and Mechanisms*. Oxford: Oxford University Press, 75–108.

——(ed.) (2002), *Linguistic Evolution through Language Acquisition: Formal and Computational Models*. Cambridge: Cambridge University Press.

——(2002), Coevolution of the language faculty and language(s) with decorrelated encodings. Paper presented at Fourth International Conference on the Evolution of Language, Harvard, 2002.

Broca, P. (1861a), Remarques sur le siège de la faculté de la parole articulée, suivies d'une observation d'aphémie (perte de parole). *Bulletin de la Société Anatomique* 36: 330–57.

——(1861b), Nouvelle observation d'aphémie produite par une lésion de la moitié postérieure des deuxième et troisième circonvolutions frontales gauches. *Bulletin de la Société Anatomique* 36: 398–407.

Brooks, A. S., D. M. Helgren, J. S. Cramer, A. Franklin, W. Hornyak, J. M. Keating, R. G. Klein, W. J. Rink, H. Schwarz, J. N. Leith Smith, K. Stewart, N. E. Todd, J. Verniers, and J. E. Yellen (1995), Dating and context of three Middle Stone Age sites with bone points in the Upper Semliki Valley, Zaire. *Science* 268: 548–53.

Browman, C. P., and L. Goldstein (1986), Towards an articulatory phonology. *Phonology Yearbook* 3: 219–52.

———(1990), Tiers in articulatory phonology, with some implicatations for casual speech. In J. Kingston and M. E. Beckman (eds.), *Papers in Laboratory Phonology I: Between the Grammar and Physics of Speech.* Cambridge: Cambridge University Press, 341–76.

———(1991), Gestural structures: distinctiveness, phonological processes, and historical change. In I. G. Mattingly and M. Studdert-Kennedy (eds.), *Modularity and the Motor Theory of Speech Perception.* Hillsdale, NJ: Erlbaum, 313–38.

———(1992), Articulatory phonology: an overview. *Phonetica* 49: 155–80.

———(1995a), Dynamics and articulatory phonology. In T. van Gelder and R. F. Port (eds.), *Mind as Motion.* Cambridge, Mass.: MIT Press, 175–93.

———(1995b), Gestural syllable position effects in American English. In F. Bell-Berti and L. J. Raphael (eds.), *Producing Speech: Contemporary issues. For Katherine Safford Harris.* Woodbury, NY: AIP Press, 19–34.

———(2000), Competing constraints on intergestural coordination and self-organization of phonological structures. *Les Cahiers de l'ICP: Bulletin de la Communication Parlée* 5: 25–34.

Brown, P. (2001), Learning to talk about motion up and down in Tzeltal: is there a language-specific bias for verb learning? In M. Bowerman and S. Levinson (eds.), *Language Acquisition and Conceptual Development.* Cambridge: Cambridge University Press.

Brown, R. (1973), *A First Language: The Early Stages.* Cambridge, Mass.: Harvard University Press.

Buchler, J. (1955), *The Writings of Charles Peirce.* New York: Dover.

Burling, R. (1993), Primate calls, human language and nonverbal communication. *Current Anthropology* 34: 25–54.

———(2000), Comprehension, production and conventionalization in the origins of language. In Knight et al., 27–39.

Bybee, J. L., R. D. Perkins, and W. Pagliuca (1994), *The Evolution of Grammar: Tense, Aspect, and Modality in the Languages of the World.* Chicago: University of Chicago Press.

———and J. Scheibmann (1999), The effect of usage on degrees of constituency: the reduction of *don't* in English. *Linguistics* 37: 575–96.

Byrd, D. (1996), Influences on articulatory timing in consonant sequences. *Journal of Phonetics* 24: 209–44.

———and E. Saltzman (1998), Intragestural dynamics of multiple phrasal boundaries. *Journal of Phonetics* 26: 173–99.

Byrne, R. W., and A. E. Russon (1998), Learning by imitation: a hierarchical approach. *Behavioral and Brain Sciences* 21(5): 667–721.

———and A. Whiten (1988), *Machiavellian Intelligence: Social Expertise and the Evo-*

*lution of Intellect in Monkeys, Apes and Humans.* Oxford: Clarendon Press.

Call, J. (2001), Chimpanzee social cognition. *Trends in Cognitive Sciences* 5: 382–7.

Callow, P. (1986), Interpreting the La Cotte sequence. In P. Callow and J. Cornford (eds.), *La Cotte de St. Brelade 1961–1978.* Norwich: Geo Books, 73–82.

Calvin, W. H. (1983), A stone's throw and its launch window: timing, precision and its implications for language and the hominid brain. *Journal of Theoretical Biology* 104: 121–35.

——(1996a), *How Brains Think: Evolving Intelligence, Then and Now.* New York: Basic Books.

——(1996b), *The Cerebral Code: Thinking a Thought in the Mosaics of the Mind.* Cambridge, Mass.: MIT Press.

——and D. Bickerton (2000), *Lingua Ex Machina: Reconciling Darwin with the Human Brain.* Cambridge, Mass.: MIT Press.

Campbell, L. (1998), *Historical Linguistics: An Introduction.* Cambridge, Mass.: MIT Press.

Cangelosi, A. (1999), Modelling the evolution of communication: from stimulus associations to grounded symbolic associations. In D. Floreano, J. D. Nicoud, and F. Mondada (eds.), *Advances in Artificial Life.* London: Springer.

————(eds). (2002), *Simulating the Evolution of Language.* London: Springer.

Caplan, D. (1987), *Neurolinguistics and Linguistic Aphasiology: An Introduction.* Cambridge: Cambridge University Press.

Carey, S. (1978), The child as word-learner. In M. Halle, J. Bresnan, and G.A. Miller (eds.), *Linguistic theory and psychological reality.* Cambridge, Mass.: MIT Press.

——and E. Bartlett (1978), Acquiring a single new word. *Papers and Reports on Child Language Development* 15: 17–29.

Carpenter, M., N. Akhtar, and M. Tomasello (1998), Sixteen-month-old infants differentially imitate intentional and accidental actions. *Infant Behavior and Development* 21: 315–30.

——K. Nagell, and M. Tomasello (1998), Social cognition, joint attention, and communicative competence from 9 to 15 months of age. *Monographs of the Society for Research in Child Development* 63.

Carré, R., B. Lindblom, and P. MacNeilage (1995), Rôle de l'acoustique dans l'évolution du conduit vocal humain. Acoustic factors in the evolution of the human vocal tract. *Comptes Rendus de l'Académie des Sciences*, Paris, vol. 30, ser. IIb, 471–6.

——and Mrayati, M. (1995), Vowel transitions, vowel systems, and the Distinctive Region Model, in C. Sorin, J. Mariani, H. Méloni, and J. Schoentgen (eds.), *Levels in Speech Communication: Relations and Interactions.* Amsterdam: Elsevier, 73–89.

Carstairs-McCarthy, A. (1999), *The Origins of Complex Language: An Inquiry into the Evolutionary Beginnings of Sentences, Syllables, and Truth.* Oxford: Oxford University Press.

Carstairs-McCarthy, A. (2000), The distinction between sentences and noun phrases: an impediment to language evolution? In Knight et al., 248–63.

Cassirer, E. (1944), *An Essay on Man. An Introduction to a Philosophy of Human Culture*. New Haven, Conn.: Yale University Press.

Cavalli-Sforza, L. L., and M. W. Feldman (1981), *Cultural transmission and evolution: a quantitative approach*. Princeton, NJ: Princeton University Press.

Cecconi, F., F. Menczer, and R. Belew (1996), Maturation and the evolution of imitative learning in artificial organisms. *Adaptive Behaviour* 4: 29–50.

Charman, T., S. Baron-Cohen, J. Swettenham, G. Baird, A. Cox, and A. Drew (2000), Testing joint attention, imitation, and play as infancy precursors to language and theory of mind. *Cognitive Development* 15(4): 481–98.

Chase, P. G., and H. L. Dibble (1987), Middle Palaeolithic symbolism. *Journal of Anthropological Archaeology* 6: 263–96.

Cheney, D., and R. Seyfarth (1990), *How Monkeys See the World: Inside the Mind of Another Species*. Chicago: University of Chicago Press.

Cheney, D. L., and R. W. Wrangham (1987), Predation. In B. B. Smuts, D. L. Cheney, R. M. Seyfarth, R. W. Wrangham, and T. T. Struhsaker (eds.), *Primate Societies*. Chicago: University of Chicago Press, 227–39.

Chiba, T., and M. Kajiyama (1941), *The Vowel: Its Nature and Structure*. Tokyo: Tokyo-Kaiseikan.

Childers, J., and M. Tomasello (2001), The role of pronouns in young children's acquisition of the English transitive construction. *Developmental Psychology* 37: 739–48.

Chomsky, N. A. (1956), Three models for the description of language. *IRE Transactions in Information Theory* 2: 113–24.

——(1957), *Syntactic Structures*. The Hague: Mouton.

——(1965), *Aspects of the Theory of Syntax*. Cambridge, Mass.: MIT Press.

——(1966), *Cartesian Linguistics: A Chapter in the History of Rationalist Thought*. New York: Harper & Row.

——(1968), *Language and Mind*. New York: Harcourt Brace.

——(1972), *Language and Mind*, 2nd edn. New York: Harcourt Brace Jovanovich.

——(1975), *Reflections on Language*. New York: Pantheon.

——(1976), On the nature of language. In H. B. Steklis, S. R. Harnard, and J. Lancaster (eds.), *Origins and Evolution of Language and Speech*. New York: New York Academy of Sciences, 46–57.

——(1980), *Rules and Representations*. New York: Columbia University Press.

——(1981a), Principles and parameters in syntactic theory. In N. Hornstein and D. Lightfoot (eds.), *Explanation in Linguistics: The Logical Problem of Language Acquisition*. London: Longman, 123–46.

——(1981b), *Government and Binding*. Dordrecht: Foris.

——(1984), *Lectures on Government and Binding*. Dordrecht: Foris.

——(1986), *Knowledge of Language*. New York: Praeger.

——(1988), *Language and Problems of Knowledge*. Cambridge, Mass.: MIT Press.

——(1994), *Language and Thought*. London: Moyer Bell.

——(1995), *The Minimalist Program*. Cambridge, Mass.: MIT Press.

——(2000), *New Horizons in the Study of Language and Mind*. Cambridge: Cambridge University Press.

——(2002), The evolution of language. Paper presented at the 4th International Conference on the Evolution of Language, Harvard University.

Christiansen, M. H. (1994), Infinite languages, finite minds: connectionism, learning and linguistic structure. Ph.D. thesis, University of Edinburgh.

——(in preparation), Cognitive constraints on word order universals: evidence from connectionist modeling and artificial grammar learning.

——and N. Chater (1999), Toward a connectionist model of recursion in human linguistic performance. *Cognitive Science* 23: 157–205.

————(eds.) (2001a), *Connectionist Psycholinguistics*. Westport, Conn.: Ablex.

————(2001b), Connectionist psycholinguistics in perspective. In Christiansen and Chater, 19–75.

————(2003), Constituency and recursion in language. In Arbib, 267–71.

——and R. Dale (forthcoming), The role of learning and development in the evolution of language: a connectionist perspective. In D. Kimbrough Oller and U. Griebel (eds.), *The Evolution of Communication Systems: A Comparative Approach*. Cambridge, Mass.: MIT Press.

————M. R. Ellefson, and C. M. Conway (2002), The role of sequential learning in language evolution: computational and experimental studies. In Cangelosi and Parisi, 165–87.

——and J. Devlin (1997), Recursive inconsistencies are hard to learn: a connectionist perspective on universal word order correlations. In M. Shafto and P. Langley (eds.), *Proceedings of the 19th Annual Cognitive Science Society Conference*. Mahwah, NJ: Erlbaum, 113–18.

——and M. Ellefson (2002), Linguistic adaptation without linguistic constraints: the role of sequential learning in language evolution. In Wray, 335–58.

——L. Kelly, R. Shillcock, and K. Greenfield (in preparation). Artificial grammar learning in agrammatism.

Christy, T. C. (1983), *Uniformitarianism in Linguistics*. Amsterdam: Benjamins.

Clahsen, H., and M. Almazen (1998), Syntax and morphology in Williams syndrome. *Cognition* 68: 167–98.

Clark, J. D. (1988), The Middle Stone Age of East Africa and the beginnings of regional identity. *Journal of World Prehistory* 2: 235–305.

Cleeremans, A. (1993), *Mechanisms of Implicit Learning: A Connectionist Model of Sequence Processing*. Cambridge, Mass.: MIT Press.

Clements, G. N. (1992), Phonological primes: features or gestures? *Phonetica* 49: 181–93.

Cleveland, J., and C. T. Snowdon (1981), The complex vocal repertoire of the adult cotton-top tamarin, *Saguinus oedipus oedipus*. *Zeitschrift für Tierpsychologie* 58: 231–70.

Clottes, J. (ed.) (2001), *La grotte Chauvet: l'art des origines*. Paris: Seuil.

Clutton-Brock, T. H., and S. D. Albon (1979), The roaring of red deer and the evolution of honest advertising. *Behaviour* 69: 145–70.

Collard, M., and B. Wood (2000), How reliable are human phylogenetic hypotheses? *Proceedings of the National Academy of the Sciences* 97: 5003–6.

Comrie, B. (1981), *Language Universals and Linguistic Typology*. Chicago: University of Chicago Press.

——(1989), *Language Universals and Linguistic Typology*, 2nd edn. Chicago: University of Chicago Press.

——(1992), Before complexity. In Hawkins and Gell-Mann (eds.), 193–211.

Condillac, E. B. de (1947 [1746]), Essai sur l'origine des connaissances humaines, ouvrage ou l'on réduit à un seul principe tout ce concerne l'entendement. In *Oeuvres Philosophiques de Condillac*. Paris: Georges Leroy.

Conway, C. M., and M. H. Christiansen (2001), Sequential learning in non-human primates. *Trends in Cognitive Sciences* 5: 539–46.

Corballis, M. C. (1983), *Human Laterality*. New York: Academic Press.

——(1991), *The Lopsided Ape*. New York: Oxford University Press.

——(1997), The genetics and evolution of handedness. *Psychological Review* 104: 714–27.

——(2002), *From Hand to Mouth: The Origins of Language*. Princeton, NJ: Princeton University Press.

Cosmides, L., and J. Tooby (1996), Are humans good intuitive statisticians after all? Rethinking some conclusions from the literature on judgement under uncertainty. *Cognition* 58: 1–73.

Cowlishaw, G. (1992), Song function in gibbons. *Behavior* 121: 131–53.

Crain, S. (1992), Language acquisition in the absence of experience. *Behavioral and Brain Sciences* 14: 597–650.

Croft, W. (1990), *Typology and Universals*. Cambridge: Cambridge University Press.

——(2000), *Explaining Language Change: An Evolutionary Approach*. London: Longman.

——(2002), *Radical Construction Grammar*. Oxford: Oxford University Press.

Crompton, A. W., R. Z. German, and A. J. Thexton (1997), Mechanisms of swallowing and airway protection in infant mammals (*Sus domesticus* and *Macaca fasicularis*). *Journal of Zoology* (London) 241: 89–102.

Crosson, B. (1992), *Subcortical Functions in Language and Memory*. New York: Guilford Press.

Crow, T. J. (1998), Sexual selection, timing and the descent of man: a theory of the genetic origins of language. *Current Psychology of Cognition* 17: 1237–77.

Crystal, D. (1987), *The Cambridge Encyclopedia of Language*. Cambridge: Cambridge University Press.

Culicover, P. W. (1997), *Principles and Parameters: An Introduction to Syntactic Theory*. Oxford: Oxford University Press.

Cummings, J. L. (1993), Frontal-subcortical circuts and human behavior. *Archive of Neurology* 50: 873–80.

——and D. F. Benson (1984), Subcortical dementia: review of an emerging concept. *Archives of Neurology* 41: 874–9.

Cunnington, R., R. Iansek, J. L. Bradshaw, and J. G. Phillips (1995), Movement-related potentials in Parkinson's disease: presence and predictability of temporal and spatial cues. *Brain* 118: 935–50.

Curtiss, S. (1989), The independence and task-specificity of language. In A. Bornstein and J. Bruner (eds.), *Interaction in Human Development*. Hillsdale, NJ: Erlbaum.

Dale, P. S., et al. (1998), Genetic influence on language delay in two-year-old children. *Nature Neuroscience* 1: 324–8.

Damasio, H. (1991), Neuroanatomical correlates of the aphasias. In M. T. Sarno (ed.), *Acquired Aphasia*, 2nd edn. New York: Academic Press.

Dang, J., and K. Honda (1996), Acoustic characteristics of the human paranasal sinuses derived from transmission characteristic measurement and morphological observation. *Journal of the Acoustical Society of America* 100: 3374–83.

Darwin, C. (1859), *On the Origin of Species*. London: John Murray.

——(1872/1998), *The Expression of the Emotions in Man and Animals*, 3rd edn. Oxford: Oxford University Press.

——(1896), *The Descent of Man and Selection in Relation to Sex*. London: William Clowes.

Davidson, I. (1990), Bilzingsleben and early marking. *Rock Art Research* 7: 52–6.

——(1991), The archaeology of language origins: a review. *Antiquity* 65: 39–48.

——(1997), The power of pictures. In M. Conkey, O. Soffer, D. Stratmann, and N. G. Jablonski (eds.), *Beyond Art: Pleistocene Image and Symbol*. San Francisco: California Academy of Sciences, 128–58.

——(1999a), First people becoming Australian. *Anthropologie* (Brno) 37(1): 125–41.

——(1999b), The game of the name. In King (ed.), 229–68.

——(2001), The requirements for human colonisation of Australia. In I. Metcalfe, J. M. B. Smith, M. Morwood, and I. Davidson (eds.), *Faunal and Floral Migration and Evolution in SE Asia-Australia*. Lisse, Netherlands: Swets & Zeitlinger, 399–408.

——(2002), The 'finished artefact fallacy': Acheulean handaxes and language origins. In Wray, 180–203.

——and W. Noble (1989), The archaeology of perception. *Current Anthropology* 30(2): 125–55.

————(1992), Why the first colonisation of the Australian region is the earliest evidence of modern human behaviour. *Archaeology in Oceania* 27: 135–42.

————(1993), Tools and language in human evolution. In Gibson and Ingold (eds.), 363–88.

————(1998), Two views on language origins. *Cambridge Archaeological Journal* 8(1): 82–8.

Dawkins, R. (1982), *The Extended Phenotype*. New York: Oxford University Press.

——(1986), *The Blind Watchmaker: Why the Evidence of Evolution Reveals a Universe Without Design*. New York: Norton.

——(1996), *Climbing Mount Improbable*. New York: Norton.

Deacon, T. (1997), *The Symbolic Species: The Coevolution of Language and the Brain*. New York: Norton.

——(2003), Multilevel selection in a complex adaptive system: the problem of language origins. In B. Weber and D. Depew (eds.), *Evolution and Learning: The Baldwin Effect Reconsidered*. Cambridge, Mass.: MIT Press.

de Boer, B. (2001), *The Origins of Vowel Systems*. Oxford: Oxford University Press.

de Graff, M. (1999), *Language Creation and Language Change: Creolization, Diachrony and Development*. Cambridge, Mass.: MIT Press

DeGusta, D., W. H. Gilbert, and S. P. Turner (1999), Hypoglossal canal size and hominid speech. *Proceedings of the National Academy of the Sciences* 96: 1800–04.

Delluc, B., and G. Delluc (1978), Les manifestations graphiques aurignaciennes sur support rocheux des environs des Eyzies (Dordogne). *Gallia Préhistoire* 21: 213–438.

Dennett, D. C. (1991), *Consciousness Explained*. New York: Little, Brown.

——(1997), *Darwin's Dangerous Idea: Evolution and the Meanings of Life*. New York: Simon & Schuster.

d'Errico, F. and A. Nowell (2000), A new look at the Berekhat Ram figurine. *Cambridge Archaeological Journal* 10(1): 123–67.

——and P. Villa (1997), Holes and grooves: the contribution of microscopy and taphonomy to the problem of art origins. *Journal of Human Evolution* 33: 1–31.

D'Esposito M., and M. P. Alexander (1995), Subcortical aphasia: distinct profiles following left putaminal hemorrhage. *Neuorology* 45: 38–41.

Deutscher, G. (1999), The different faces of uniformitarianism. Paper read at the 14th International Conference on Historical Linguistics, Vancouver.

——(2000), *Syntactic Change in Akkadian: The Evolution of Sentential Complementation*. Oxford: Oxford University Press.

de Waal, F. B. M. (1982), *Chimpanzee Politics*. London: Cape.

——(1988), The communicative repertoire of captive bonobos (*Pan paniscus*) compared to that of chimpanzees. *Behaviour* 106: 183–251.

——(1989), *Peacemaking among Primates*. Cambridge, Mass.: Harvard University Press.

Diamond, A. S. (1959), *The History and Origin of Language*. London: Methuen.

Dibble, H. L. (1989), The implications of stone tool types for the presence of language during the Lower and Middle Paleolithic. In P. A. Mellars and C. B. Stringer (eds.), *The Human Revolution*. Princeton, NJ: Princeton University Press, 415–31.

Diessel, H., and M. Tomasello (2000), The development of relative constructions in early child speech. *Cognitive Linguistics* 11: 131–52.

———(2001), The acquisition of finite complement clauses in English: a corpus-based analysis. *Cognitive Linguistics* 12: 97–141.

Dik, S. C. (1997), *The Theory of Functional Grammar*, ed. K. Hengeveld (2 vols.). Berlin: Mouton de Gruyter.

Dinnsen, D. (1992), Variation in developing and fully developed phonetic inventories. In Ferguson et al., 423–35.

Dixon, R. M. W. (1997), *The Rise and Fall of Languages*. Cambridge: Cambridge University Press.

Donald, M. (1991), *Origins of the Modern Mind*. Cambridge, Mass.: Harvard University Press.

Doupe, A. J., and P. K. Kuhl (1999), Birdsong and human speech: common themes and mechanisms. *Annual Review of Neuroscience* 22: 567–631.

Dronkers, N. F., J. K. Shapiro, B. Redfern, and R. T. Knight (1992), The role of Broca's area in Broca's aphasia. *Journal of Clinical and Experimental Neuropsychology* 14: 52–3.

Dryer, M. (1992), The Greenbergian word order correlations. *Language* 68: 81–138.

——(1997), Are grammatical relations universal? In J. Bybee, J. Haiman, and S. Thompson (eds.), *Essays on Language Function and Language Type*. Amsterdam: Benjamins.

Dunbar, R. I. M. (1993), The co-evolution of neocortical size, group size and language in humans. *Behavioral and Brain Sciences* 16: 681–735.

——(1996/1998), *Grooming, Gossip and the Evolution of Language*. London: Faber & Faber/Cambridge, Mass.: Harvard University Press.

——(1999), Culture, honesty and the freerider problem. In R. I. M. Dunbar, C. Knight, and C. Power (eds.), *The Evolution of Culture*. Edinburgh: Edinburgh University Press, 194–213.

——and P. Dunbar (1975), Social dynamics of gelada baboons. In F. S. Szalay (ed.), *Contributions to Primatology*, vol. 6. New York: Karger.

——N. D. C. Duncan, and D. Nettle (1995), Size and structure of freely forming conversational groups. *Human Nature* 6: 67–78.

Durham, W. (1991), *Coevolution, Genes, Culture and Human Diversity*. Palo Alto, Calif.: Stanford University Press.

Durie, M., and M. Ross (eds.) (1996), *The Comparative Method Reviewed: Regularity and Irregularity in Language Change*. Oxford: Oxford University Press.

Elman, J. (1990), Finding structure in time. *Cognitive Science* 14: 179–211.

Elman, J. L., E. A. Bates, M. H. Johnson, A. Karmiloff-Smith, D. Parisi, and K. Plunkett (1996), *Rethinking Innateness*. Cambridge, Mass.: MIT Press.

Emery, N. J., E. N. Lorincz, D. I. Perrett, M. W. Oram, and C. I. Baker (1997), Gaze following and joint attention in rhesus monkeys (*Macaca mulatta*). *Journal of Comparative Psychology* 111(3): 286–93.

Enard, W., M. Przeworski, S. E. Fisher, C. S. Lai, V. Wiebe, T. Kitano, A. P. Monaco, and S. Pääbo (2002), Molecular evolution of FOXP2, a gene involved in speech and language. *Nature* 418: 869–72.

Enquist, M., and O. Leimar (1993), The evolution of cooperation in mobile organisms. *Animal Behaviour* 45: 747–57.

Epstein, S. D., E. M. Groat, R. Kawashima, and H. Kitahara (1998), *A Derivational Approach to Syntactic Relations*. New York: Oxford University Press.

Fadiga, L., L. Craighero, G. Buccino, and G. Rizzolatti (2002), Speech listening specifically modulates the excitability of tongue muscles: a TMS study. *European Journal of Neuroscience* 15: 399–402.

Fagg, A. H., and M. A. Arbib (1998), Modelling parietal-premotor interactions in primate control of grasping. *Neural Networks* 11: 1277–1303.

Falk, D. (1987), Hominid paleoneurology. *Annual Review of Anthropology* 16: 13–30.

Fant, G. (1960), *Acoustic Theory of Speech Production*. The Hague: Mouton.

Fauconnier, G. (1985), *Mental Spaces*. Cambridge, Mass.: MIT Press. (Rev. edn. Cambridge University Press, 1994.)

——and M. Turner (2002), *The Way We Think: Conceptual Blendings and the Mind's Hidden Complexities*. New York: Basic Books.

Ferguson, C. A., and C. B. Farwell (1975), Words and sounds in early language acquisition. *Language* 51: 419–39.

——L. Menn, and C. Stoel-Gammon (eds.) (1992), *Phonological Development: Models, Research, Implications*. Timonium, Md.: York Press.

Fisher, R. A. (1930), *The Genetical Theory of Natural Selection*. Oxford: Clarendon Press.

Fitch, W. T. (1994), Vocal tract length and the evolution of language. PhD dissertation, Brown University.

——(1997), Vocal tract length and formant frequency dispersion correlate with body size in rhesus macaques. *Journal of the Acoustical Society of America* 102: 1213–22.

——(1999), Acoustic exaggeration of size in birds by tracheal elongation: comparative and theoretical analyses. *Journal of Zoology* 248: 31–49.

——(2000a), The evolution of speech: a comparative review. *Trends in Cognitive Sciences* 4: 258–67.

——(2000b), The phonetic potential of nonhuman vocal tracts: Comparative cineradiographic observations of vocalizing animals. *Phonetica* 57: 205–18.

——(2000c), Skull dimensions in relation to body size in nonhuman mammals: the causal bases for acoustic allometry. *Zoology* 103: 40–58.

——and J. Giedd (1999), Morphology and development of the human vocal tract: a study using magnetic resonance imaging. *Journal of the Acoustical Society of America* 106: 1511–22.

——and M. D. Hauser (1995), Vocal production in nonhuman primates: acoustics, physiology and functional constraints on honest advertisement. *American Journal of Primatology* 37: 191–219.

————(2003), Unpacking 'honesty': vertebrate vocal production and the evolution of acoustic signals. In A. Simmons, A. N. Popper, and R. R. Fay (eds.), *Acoustic Communication*. New York: Springer, 65–137.

——and J. P. Kelley (2000), Perception of vocal tract resonances by whooping cranes, *Grus Americana*. *Ethology* 106: 559–74.

——and D. Reby (2001), The descended larynx is not uniquely human. *Proceedings of the Royal Society, Biological Sciences*, 268: 1669–75.

Flowers, K. A., and C. Robertson (1985), The effects of Parkinson's disease on the ability to maintain a mental set. *Journal of Neurology, Neurosurgery, and Psychiatry* 48: 517–29.

Foley, R. A. (1987), Hominid species and stone-tool assemblages. *Antiquity* 61: 380–92.

——and M. M. Lahr (1997), Mode 3 technologies and the evolution of modern humans. *Cambridge Archaeological Journal* 7(1): 3–36.

Formicola, V., A. Pontrandolfi, and J. Svoboda (2001), The Upper Palaeolithic triple burial of Dolni Vestonice. *American Journal of Physical Anthropology* 115: 372–9.

Fowler, C. A. (1980), Coarticulation and theories of extrinsic timing. *Journal of Phonetics* 8: 113–33.

Franco, F., and G. E. Butterworth (1996), Pointing and social awareness: declaring and requesting in the second year of life. *Journal of Child Language* 23: 307–36.

Frege, G. (1892), *Über Sinn und Bedeutung*. Trans. as *On Sense and Reference*. In P. Geach and M. Black (eds.), *Translations from the Philosophical Writings of Gottlob Frege*. Oxford: Blackwell, 1952.

Frey, R., and R. R. Hofmann (2000), Larynx and vocalization of the Takin (Budorcas taxicolor Hodgson, 1850 - Mammalia, Bovidae). *Zoologischer Anzeiger* 239: 197–214.

Fuster, J. M. (1989), *The Prefrontal Cortex: Anatomy, Physiology, and Neuropsychology of the Frontal Lobe*, 2nd edn. New York: Raven Press.

Gabunia, L., A. Vekua, and D. Lordkipanidze (2000), The environmental contexts of early human occupation of Georgia (Transcaucasia). *Journal of Human Evolution* 38(6): 785–802.

Gafos, A. (2002), A grammar of gestural coordination. *Natural Language and Linguistic Theory* 20: 269–337.

Gamble, C. (1994), *Timewalkers: The Prehistory of Global Colonization*, Cambridge, Mass.: Harvard University Press.

Gardner, R. A., and B. T. Gardner (1969), Teaching sign language to a chimpanzee. *Science* 165: 664–72.

————(1984), A vocabulary test for chimpanzees (Pan troglodytes). *Journal of Comparative Psychology* 4: 381–404.

————(1994), Development of phrases in the utterances of children and cross-fostered chimpanzees. In R. A. Gardner, B. T. Gardner, B. Chiarelli, and R. Plooj (eds.), *The Ethological Roots of Culture*. Dordrecht: Kluwer Academic, 223–55.

Gargett, R. H. (1999), Middle Palaeolithic burial is not a dead issue: the view from Qafzeh, Saint-Césaire, Kebara, Amud, and Dederiyeh. *Journal of Human Evolution* 37: 27–90.

Gautier, J. P. (1971), Etude morphologique et fonctionnelle des annexes extra-laryngées des cercopithecinae: liaison avec les cris d'espacement. *Biologica Gabonica* 7: 230–67.

Gazdar, G., E. Klein, G. Pullum, and I. Sag (1985), *Generalized Phrase Structure Grammar*. Oxford: Blackwell.

Geman, S., E. Bienenstock, and R. Doursat (1992), Neural networks and the bias/variance dilemma. *Neural Computation* 4: 1–58.

Gentner, D., and A. Markman (1997), Structure mapping in analogy and similarity. *American Psychologist* 52: 45–56.

Gibbs, S., M. Collard, and B. Wood (2000), Soft-tissue characters in higher primate phylogenetics. *Proceedings of the National Academy of the Sciences* 97(20): 11130–32.

Gibson, E., and K. Wexler (1994), Triggers. *Linguistic Inquiry* 25: 407–54.

Gibson, K. R. (1996), The biocultural human brain, seasonal migrations, and the emergence of the Upper Palaeolithic. In P. Mellars and K. Gibson (eds.), *Modelling the Early Human Mind*. Cambridge: McDonald Institute for Archaeological Research, 33–46.

——and T. Ingold (eds.) (1993), *Tools, Language and Cognition in Human Evolution*. Cambridge: Cambridge University Press.

——and S. Jessee (1999), Language evolution and expansions of multiple neurological processing areas. In King (ed.), 189–227.

Gick, B. (1999), A gesture-based account of intrusive consonants in English. *Phonology* 16: 29–54.

——(in press), Articulatory correlates of ambisyllabicity in English glides and liquids. In J. Local, R. Ogden, and P. Temple (eds.), *Papers in Laboratory Phonology 6: Constraints on Phonetic Interpretation*. Cambridge: Cambridge University Press.

Gillespie, R. (1998), Alternative timescales: a critical review of Willandra Lakes dating. *Archaeology in Oceania* 33(3): 169–82.

Giorgi, A., and G. Longobardi (1991), *The syntax of noun phrases: configuration, parameters and empty categories*. Cambridge: Cambridge University Press.

Givón, T. (1979), *On Understanding Grammar*. New York: Academic Press.

——(1995), *Functionalism and Grammar*. Amsterdam: Benjamins.

Gold, E. M. (1967), Language identification in the limit. *Information and Control* 10: 447–74.

Goldberg, A. (1995), *Constructions: A Construction Grammar Approach to Argument Structure*. Chicago: University of Chicago Press.

Goldberg, P. (2000), Micromorphology and site formation at Die Kelders Cave I, South Africa. *Journal of Human Evolution* 38(1): 43–90.

Goldin-Meadow, S., and D. McNeill (1999), The role of gesture and mimetic representation in making language the province of speech. In M. C. Corballis and S. E G. Lea (eds.), *The Descent of Mind*. Oxford: Oxford University Press, 155–72.

Goldstein, K. (1948), Language and Language Disturbances. New York: Grune & Stratton.

Goller, F., and O. N. Larsen (1997), *In situ* biomechanics of the syrinx and sound generation in pigeons. *Journal of Experimental Biology* 200: 2165–76.

Gómez, J. C., E. Sarría, and J. Tamarit (1993), The comparative study of early communication and theories of mind. In S. Baron-Cohen, H. Tager-Flusberg, and D. Cohen (eds.), *Understanding Other Minds*. Oxford: Oxford University Press.

Goodall, J. (1986), *The Chimpanzees of Gombe: Patterns of Behavior*. Cambridge, Mass.: Harvard University Press.

Goodman, N. (1978), *Ways of Worldmaking*. Indianapolis: Hackett Publishing.

Gopnik, M., and M. Crago (1991), Familial segregation of a developmental language disorder. *Cognition* 39: 1–50.

Gordon, P. (1985), Level-ordering in lexical development. *Cognition* 21: 73–93.

Goren-Inbar, N. (1986), A figurine from the Acheulian site of Berekhat Ram. *Mitukefat Haeven* 19: 7–12.

Gorman, A.C., (2000), The archaeology of body modification. PhD thesis, University of New England, Armidale, NSW.

Gould, S. J. (1980), *The Panda's Thumb: More Reflections in Natural History*. New York: Norton.

——(1991), Exaptation: a crucial tool for an evolutionary psychology. *Journal of Social Issues* 47: 43–65.

——(1997), Darwinian fundamentalism. *New York Review of Books* (12 June): 34–52.

Gowlett, J. A. J. (1986), Culture and conceptualisation: the Oldowan-Acheulian gradient. In G. N. Bailey and P. Callow (eds.), *Stone Age Prehistory*. Cambridge: Cambridge University Press, 243–60.

Grafman J. (1989), Plans, actions and mental sets: the role of the frontal lobes. In E. Perecman (ed.), *Integrating Theory and Practice in Clinical Neuropsychology*. Hillsdale, NJ: Erlbaum.

Graybiel, A. M. (1995), Building action repertoires: memory and learning functions of the basal ganglia. *Current Opinion in Neurobiology* 5: 733–41.

——(1997), The basal ganglia and cognitive pattern generators. *Schizophrenia Bulletin* 23: 459–69.

Graziano, M. S. A., G. S. Yap, and C. G. Gross (1994), Coding of visual space by premotor neurons. *Science* 266: 1054–7.

Greenberg, B. D., D. L. Murphy, and S. A. Rasmussen (2000), Neuroanatomically based approaches to obsessive-compulsive disorder: neurosurgery and transcranial magnetic stimulation. *Psychiatric Clinics of North America* 23: 671–85.

Greenberg, J. H. (1963), *Universals of Language*. Cambridge, Mass.: MIT Press.

——(1987), *Language in the Americas*. Stanford, Calif.: Stanford University Press.

——C. A. Ferguson, and E. A. Moravcsik (1978), *Universals of Human Language* (4 vols.). Stanford, Calif.: Stanford University Press.

Greenfield, P. M., and E. S. Savage-Rumbaugh (1993), Comparing communicative competence in child and chimp. *Journal of Child Language* 20: 1–26.

Grice, H. P. (1975), Logic and conversation. In P. Cole (ed.), *Syntax and Semantics*, vol. 3. New York: Academic Press, 41–58.

Grossman, M. (1980), A central processor for hierarchically structured material: evidence from Broca's aphasia. *Neuropsychologia* 18: 299–308.

——S. Carvell, S. Gollomp, M. B. Stern, G. Vernon, and H. I. Hurtig (1991), Sentence comprehension and praxis deficits in Parkinson's disease. *Neurology* 41: 1620–28.

————————M. Reivich, D. Morrison, A. Alavi, and H. I. Hurtig (1993), Cognitive and physiological substrates of impaired sentence processing in Parkinson's Disease. *Journal of Cognitive Neuroscience* 5: 480–98.

Groves, C. P. (1989), *A Theory of Human and Primate Evolution*. Oxford: Clarendon Press.

Hahn, J. (1986), *Kraft und Aggression*. Tübingen: Institut für Urgeschichte der Universität Tübingen.

Hailman, J. P., and M. S. Ficken (1987), Combinatorial animal communication with computable syntax: chick-a-dee calling qualifies as 'language' by structural linguistics. *Animal Behaviour* 34: 1899–1901.

Haiman, J. (1985), *Natural Syntax: Iconicity and Erosion*. Cambridge: Cambridge University Press.

——(1994), Ritualization and the development of language. In Pagliuca (ed.), 3–28.

Hanks, W. (1992), The indexical ground of deictic reference. In A. Duranti and C. Goodwin, *Rethinking Context: Language as an Interactive Phenomenon*. Cambridge: Cambridge University Press, 43–76.

Hare, B., J. Call, and M. Tomasello (2001), Do chimpanzees know what conspecifics know? *Animal Behaviour* 61(1): 139–51.

Harnad, S. (1987), *Categorical Perception: The Groundwork of Cognition*. Cambridge: Cambridge University Press.

Harnad, S. R., S. D. Steklis, and J. Lancaster (eds.) (1976), *Origins and Evolution of Language and Speech*. New York: New York Academy of Sciences.

Harrington, D. L., and L. Haaland (1991), Sequencing in Parkinson's Disease: abnormalities in programming and controlling movement. *Brain* 114: 99–115.

Harris, R. (1996), *The Origin of Language: Key Issues*. Bristol: Thoemmes.

Harris, Z. S. (1955), From phoneme to morpheme. *Language* 31: 190–222.

Harrison, D. F. N. (1995), *The Anatomy and Physiology of the Mammalian Larynx*. New York: Cambridge University Press.

Harrison, M. A. (1978), *Introduction to Formal Language Theory*. Reading, Mass.: Addison-Wesley.

Harvey, I. (1993), The puzzle of the persistent question marks: a case study of genetic drift. In S. Forrest (ed.), *Genetic Algorithms: Proceedings of the 5th International Conference*. San Mateo, Calif.: Morgan Kaufmann.

Hashimoto T., and T. Ikegami (1996), Emergence of net-grammar in communicating agents. *BioSystems* 38: 1–14.

Hauser, M. D. (1996), *The Evolution of Communication*. Cambridge, Mass.: MIT Press.

——(1997), Artifactual kinds and functional design features: what a primate understands without language. *Cognition* 64: 285–308.

——(2000), *Wild Minds: What Animals Really Think*. New York: Holt.

——(2002), What's so special about speech? In E. Dupoux (ed.), *Language, Brain and Cognitive Development: Essays in Honor of Jacques Mehler*. Cambridge, Mass.: MIT Press.

——S. Carey, and L. B. Hauser (2000), Spontaneous number representation in semi-free-ranging rhesus monkeys. *Proceedings of the Royal Society* 267: 829–33.

——N. Chomsky, and W. T. Fitch (2002). The language faculty: what is it, who has it, and how did it evolve? *Science* 298: 1569–79.

——E. L. Newport, and R. N. Aslin (2001), Segmenting a continuous acoustic speech stream: serial learning in cotton-top tamarin monkeys. *Cognition* 78: B53–64.

Hawkins, J. A. (1994), *A Performance Theory of Order and Constituency*. Cambridge: Cambridge University Press.

——and Gell-Mann, M. (1992), *The Evolution of Human Languages*. Reading, Mass.: Addison-Wesley.

Hazlehurst B., and E. Hutchins (1998), The emergence of propositions from the coordination of talk and action in a shared world. *Language and Cognitive Processes* 13: 373–424.

Heine, B. (1991), *Grammaticalization*. Chicago: University of Chicago Press.

——U. Claudi, and F. Hünnemeyer (1991), *Grammaticalization: A Conceptual Framework*. Chicago: University of Chicago Press.

——and T. Kuteva (2002a), On the evolution of grammatical forms. In Wray, 376–97.

————(2002b), *World Lexicon of Grammaticalization*. Cambridge: Cambridge University Press.

Hellwag, C. (1781), De Formatione Loquelae. Dissertation, Tübingen.

Henshilwood, C. S., F. d'Errico, R. Yates, Z. Jacobs, C. Tribolo, G. A. T. Duller, N. Mercier, J. C. Sealy, H. Valladas, I. Watts, and A. Wintle (2002), Emergence of modern human behavior: Middle Stone Age engravings from South Africa. *Science* 295: 1278–80

Henton, C. (1992), The abnormality of male speech. In G. Wolf (ed.), *New Departures in Linguistics*. New York: Garland.

Herbert, R. K. (1990), The relative markedness of click sounds: evidence from language change, acquisition, and avoidance. *Anthropological Linguistics* 32: 120–38.

Hewes, G. W. (1973), Primate communication and the gestural origin of language. *Current Anthropology* 14: 5–24.

Hienz, R. D., M. B. Sachs, and J. M. Sinnott (1981), Discrimination of steady-state vowels by blackbirds and pigeons. *Journal of the Acoustical Society of America* 70: 699–706.

Hill, W. C., and A. H. Booth (1957), Voice and larynx in African and Asiatic Colobidae. *Journal of the Bombay Natural History Society* 54: 309–21.

Hillenbrand, J. L., A. Getty, M. J. Clark, and K. Wheeler (1995), Acoustic characteristics of American English vowels. *Journal of the Acoustical Society of America* 97: 3099–3111.

Hinton, G. E., and S. J. Nowlan (1987), How learning can guide evolution. *Complex Systems* 1: 495–502.

Hockett, C. F. (1958), *A Course in Modern Linguistics*. New York: Macmillan.

——(1960), The origin of speech. *Scientific American* 203: 88–111.

——and R. Ascher (1964), The human revolution. *Current Anthropology* 5(3): 135–68.

Hoehn, M. M., and M. D. Yahr (1967), Parkinsonism: onset, progression and mortality. *Neurology* 17: 427–42.

Holloway, R. L. (1969), Culture: A human domain. *Current Anthropology* 10: 395–413.

——(1983), Human brain evolution: a search for units, models and synthesis. *Canadian Journal of Anthropology* 3: 215–30.

Hombert, J.-M., and E. Marsico (1996), Do vowel systems increase in complexity? Paper presented at First Evolution of Human Language Conference, Edinburgh.

Hopkins, W. D. (1996), Chimpanzee handedness revisited: 55 years since Finch (1941). *Psychonomic Bulletin and Review* 3: 449–57.

Hopper, P. J. and E. C. Traugott (1993), *Grammaticalization*. Cambridge: Cambridge University Press.

Horning, J. (1969), A study of grammatical inference. PhD, Stanford University.

Humboldt, W. von (1836/1999), *Über die Verschiedenheit des Menschlichen Sprachbaues*. Berlin. Trans. as *On Language* by P. Heath, ed. M. Losonsky. Cambridge: Cambridge University Press.

Hunt, R. H., and R. N. Aslin (1998), Statistical learning of visuomotor sequences: implicit acquisition of sub-patterns. In *Proceedings of the 20th Annual Conference of the Cognitive Science Society*. Hillsdale, NJ: Erlbaum.

Hurford, J. R. (1989), Biological evolution of the Saussurean sign as a component of the language acquisition device. *Lingua* 77: 187–222.

——(1991), The evolution of critical period for language acquisition. *Cognition* 40: 159–201.

——(2000a), The emergence of syntax. In Knight et al., 219–30.

——(2000b), Social transmission favours linguistic generalization. In Knight et al., 324–52.

——(2002b), Expression/induction models of language evolution: dimensions and issues. In Briscoe, 301–44.

——(2003), Language beyond our grasp: what mirror neurons can, and cannot, do for language evolution. In D. K. Oller, U. Griebel, and K. Plunkett (eds.), *The Evolution of Communication Systems: A Comparative Approach*. Cambridge, Mass.: MIT Press.

——(2003), The neural basis of predicate-argument structure. *Behavioral and Brain Sciences*.

——and W. T. Fitch (eds.) (2002), *Proceedings of the Fourth International Conference on the Evolution of Language*. Cambridge, Mass.: Department of Psychology, Harvard University.

——M. Studdert-Kennedy, and C. Knight (eds.) (1998), *Approaches to the Evolution of Language: Social and Cognitive Bases.* New York: Cambridge University Press.

Hurst, J. A., et al. (1990), An extended family with a dominantly inherited speech disorder. *Developmental Medicine and Child Neurology* 32: 347–55.

Indefrey. P., C. M. Brown, F. Hellwig, K. Amunts, H. Herzog, R. J. Seitz, and P. Hagoort (2001), A neural correlate of syntactic encoding during speech production. *Proceedings of the National Academy of Sciences* 98: 5933–6.

Ingman, M., H. Kaessmann, S. Pääbo, and U. Gyllensten (2000), Mitochondrial genome variation and the origin of modern humans. *Nature* 408: 708–13.

Ingold, T. (1993), Technology, language, intelligence: a reconsideration of basic concepts. In Gibson and Ingold, 449–72.

Ingram, D. (1989), *First Language Acquisition: Method, Description, and Explanation.* New York: Cambridge University Press.

——(1992), Early phonological acquisition: a cross-linguistic perspective. In Ferguson et al., 423–35.

Itakura, S. (1996), An exploratory study of gaze-monitoring in nonhuman primates. *Japanese Psychological Research* 38(3): 174–80.

Jablonka, E., and M. Lamb (1995), *Epigenetic Inheritance and Evolution.* Oxford: Oxford University Press.

Jackendoff, R. (1994), *Patterns in the Mind: Language and Human Nature.* New York: Basic Books.

——(1997), *The Architecture of the Language Faculty.* Cambridge, Mass.: MIT Press.

——(1999), Possible stages in the evolution of the language capacity. *Trends in Cognitive Science* 3: 272–9.

——(2002), *Foundations of Language: Brain, Meaning, Grammar, Evolution.* Oxford: Oxford University Press.

Jacob, F. (1977), The linguistic model in biology. In D. Armstrong and C. H. van Schoonefeld (eds.), *Roman Jakobson: Echoes of his Scholarship.* Lisse: de Ridder, 185–92.

Jain, D., J. Osherson, J. Royer, and A. Sharma (1998), *Systems that Learn.* Cambridge, Mass.: MIT Press.

Jakobson, R. (1940), Kindersprache, Aphasie und allgemeine Lautgesetze. In *Selected Writings,* trans. A. R. Keiler. The Hague: Mouton.

——(1968), *Child Language, Aphasia, and Phonological Universals.* The Hague: Mouton.

——(1970), Linguistics. In *Main Trends of Research in the Social and Human Sciences,* vol 1. Paris/The Hague: UNESCO/Mouton, 437–40.

——C. G. M. Fant, and M. Halle (1951/1963), *Preliminaries to Speech Analysis.* Cambridge, Mass.: MIT Press.

Janik, V. M, and P. B. Slater (1997), Vocal learning in mammals. *Advances in the Study of Behavior* 26: 59–99.

Jelinek, A. (1990), The Amudian in the context of the Mugharan Tradition in Tabun Cave (Mount Carmel), Israel. In P. A. Mellars (ed.), *The Emergence of Modern Humans.* Edinburgh: Edinburgh University Press, 81–90.

Jellinger K. (1990), New developments in the pathology of Parkinson's disease. In M. B. Streifler, A. D. Korezyn, J. Melamed, and M. B. H. Youdim (eds.), *Advances in Neurology*, vol. 53: *Parkinson's Disease: Anatomy, Pathology and Therapy*. New York: Raven Press, 1–15.

Jenkins, L. (2000), *Biolinguistics: Exploring the Biology of Language*. Cambridge: Cambridge University Press.

Jolly, A. (1985), *The Evolution of Primate Behavior*, 2nd edn. New York: Macmillan.

Jones, R., and I. Johnson (1985), Deaf Adder Gorge: Lindner Site, Nauwalabila I. In R. Jones (ed.), *Archaeological Research in Kakadu National Park*. Canberra: Australian National Parks and Wildlife Service, 165–227.

Joshi, A., K. Vijay-Shanker, and D. Weir (1991), The convergence of mildly context-sensitive grammar formalisms. In P. Sells, S. Shieber, and T. Wasow (eds.), *Foundational Issues in Natural Language Processing*. Cambridge, Mass.: MIT Press, 31–82.

——L. Levy, and M. Takahashi (1975), Tree adjunct grammars. *Journal of Computer and System Sciences* 10(1): 136–63.

Jusczyk, P. W. (1997), *The Discovery of Spoken Language*. Cambridge, Mass.: MIT Press.

Kahane, J. (1978), A morphological study of the human prepubertal and pubertal larynx. *American Journal of Anatomy* 151: 11–20.

Kalmár, I. (1985), Are there really no primitive languages? In D. R. Olson, N. Torrance, and A. Hildyard (eds.), *Literacy, Language, and Learning: The Nature and Consequences of Reading and Writing*. Cambridge: Cambridge University Press.

Katz, W. F. (1988), Anticipatory coarticulation in aphasia: acoustic and perceptual data. *Brain and Language* 35: 340–68.

Kauffman, S. A. (1993), *The Origins of Order: Self Organization and Selection in Evolution*. Oxford: Oxford University Press.

——(1995), *At Home in the Universe: The Search for Laws of Self-Organization and Complexity*. Oxford: Oxford University Press.

Kay, R. F., M. Cartmill, and M. Barlow (1998), The hypoglossal canal and the origin of human vocal behaviour. *Proceedings of the National Academy of Sciences USA*, 95: 5417–19.

Keating, P. A. (1990), The window model of coarticulation: articulatory evidence. In J. Kingston and M. E. Beckman (eds.), *Papers in Laboratory Phonology I: Between the Grammar and Physics of Speech*. Cambridge: Cambridge University Press, 451–70.

Keeley, L. H., and N. Toth (1981), Microwear polishes on early stone tools from Koobi Fora, Kenya. *Nature* 293: 464–5.

Kelemen, G. (1969), Anatomy of the larynx and the anatomical basis of vocal performance. In G. Bourne (ed.) *The Chimpanzee*, vol. 1. Basel: Karger, 165–87.

Keller, R. (1994), *On Language Change: The Invisible Hand in Language*. London: Routledge.

Kennedy, G. E., and N. A. Faumuina (2001), KMH2 and the comparative morphology of the hyoid. *Yearbook of Physical Anthropology* 44: 89.

Keverne, E. B., N. Martensz, and B. Tuite (1989), Beta-endorphin concentrations in cerebrospinal fluid of monkeys are influenced by grooming relationships. *Psychoneuroendocrinology* 14: 155–61

Kim, J. J., et al. (1994), Sensitivity of children's inflection to morphological structure. *Journal of Child Language* 21: 173–209.

Kimura, D. (1983), Sex differences in cerebral organization for speech and praxic functions. *Canadian Journal of Psychology* 37: 19–35.

——(1993), *Neuromotor Mechanisms in Human Communication*. Oxford: Oxford University Press.

——T. Aosaki, and A. Graybiel (1993), Role of basal ganglia in the acquisition and initiation of learned movement. In N. Mano, I. Hamada, and M. R. DeLong (eds.), *Role of the Cerebellum and Basal Ganglia in Voluntary Movements*. Amsterdam: Elsevier.

King, B. J. (ed.) (1999), *The Origins of Language: What Nonhuman Primates Can Tell Us*. Sante Fe, NM: School of American Research Press.

Kiparsky, P. (1976), Historical linguistics and the origins of language. In Harnad et al. (eds.), 97–103.

Kirby, S. (1998a), Fitness and the selective adaptation of language. In Hurford et al., 359–83.

——(1999), *Function, Selection and Innateness: The Emergence of Language Universals*. Oxford: Oxford University Press.

——(2000), Syntax without natural selection: how compositionality emerges from vocabulary in a population of learners. In Knight et al., 303–23.

——(2001), Spontaneous evolution of linguistic structure: an iterated learning model of the emergence of regularity and irregularity. *IEEE Journal of Evolutionary Computation* 5(2): 102–10.

——(2002a), Natural language from artificial life. *Artificial Life* 8: 185–215.

——(2002b), Learning, bottlenecks and the evolution of recursive syntax. In Briscoe, 173–204.

——and J. Hurford (1997), Learning, culture, and evolution in the origin of linguistic constraints. In P. Husbands and H. Inman (eds.), *Proceedings of the Fourth European Conference on Artificial Life*. Cambridge, Mass.: MIT Press, 493–502.

————(2002), The emergence of linguistic structure: an overview of the iterated learning model. In Cangelosi and Parisi, 121–47.

——K. Smith, and H. Brighton (2003), Language evolves to aid its own survival. In preparation.

Klima, E., and U. Bellugi (1979), *Signs of Language*. Cambridge, Mass.: Harvard University Press.

Kluender, K. R., A. J. Lotto, L. L. Holt, and S. L. Bloedel (1998), Role of experience for language-specific functional mappings of vowel sounds. *Journal of the Acoustical Society of America* 104: 3568–82.

Knight, C. (1998), Ritual/speech coevolution: a solution to the problem of deception. In Hurford et al., 68–91.

Knight, C. (2000), Play as precursor of phonology and syntax. In Knight et al., 9 119.

——M. Studdert-Kennedy, and J. R. Hurford (eds.) (2000), *The Evolutionary Eme gence of Language: Social Function and the Origins of Linguistic Form*. Car bridge: Cambridge University Press.

Kohler, K. (1998), The development of sound systems in human language. In Hu ford et al., 265–78.

Komarova, N. L., P. Niyogi, and M. A. Nowak (2001), The evolutionary dynamics (grammar acquisition. *Journal of Theoretical Biology* 209: 43–59

——and M. A. Nowak (2001), Evolutionary dynamics of the lexical matrix. *Bullet* of *Mathematical Biology* 63: 451–85.

———(2003), Linguistic coherence in finite populations. *Journal of Theoretic Biology* 221: 445–57.

——and I. Rivin (2003), Mathematics of learning, *Electr. Ann. of the AMS*, subm ted to Proceedings of the Royal Society A; also Arxiv math. PR/0105235.

Kreitman, M. (2000), Methods to detect selection in populations with application to the human. *Annual Review of Genomics and Human Genetics*, 539–59.

Kroch, A. (1989), Reflexes of grammar in patterns of language change. *Languag Variation and Change* 1: 199–244.

Krug, M. (1998), String frequency: a cognitive motivating factor in coalescenc language processing, and language change. *Journal of English Linguistics* 26: 28 320.

Krushinsky, L. V. (1965), Solution of elementary logical problems by animals on th basis of extrapolation. *Progress in Brain Research* 17: 280–308.

Kugler, P. N., and M. T. Turvey (1987), *Information, Natural Law, and the Se Assembly of Rhythmic Movement*. Hillsdale, NJ: Erlbaum.

Kuhl, P. K. (1989), On babies, birds, modules, and mechanisms: a comparative a proach to the acquisition of vocal communication. In R. J. Dooling and S. I Hulse (eds.) *The Comparative Psychology of Audition*. Hillsdale, NJ: Erlbau 379–422.

——(1991), Human adults and human infants show a 'perceptual magnet effect' fc the prototypes of speech categories, monkeys do not. *Perception and Psychoph) ics* 50: 93–107.

——(2000), Language, mind, and brain: experience alters perception. In M. Ga zaniga (ed.), *The New Cognitive Neurosciences*, 2nd edn. Cambridge, Mass.: MI Press, 99–118.

Kuhn, S. L., M. C. Stiner, D. S. Reese, and E. Gulec (2001), Ornaments of the ear est Upper Paleolithic. *Proceedings of the National Academy of Sciences* 98(13 7641–6.

Labov, W. (1969), The logic of nonstandard English. *Georgetown Monographs o Language and Linguistics* 22: 1–31.

Lai, C. S. L, S. E. Fisher, J. A. Hurst, F. Vargha-Khadem, and A. P. Monaco (2001), forkhead-domain gene is mutated in a severe speech and language disorder. *N ture* 413: 519–23.

Laitman, J. T., and J. S. Reidenberg (1988), Advances in understanding the relationship between the skull base and larynx with comments on the origins of speech. *Journal of Human Evolution* 3: 99–109.

Lakoff, George (1987), *Women, Fire, and Dangerous Things: What Categories Reveal about the Mind.* Chicago: University of Chicago Press.

Langacker, R. (1987a), *Concept, Image, and Symbol: The Cognitive Basis for Grammar.* New York: Mouton de Gruyter.

——(1987b), *Foundations of Cognitive Grammar,* volume i. Stanford, Calif.: Stanford University Press.

——(1988), A usage-based model. In B. Rudzka-Ostyn (ed.), *Topics in Cognitive Linguistics.* Amsterdam: Benjamins.

——(1991), *Foundations of Cognitive Grammar,* volume ii. Stanford, Calif.: Stanford University Press.

Lange, K. W., T. W. Robbins, C. D. Marsden, M. James, A. M. Owen, and G. M. Paul (1992), L-Dopa withdrawal in Parkinson's disease selectively impairs cognitive performance in tests sensitive to frontal lobe dysfunction. *Psychopharmacology* 107: 394–404.

Langer, S. (1957), *Philosophy in a New Key.* Cambridge, Mass.: Harvard University Press.

Langton, S. R. H., R. J. Watt, and V. Bruce (2000), Do the eyes have it? Cues to the direction of social attention. *Trends in Cognitive Sciences* 4(2): 50–9.

Lashley, K. S. (1951), The problem of serial order in behaviour. In L. A. Jeffress (ed.), *Cerebral Mechanisms in Behaviour: The Hixon Symposium..* New York: Wiley.

Laver, J. (1994), *Principles of Phonetics.* Cambridge: Cambridge University Press.

Lawrence, S., C. L. Giles, and S. Fong (1996), Can recurrrent neural networks learn natural language grammars? *Proceedings of the International Conference on Neural Networks (ICNN96),* Washington, DC, 1853–8.

Leavens, D. A., W. D. Hopkins, and K. A. Bard (1996), Indexical and referential pointing in chimpanzees (*Pan troglodytes*). *Journal of Comparative Psychology,* 110(4): 346–53.

Lee, R. B. (1979), *The !Kung San: Men, Women, and Work in a Foraging Society.* Cambridge: Cambridge University Press.

Leech, G. (1974), *Semantics.* New York: Pelican.

Leigh, S. R. (1992), Cranial capacity evolution in *Homo erectus* and early *Homo sapiens. American Journal of Physical Anthropology* 87: 1–13.

Lenneberg, E. H. (1967), *Biological Foundations of Language.* New York: Wiley.

Leonard, L. B. (1998), *Children with Specific Language Impairment.* Cambridge, Mass.: MIT Press.

Li, M., and P. Vitanyi (1997), *An Introduction to Kolmogorov Complexity and its Applications.* Berlin: Springer.

Liberman, A. M. (1957), Some results of research on speech perception. *Journal of the Acoustical Society of America* 29: 117–23.

——(1996), *Speech: A Special Code.* Cambridge, Mass.: MIT Press.

Liberman, A. M., F. S. Cooper, D. P. Shankweiler, and M. Studdert-Kennedy (1967), Perception of the speech code. *Psychological Review* 74: 431–61.

——and I. G. Mattingley (1985), The motor theory of speech perception revised. *Cognition* 21: 1–36.

————(1988), Specialized perceiving systems for speech and other biologically significant sounds. In G. M. Edelman, W. E. Gail, and W. M. Cowan (eds.), *Auditory Function*. New York: Wiley, 775–93.

——F. S. Cooper, D. P. Shankweiler, and M. Studdert-Kennedy (1967), Perception of the speech code. *Psychological Review* 74: 431–61.

Lichtheim, L. (1885), On aphasia. *Brain* 7: 433–84.

Lieberman, D. E. (1998), Sphenoid shortening and the evolution of modern human cranial shape. *Nature* 393: 158–62.

——and R. C. McCarthy (1999), The ontogeny of cranial base angulation in humans and chimpanzees and its implications for reconstructing pharyngeal dimensions. *Journal of Human Evolution* 36: 487–517.

————K. M. Hiiemae, and J. B. Palmer (2001), Ontogeny of postnatal hyoid and larynx descent in humans. *Archives of Oral Biology* 2001: 117–28.

——and J. J. Shea (1994), Behavioral differences between archaic and modern humans in the Levantine Mousterian. *American Anthropologist* 96(2): 300–32.

Lieberman, P. (1968), Primate vocalization and human linguistic ability. *Journal of the Acoustical Society of America* 44: 1574–84.

——(1975), *On the Origins of Language: An Introduction to the Evolution of Human Speech*. New York: Macmillan.

——(1984), *The Biology and Evolution of Language*. Cambridge, Mass.: Harvard University Press.

——(1985), On the evolution of human syntactic ability: its pre-adaptive bases, motor control and speech. *Journal of Human Evolution* 14: 657–68

——(1991), *Uniquely Human: The Evolution of Speech, Thought, and Selfless Behavior*. Cambridge, Mass.: Harvard University Press.

——(1998), *Eve Spoke: Human Language and Human Evolution*. New York: Norton.

——(2000), *Human Language and Our Reptilian Brain: The Subcortical Bases of Speech, Syntax and Thought*. Cambridge, Mass.: Harvard University Press.

——(2002), On the nature and evolution of the neural bases of human language. *Yearbook of Physical Anthropology* 45: 36–62.

——and E. S. Crelin (1971), On the speech of Neanderthal man. *Linguistic Inquiry* 2: 203–22.

——J. Friedman, and L. S. Feldman (1990), Syntactic deficits in Parkinson's disease *Journal of Nervous and Mental Disease* 178: 360–65.

——E. T. Kako, J. Friedman, G. Tajchman, L. S. Feldman, and E. B. Jiminez (1992) Speech production, syntax comprehension, and cognitive deficits in Parkinson's disease. *Brain and Language* 43: 169–89.

——B. G. Kanki, and A. Protopapas (1995), Speech production and cognitive decrements on Mount Everest. *Aviation, Space and Environmental Medicine* 66 857–64.

————A. Protopapas, E. Reed, and J. W. Youngs (1994), Cognitive defects at altitude. *Nature* 372: 325.

————and D. H. Klatt (1972), Phonetic ability and related anatomy of the newborn, adult human, Neanderthal man, and the chimpanzee. *American Anthropologist* 74: 287–307.

——D. H. Klatt, and W. H. Wilson (1969), Vocal tract limitations on the vowel repertoires of rhesus monkeys and other nonhuman primates. *Science* 164: 1185–7.

——J. T. Laitman, J. S. Reidenberg, and P. J. Gannon (1992), The anatomy, physiology, acoustics and perception of speech. *Journal of Human Evolution* 23: 447–67.

.ieven, E., J. Pine, and G. Baldwin (1997), Lexically-based learning and early grammatical development. *Journal of Child Language* 24: 187–220.

.ightfoot, D. (1991a), *How to Set Parameters: Arguments from Language Change.* Cambridge, Mass.: MIT Press.

——(1991b), Subjacency and sex. *Language and Communication* 11: 67–9.

——(1999), *The Development of Language: Acquisition, Change, and Evolution.* Oxford: Blackwell.

——(2000), The spandrels of the linguistic genotype. In Knight et al., 231–47.

——and S. Anderson (2002), *The Language Organ.* New York: Cambridge University Press.

.indblom, B. (1992), Phonological units as adaptive emergents of lexical development. In Ferguson et al., 131–63.

——(1998), Systemic constraints and adaptive change in the formation of sound structure. In Hurford et al., 242–64.

——(2000), Developmental origins of adult phonology: the interplay between phonetic emergents and the evolutionary adaptations of sound patterns. *Phonetica* 57: 297–314.

——B. MacNeilage, and M. Studdert-Kennedy (1984), Self-organizing processes and the explanation of phonological universals. In B. Butterworth, B. Comrie, and Östen Dahl (eds.), *Explanations for Language Universals.* Berlin: Mouton, 181–203.

——and I. Maddieson (1988), Phonetic universals in consonant systems. In L. M. Hyman and C. N. Li (eds.), *Language, Speech and Mind.* London: Routledge, 62–78.

.isker, L., and A. S. Abramson (1964), A cross language study of voicing in initial stops: acoustical measurements. *Word* 20: 384–442.

.ivingstone, D., and C. Fyfe, (2000), Modelling language–physiology coevolution. In Knight et al., 199–218.

.ock, A., and C. Peters (eds.) (1996), *Handbook of Human Symbolic Evolution.* Oxford: Clarendon Press.

.oritz, D. (1999), *How the Brain Evolved Language.* Oxford: Oxford University Press.

.otto, A. J., K. R. Kluender, and L. L. Holt (1998), Depolarizing the perceptual magnet effect. *Journal of the Acoustical Society of America* 103: 3648–55.

Lupyan, G., and M. H. Christiansen (2002), Case, word order, and language learnability: insights from connectionist modeling. In *Proceedings of the 24th Annual Conference of the Cognitive Science Society*. Mahwah, NJ: Erlbaum, 596–601.

Lyn, H., and S. Savage-Rumbaugh (2000), Observational word learning in two bonobos (*Pan paniscus*): ostensive and non-ostensive contexts. *Language and Communication* 20(3): 255–73.

McBrearty, S. and A. S. Brooks (2000), The revolution that wasn't. *Journal of Human Evolution* 39(5): 453–563.

McCarthy, J. J. (1988), Feature geometry and dependency: a review. *Phonetica* 45: 84–108.

McComb, K. E. (1991), Female choice for high roaring rates in red deer, *Cervus elaphus*. *Animal Behaviour* 41: 79–88.

McGrew, W. C. (1992), *Chimpanzee material culture*. Cambridge: Cambridge University Press.

——and L. F. Marchant (1997), On the other hand: current issues in a meta-analysis of the behavioural laterality of hand function in nonhuman primates. *Yearbook of Physical Anthropology* 40: 201–32.

McGurk, H., and J. MacDonald (1976), Hearing lips and seeing voices. *Nature* 264: 746–8.

Mackay, D. (1999), Rate of information acquisition by a species subjected to natural selection. http://wol.ra.phy.cam.ac.uk/mackay

MacLarnon, A., and G. Hewitt (1999), The evolution of human speech: the role of enhanced breathing control. *American Journal of Physical Anthropology* 109: 341–63.

McLeod, P., K. Plunkett, and E. T. Rolls (1998), *Introduction to Connectionist Modelling of Cognitive Processes*. Oxford: Oxford University Press.

MacNeilage, P. F. (1998a), Evolution of the mechanism of language output: comparative neurobiology of vocal and manual communication. In Hurford et al. 222–41.

——(1998b), The frame/content theory of evolution of speech production. *Behavioral and Brain Sciences* 21: 499–546.

——and B. L. Davis (2000), On the origin of internal structure of word forms. *Science* 288: 527–31.

McNeill, D. (1992), *Hand and Mind: What Gestures Reveal about Thought*. Chicago: University of Chicago Press.

——(1985), So you think gestures are nonverbal? *Psychological Review*, 92: 350–71.

McPherron, S. P. (2000), Handaxes as a measure of the mental capabilities of early hominids. *Journal of Archaeological Science* 27(8): 655–63.

McWhorter, J. (2002), *The Power of Babel: A Natural History of Language*. New York: W. H. Freeman.

Manuel, S. Y. (1990), The role of contrast in limiting vowel-to-vowel coarticulation in different languages. *Journal of the Acoustical Society of America* 88: 1286–98.

Manzini, R., and K. Wexler (1987), Parameters, binding theory, and learnabilit

*Linguistic Inquiry* 18: 413–44.

Maratsos, M., R. Gudeman, P. Gerard-Ngo, and G. DeHart (1987), A study in novel word learning: the productivity of the causative. In B. MacWhinney (ed.), *Mechanisms of Language Acquisition*. Hillsdale, NJ: Erlbaum.

Marchman, V., and E. Bates (1994), Continuity in lexical and morphological development: a test of the critical mass hypothesis. *Journal of Child Language* 21: 339–66.

Marean, C. W., and S. Y. Kim (1998), Mousterian large-mammal remains from Kobeh Cave. *Current Anthropology* 39 (suppl. S): 113.

Marie, P. (1926), *Travaux et mémoires*. Paris: Masson.

Marler, P. (1976), Social organization, communication and graded signals: the chimpanzee and the gorilla. In P. P. G. Bateson and R. A. Hinde (eds.), *Growing Points in Ethology*. Cambridge: Cambridge University Press, 239–80.

——(1977), The structure of animal communication sounds. In T. H. Bullock (ed.), *Recognition of Complex Acoustic Signals*. Berlin: Dahlem Konferenzen.

Marsden, C. D., and J. A. Obeso (1994), The functions of the basal ganglia and the paradox of sterotaxic surgery in Parkinson's disease. *Brain* 117: 877–97.

Marshack, A. (1997), The Berekhat Ram figurine. *Antiquity* 71(272): 327–37.

Martin, A., C. L. Wiggs, L. G. Ungerleider, and J. V. Haxby (1996), Neural correlates of category-specific knowledge. *Nature* 379: 649–52.

Martin, R., and J. Uriagereka (2000), Introduction: some possible foundations of the Minimalist Program. In R. Martin, D. Michaels, and J. Uriagereka (eds.), *Step by Step: Essays on Minimalist Syntax in Honor of Howard Lasnik*. Cambridge, Mass.: MIT Press, 1–29.

Matthews, G. H. (1961), *Hidatsa Syntax*. Cambridge, Mass.: MIT Press.

Mayley, G. (1996), Landscapes, learning costs and genetic assimilation. In P. Turney, D. Whitely, and R. Anderson (eds.), *Evolution, Learning and Instinct: 100 Years of the Baldwin Effect*. Cambridge, Mass.: MIT Press.

Maynard Smith, J. (1982), *Evolution and the Theory of Games*. New York: Cambridge University Press.

——(1988), *An Introduction to the Mathematical Theory of Evolution*. New York.

——(1998), *Evolutionary Genetics*, 2nd edn. New York: Oxford University Press.

——and Szathmáry, E. (1995), *The Major Transitions in Evolution*. Oxford: Oxford University Press.

Mayr, E. (1982), *The Growth of Biological Thought*. Cambridge, Mass.: Harvard University Press.

Mega, M. S., and M. F. Alexander (1994), Subcortical aphasia: the core profile of capsulostriatal infarction. *Neurology* 44: 1824–9.

Meier, R. P., and E. L. Newport (1990), Out of the hands of babes: On a possible sign language advantage in language acquisition. *Language* 66: 1–23.

Mellars, P. A. (1973), The character of the Middle-Upper Palaeolithic transition in south-west France. In C. Renfrew (ed.), *The Explanation of Culture Change*. London: Duckworth, 255–76.

Mellars, P. A. (1989), Technological changes across the Middle-Upper Palaeolithic transition. In P. A. Mellars and C. B. Stringer (eds.), *The Human Revolution*. Princeton, NJ: Princeton University Press, 338–65.

Meltzoff, A. (1988), Infant imitation after a one week delay: long term memory for novel acts and multiple stimuli. *Developmental Psychology* 24: 470–76.

——(1995), Understanding the intentions of others: re-enactment of intended acts by 18-month-old children. *Developmental Psychology* 31: 838–50.

——and K. Moore (1997), Explaining facial imitation: a theoretical model. *Early Development and Parenting* 6: 179–92.

Menn, L. (1983), Development of articulatory, phonetic and phonological capabilities. In B. Butterworth (ed.), *Language Production*, vol. 2. London: Academic Press, 3–30.

Mercier, N., H. Valladas, G. Valladas, J.-L. Reyss, J.-L., A. Jelinek, L. Meignen, and J.-L. Joron (1995), TL dates of burnt flints from Jelinek's excavations at Tabun and their implications. *Journal of Archaeological Science* 22: 495–509.

Mergell, P., W. T. Fitch, and H. Herzel (1999), Modeling the role of non-human vocal membranes in phonation. *Journal of the Acoustical Society of America* 105: 2020–28.

Mesulam, M. M. (1990), Large-scale neurocognitive networks and distributed processing for attention, language, and memory. *Annals of Neurology* 28: 597–613.

Middleton. F. A., and P. L. Strick (1994), Anatomical evidence for cerebellar and basal ganglia involvement in higher cognition. *Science* 266: 458–61.

Miller, G. A. (1991), *The Science of Words*. New York: W. H. Freeman.

Miller, G. F. (1999), Sexual selection for cultural displays. In R. Dunbar, C. Knight and C. Power (eds) *The Evolution of Culture*. Edinburgh: Edinburgh University Press, 71–91.

——(2000), *The Mating Mind: How Sexual Choice Shaped the Evolution of Human Nature*. New York: Doubleday.

Miller, G. H., P. B. Beaumont, H. J. Deacon, A. S. Brooks, P. E. Hare, and A. J. T. Jull (1999), Earliest modern humans in southern Africa dated by isoleucine epimerization in ostrich eggshell. *Quaternary Science Reviews* 18(13): 1537–48.

Milroy, J. (1992), *Linguistic Variation and Change: On the Historical Sociolinguistics of English*. Oxford: Blackwell.

Mitani, J. C., and P. Marler (1989), A phonological analysis of male gibbon singing behavior. *Behaviour* 109: 20–45.

Mitchell, T. (1997), *Machine Learning*. New York: McGraw-Hill.

Mithun, M. (1984), How to avoid subordination. *Berkeley Linguistics Society* 10: 493–523.

Moore, C., and P. Dunham (1995), *Joint Attention: Its origins and Role in Development*. Hillsdale, NJ: Erlbaum.

Morales, M., P. Mundy, C. E. F. Delgado, M. Yale, R. Neal, and H. K. Schwartz (2000) Gaze following, temperament and language development in 6-month-olds: a replication and extension. *Infant Behavior and Development* 23(2): 231–6.

Morris, C. (1938), *Foundations of the Theory of Signs*. Chicago: University of Chicago Press.

——(1964), *Signification and Significance*. Cambridge, Mass.: MIT Press.

Morris R. G., J. J. Downes, B. J. Sahakian, J. L. Evenden, A. Heald, and T. W. Robbins (1988), Planning and spatial working memory in Parkinson's disease. *Journal of Neurology, Neurosurgery, and Psychiatry* 51: 757–66.

Morse, K. (1993), Shell beads from Mandu Mandu rockshelter, Cape Range Peninsula, Western Australia, before 30,000 bp. *Antiquity* 67: 877–83.

Muggleton, S. (1996), Learning from positive data. In *Proceedings of the 6th Inductive Logic Programming Workshop, Stockholm*.

Muller, J. (1848), *The Physiology of the Senses, Voice and Muscular Motion with the Mental Faculties*, trans. W. Baly. London: Walton & Maberly.

Myowa-Yamakoshi, M., and T. Matsuzawa (1999), Factors influencing imitation of manipulatory actions in chimpanzees (*Pan troglodytes*). *Journal of Comparative Psychology* 113: 128–36.

Naeser, M. A., M. P. Alexander, N. Helms-Estabrooks, H. L. Levine, S. A. Laughlin, and N. Geschwind (1982), Aphasia with predomininantly subcortical lesion sites: description of three capsular/putaminal aphasia syndromes. *Archives of Neurology* 39: 2–14.

Natsopoulos, D., G. Grouios, S. Bostantzopoulou, G. Mentenopoulos, Z. Katsarou, and J. Logothetis (1993), Algorithmic and heuristic strategies in comprehension of complement clauses by patients with Parkinson's disease. *Neuropsychologia* 31: 951–64.

Nearey, T. (1979), *Phonetic Features for Vowels*. Bloomington: Indiana University Linguistics Club.

Negus, V. E. (1949), *The Comparative Anatomy and Physiology of the Larynx*. New York: Hafner.

Neidle, C., J. Kegl, D. MacLaughlin, B. Bahan, and R. G. Lee (2000), *The Syntax of American Sign Language*. Cambridge, Mass.: MIT Press.

Nettle, D. (1999a), Is the rate of linguistic change constant? *Lingua* 108: 119–36.

——(1999b), *Linguistic Diversity*. Oxford: Oxford University Press.

——and R. I. M. Dunbar (1997), Social markers and the evolution of reciprocal exchange. *Current Anthropology* 38: 93–8.

Neville, H. J., D. Bavelier, D. Corina, J. Rauschecker, A. Karni, A. Lalwani, et al. (1997), Cerebral organization for deaf and hearing subjects: biological constraints and effects of experience. *Proceedings of the National Academy of Sciences* 95: 922–9.

Newmeyer, F. J. (1990), Speaker–hearer asymmetry as a factor in language evolution: a functional explanation for formal principles of grammar. *Berkeley Linguistics Society* 16: 241–7.

——(1991), Functional explanation in linguistics and the origins of language. *Language and Communication* 11: 3–28.

——(1998), *Language Form and Language Function*. Cambridge, Mass.: MIT Press.

Newmeyer, F. J. (2000), On the reconstruction of 'proto-world' word order. In Knight et al., 372–90.

——(2002), Uniformitarian assumptions and language evolution research. In Wray, 359–75.

Newport, E. (1999), Reduced input in the acquisition of signed languages: contributions to the study of creolization. In de Graff (ed.), 161–78.

Nichols, J. (1992), *Linguistic Diversity in Space and Time*. Chicago: University of Chicago Press.

Nishitani, N., and R. Hari (2000), Dynamics of cortical representation for action. *Proceedings of the National Academy of Sciences USA*, 97: 913–18.

Niyogi, P. (1999), *The Informational Complexity of Learning from Examples*. Dordrecht: Kluwer.

——and R. C. Berwick (1997a), Evolutionary consequences of language learning. *Linguistics and Philosophy* 20: 697–719.

————(1997b), A dynamical systems model for language change. *Complex Systems* 11: 161–204.

Noble, W., and I. Davidson (1996), *Human Evolution, Language and Mind: A Psychological and Archaeological Inquiry*. Cambridge: Cambridge University Press.

————(1997), Reply to Mithen. *Cambridge Archaeological Journal* 7(2): 279–86.

————(2001), Discovering the symbolic potential of communicative signs: the origins of speaking a language. In A. Nowell (ed.), *In the Mind's Eye: Multidisciplinary Perspectives on the Evolution of Human Cognition*. Ann Arbor, Mich.: International Monographs in Prehistory, 187–200.

Nottebohm, F. (1976), Vocal tract and brain: a search for evolutionary bottlenecks. *Annals of the New York Academy of Sciences* 280: 643–9.

——(1999), The anatomy and timing of vocal learning in birds. In M. D. Hauser and M. Konishi (eds.), *The Design of Animal Communication*. Cambridge, Mass.: MIT Press, 63–110.

Nowak, M. A., and N. L. Komarova (2001), Towards an evolutionary theory of language. *Trends in Cognitive Sciences* 5: 288–95.

————and D. Krakauer (1999), The evolutionary language game. *Journal of Theoretical Biology* 200: 147–62.

————and P. Nyogi (2002), Computational and evolutionary aspects of language. *Nature* 417: 611–17.

——————(2001), Evolution of universal grammar. *Science* 291: 114–18.

——and D. C. Krakauer (1999), The evolution of language. *Proceedings of the National Academy of Sciences USA* 96: 8028–33.

————and A. Dress (1999), An error limit for the evolution of language. *Proceedings of the Royal Society of London* B, 266: 2131–6.

——J. B. Plotkin, and V. A. A. Jansen (2000), The evolution of syntactic communication. *Nature* 404: 495–8.

Nowell, A. (ed.) (2001), *The Archaeology of Mind*. Ann Arbor, Mich.: International Monographs in Prehistory.

Nowicki, S., and R. R. Capranica (1986), Bilateral syringeal coupling during phonation of a songbird. *Journal of Neuroscience* 6: 3593–3610.

Nunberg, G. (1993), Indexicality and deixis. *Linguistics and Philosophy* 68: 1–43.

Ohala, J. J. (1983), The origin of sound patterns in vocal tract constraints. In P. F. MacNeilage (ed.), *The Production of Speech*. New York: Springer, 189–216.

——(1984), An ethological perspective on common cross-language utilization of F0 of voice. *Phonetica* 41: 1–16.

Oliphant, M. (2002a), Rethinking the language bottleneck: why don't animals learn to communicate? In K. Dautenhahn and C. L. Nehaniv (eds.), *Imitation in Animals and Artifacts: Complex Adaptive Systems*. Cambridge, Mass.: MIT Press, 311–25.

——(2002b), Learned systems of arbitrary reference: the foundation of human linguistic uniqueness. In Briscoe, 23–52.

Oller, D. K. (2000), *The Emergence of the Speech Capacity*. Mahwah, NJ: Erlbaum.

Olmsted, D. L. (1971), *Out of the Mouths of Babes: Earliest Stages in Language Learning*. The Hague: Mouton.

Orlov, T., V. Yakoviev, S. Hochstein, and E. Zohary (2000), Macaque monkeys categorize images by their ordinal number. *Nature* 404: 77–80.

Osherson, D., M. Stob, and S. Weinstein (1986), *Systems That Learn*. Cambridge, Mass.: MIT Press.

O'Sullivan, P. B., M. Morwood, D. Hobbs, F. Aziz, F. A. Suminto, M. Situmorang, A. Raza, and R. Maas (2001), Archaeological implications of the geology and chronology of the Soa basin, Flores, Indonesia. *Geology* 29(7): 607–10.

Owings, D., and E. Morton (1998), *Animal Vocal Communication: A New Approach*. Cambridge: Cambridge University Press.

Owren, M. J., and R. Bernacki (1988), The acoustic features of vervet monkey (*Cercopithecus aethiops*) alarm calls. *Journal of the Acoustical Society of America* 83: 1927–35.

——and D. Rendall (1997), An affect-conditioning model of nonhuman primate vocal signaling. In D. H. Owings, M. D. Beecher, and N. S. Thompson (eds.), *Perspectives in Ethology*, vol. 12: *Communication*. New York: Plenum, 299–346.

————(2001), Sound on the rebound: bringing form and function back to the forefront in understanding nonhuman primate vocal signaling. *Evolutionary Anthropology* 10(2): 58–71.

Oztop, E., and M. A. Arbib (2002), Schema design and implementation of the grasp-related mirror neurone system. *Biological Cybernetics* 87: 116–40.

——N. Bradley, and M. A. Arbib (2003), Learning to grasp I: The Infant Learning to Grasp Model (ILGM) (to appear).

Pagliuca, W. (ed.) (1994), *Perspectives on Grammaticalization*. Amsterdam: Benjamins.

Parent, A. (1986), *Comparative Neurobiology of the Basal Ganglia*. New York: Wiley.

Peirce, C. (1955), *Philosophical Writings of Peirce*, ed. J. Buchler. New York: Dover, Ch. 7.

Pepperberg, I. M. (2000), *The Alex Studies: Cognitive and Communicative Abilities of Grey Parrots*. Cambridge, Mass,: Harvard University Press.

Perkins, R. D. (1992), *Deixis, Grammar, and Culture*. Amsterdam: Benjamins.

Peters, G., and M. Hast (1994), Hyoid structure, laryngeal anatomy, and vocalization in felids. *Zeitschrift für Säugetierkunde* 59: 87–104.

Peterson, G. E., and H. L. Barney (1952), Control methods used in a study of the vowels. *Journal of the Acoustical Society of America* 24: 175–84.

Petitto, L. A., and P. Marentette (1991), Babbling in the manual mode: evidence for the ontogeny of language. *Science* 251: 1493–6.

Pettitt, P. B., and N. O. Bader (2000), Direct AMS radiocarbon dates for the Sungir mid-Upper Palaeolithic burials. *Antiquity* 74(284): 269.

Pfeiffer, J. E. (1985), *The Emergence of Humankind*. New York: Harper & Row.

Piaget, J. (1945/1962), *Play, Dreams, and Imitation*. New York: Norton.

Piatelli-Palmarini, M. (1989), Evolution, selection, and cognition: From 'learning' to parameter setting in biology and the study of language. *Cognition* 31: 1–44.

Pickett, E. R., E. Kuniholm, A. Protopapas, J. Friedman, and P. Lieberman (1998), Selective speech motor, syntax, and cognitive deficits associated with bilateral damage to the putamen and the head of the caudate nucleus: a case study. *Neuropsychologia* 36: 173–88.

Pierrehumbert, J. B. (1990), Phonological and phonetic representation. *Journal of Phonetics* 18: 375–94.

Pine, J., E. Lieven, and G. Rowland (1998), Comparing different models of the development of the English verb category. *Linguistics* 36: 4–40.

Pinker, S. (1979), Formal models of language learning. *Cognition* 7: 217–83.

——(1984), *Language Learnability and Language Development*. Cambridge, Mass.: Harvard University Press.

——(1989), *Learnability and Cognition: The Acquisition of Argument Structure*. Cambridge, Mass.: MIT Press.

——(1994), *The Language Instinct*. New York: HarperCollins.

——(1997), *How the Mind Works*. New York: Norton.

——(1999), *Words and Rules: The Ingredients of Language*. New York: HarperCollins.

——(2002), *The Blank Slate: The Modern Denial of Human Nature*. New York: Viking.

——and P. Bloom (1990), Natural language and natural selection. *Behavioral and Brain Sciences* 13: 707–84.

——D. S. Lebeaux, and L. A. Frost (1987), Productivity and constraints in the acquisition of the passive. *Cognition* 26: 195–267.

——and A. Prince (1988), On language and connectionism: analysis of a parallel distributed processing model of language acquisition. *Cognition* 28: 73–193.

Pizutto, E., and C. Caselli (1992), The acquisition of Italian morphology. *Journal of Child Language* 19: 491–557.

———(1994), The acquisition of Italian verb morphology in a cross-linguistic perspective. In Y. Levy (ed.), *Other Children. Other Languages*. Hillsdale, NJ: Erlbaum.

Pocock, R. I. (1916), On the hyoidean apparatus of the lion (*F. leo*) and related species of *Felidae. Annals and Magazine of Natural History* 8: 222–9.

Pollack, R. (1994), *Signs of Life*. Boston: Houghton-Mifflin.

Pollard, C., and I. A. Sag (1994), *Head-Driven Phrase Structure Grammar*. Chicago: University of Chicago Press.

Povinelli, D. (1999), *Folk Physics for Apes*. Oxford: Oxford University Press.

Power, C. (1998), Old wives' tales: the gossip hypothesis and the reliability of cheap signals. In Hurford et al., 111–29.

Premack, D. (1986), 'Gavagai!' or the future history of the animal language controversy. *Cognition* 19: 207–96.

Prince, A., and P. Smolensky (1997), Optimality: from neural networks to universal grammar. *Science* 275: 1604–10.

Przeworski, M., R. R. Hudson, and A. Di Rienzo (2000), Adjusting the focus on human variation. *Trends in Genetics* 16: 296–302.

Pullum, G. (1983), How many possible human languages are there? *Linguistic Inquiry* 14: 447–67.

——and G. Gazdar (1982), Natural languages and context free languages. *Linguistics and Philosophy* 4: 471–504.

——and B. C. Scholz (2002), Empirical assessment of stimulus poverty arguments. *Linguistic Review* 19(1–2).

Quine, W. V. O. (1960), *Word and Object*. Cambridge, Mass.: MIT Press.

——(1961), Two dogmas of empiricism. In Quine, *From a Logical Point of View*. Cambridge, Mass.: Harvard University Press.

Radford, A. (1997), *Syntax: A Minimalist Introduction*. Cambridge: Cambridge University Press.

Raemaekers, J. J., P. M. Raemaekers, and E. H. Haimoff (1984), Loud calls of the gibbons (*Hylobates lar*): repertoire, organization and context. *Behavior* 91: 146–89.

Ragir, S. (2002), Constraints on communities with indigenous sign languages: clues to the dynamics of language origins. In Wray, 272–96.

Rainey, A. (1991), Some Australian bifaces. *Lithics* 12: 33–6.

Ramachandran, V. S., and S. Blakeslee, S. (1998), *Phantoms in the Brain*. New York: Morrow.

Ramus, F., M. D. Hauser, C. T. Miller, D. Morris, and J. Mehler (2000), Language discrimination by human newborns and cotton-top tamarins. *Science* 288: 349–51.

Recanzone, G. H. (2000), Cerebral cortical plasticity: perception and skill acquisition. In M. Gazzaniga (ed.), *The New Cognitive Neurosciences*. Cambridge, Mass.: MIT Press, 237–47.

Rendall, D. (1996), Social communication and vocal recognition in free-ranging rhesus monkeys (*Macaca mulatta*). Ph.D., University of California, Davis.

Rendall, D., M. J. Owren, and P. S. Rodman (1998), The role of vocal tract filtering in identity cueing in rhesus monkey (*Macaca mulatta*) vocalizations. *Journal of the Acoustical Society of America* 103: 602–14.

Renfrew, C., and P. Bahn (2000), *Archaeology*. London: Thames & Hudson.

Renshaw, E. (1991), *Modelling Biological Populations in Space and Time*. Cambridge: Cambridge University Press.

Richards, R. (1987), *Darwin and the Emergence of Evolutionary Theories of Mind and Behaviour*. Chicago: University of Chicago Press.

Richman, B. (1976), Some vocal distinctive features used by gelada monkeys. *Journal of the Acoustical Society of America* 60: 718–24.

Ridley, M. (1986), *The Problems of Evolution*. New York: Oxford University Press.

——(1990), Reply to Pinker and Bloom. *Behavioral and Brain Sciences* 13: 756.

——(1993), *The Red Queen: Sex and the Evolution of Human Nature*. New York: Macmillan.

Riede, T., and W. T. Fitch (1999), Vocal tract length and acoustics of vocalization in the domestic dog, *Canis familiaris*. *Journal of Experimental Biology* 202: 2859–69.

Rigaud, J. P., J. F. Simek, and T. Ge (1995), Mousterian fires from Grotte XVI (Dordogne, France). *Antiquity* 69(266): 902–12.

Rissanen, J. (1989), *Stochastic Complexity in Statistical Inquiry*. Singapore: World Scientific.

Ristad, E., and J. Rissanen (1994), Language acquisition in the MDL framework. In Ristad (ed.), *Language Computation*. Philadelphia: American Mathematical Society.

Rizzolatti, G., and M. A. Arbib (1998), Language within our grasp. *Trends in Neurosciences* 21(5): 188–94.

——L. Fadiga, V. Gallese, and L. Fogassi (1995), Premotor cortex and the recognition of motor actions. *Cognitive Brain Research*, 3: 131–41.

———M. Matelli, V. Bettinardi, D. Perani, and F. Fazio (1996), Localisation of grasp representations in humans by positron emission tomography, 1: Observation versus execution. *Experimental Brain Research* 111: 246–52.

Roberts, I. (2001), Language change and learnability. In S. Bertolo (ed.), *Language Acquisition and Learnability*. Cambridge: Cambridge University Press, 81–125.

Roberts, M. B. (1986), Excavations of the Lower Palaeolithic site at Amey's Eartham pit, Boxgrove, West Sussex. *Proceedings of the Prehistoric Society* 52: 215–45.

Roberts, R. G., R. Jones, and M. A. Smith (1993), Optical dating at Deaf Adder Gorge, Northern Territory, indicates human occupation between 53,000 and 60,000 years ago. *Australian Archaeology* 37: 58–9.

Robins, R. H. (1979), *A Short History of Linguistics*, 2nd edn. London: Longman.

Roca, I., and W. Johnson (1999), *A Course in Phonology*. Oxford: Blackwell.

Rossen, M., et al. (1996), Interaction between language and cognition: evidence from Williams syndrome. In J. H. Beitchman, N. J. Cohen, M. M. Konstantareas and R. Tannock (eds.), *Language, Learning, and Behavior Disorders*. New York: Cambridge University Press.

Rubino, R., and J. Pine (1998), Subject–verb agreement in Brazilian Portugese: what low error rates hide. *Journal of Child Language* 25: 35–60.

Ruhlen, M. (1994), *The Origin of Language: Tracing the Evolution of the Mother Tongue*. New York: Wiley.

Ryalls, J. (1986), An acoustic study of vowel production in aphasia. *Brain and Language* 29: 48–67.

Sacks, O. (1991), *Seeing Voices*. London: Picador.

Sadock, J. M. (1991), *Autolexical syntax: a theory of parallel grammatical representations*. Chicago: University of Chicago Press.

Saffran, J. R., R. N. Aslin, and E. L. Newport (1996), Statistical learning by 8-month-old infants. *Science* 274: 1926–8.

——E. Johnson, R. N. Aslin, and E. Newport (1999), Statistical learning of tone sequences by human infants and adults. *Cognition* 70: 27–52.

Saltzman, E. L., and K. G. Munhall (1989), A dynamical approach to gestural patterning in speech production. *Ecological Psychology* 1: 333–82.

Sampson, G. (1989), Language acquisition: growth or learning? *Philosophical Papers* 1(3): 203–40.

——(1999), *Educating Eve: The Language Instinct Debate*. London: Cassell Academic.

Sanford, A. J., and P. Sturt (2002), Depth of processing in language comprehension: not noticing the evidence. *Trends in Cognitive Sciences* 6: 382–6.

Santos, L. R., M. D. Hauser, and E. S. Spelke (2002), Recognition and categorization of biologically significant objects in rhesus monkeys (*Macaca mulatta*): The domain of food. *Cognition* 82: 127–55.

Sasaki, C. T., P.A. Levine, J. T. Laitman, and E.S. Crelin (1977), Postnatal descent of the epiglottis in man. *Archives of Otolaryngology* 103: 169–71.

Savage-Rumbaugh, S. (1986), *Ape Language: From Conditioned Response to Symbol*. New York: Columbia University Press.

——and R. Lewin (1994), *Kanzi*. New York: McGraw-Hill.

——K. McDonald, R. A. Sevcik, W. D. Hopkins, and E. Rubert (1986), Spontaneous symbol acquisition and communicative use by pygmy chimpanzees (*Pan paniscus*). *Journal of Experimental Psychology: General* 115(3), 211–35.

——J. Murphy, R. A. Sevcik, K. E. Brakke, S. L. Williams, and D. M. Rumbaugh (1993), Language comprehension in ape and child. *Monographs of the Society for Research in Child Development* 58: 1–221.

——and D. Rumbaugh (1993), The emergence of language. In Gibson and Ingold (eds.), 86–100.

——S. Shanker, and T. J. Taylor (1998), *Apes, Language, and the Human Mind*. New York: Oxford University Press.

Schick, K. D. (1994), The Movius Line reconsidered. In R.S. Coruccini and R. L. Ciochon (eds.), *Integrative Paths to the Past*. Englewood Cliffs, NJ: Prentice-Hall, 569–96.

Schneider, R., H.-J. Kuhn, and G. Kelemen (1967), Der Larynx des männlichen *Hypsignathus monstrosus* Allen, 1861 (*Pteropodidae, Megachiroptera, Mammalia*). *Zeitschrift für wissenschaftliche Zoologie* 175: 1–53.

Schön Ybarra, M. (1988), Morphological adaptations for loud phonation in the vocal organ of howling monkeys. *Primate Report* 22: 19–24.

——(1995), A comparative approach to the nonhuman primate vocal tract: implications for sound production. In E. Zimmerman and J. D. Newman (eds.), *Current Topics in Primate Vocal Communication*. New York: Plenum Press, 185–98.

Schrödinger, E. (1944), *What is Life?* Cambridge: Cambridge University Press.

Sebeok, T. (1990), *Essays in Zoosemiotics*. Toronto: University of Toronto Press.

Senecail, B. (1979), *L'Os hyoïde: introduction anatomique a l'étude de certains mécanismes de la phonation*. Paris: Faculté de Médicine de Paris.

Senghas, A., and M. Coppola (2001), Children creating language: how Nicaraguan sign language acquired a spatial grammar. *Psychological Science* 12: 323–8.

Shallice, T. (1988), *From Neuropsychology to Mental Structure*. Cambridge: Cambridge University Press.

Shieber, S. M. (1985), Evidence against the context-freeness of natural language. *Linguistics and Philosophy* 8: 333–43.

Siegal, M., R. Varley, and S. C. Want (2001), Mind over grammar: reasoning in aphasia and development. *Trends in Cognitive Sciences* 5: 296–301.

Singer, R., and J. Wymer (1982), *The Middle Stone Age at Klasies River Mouth in South Africa*. Chicago: University of Chicago Press.

Sinha, C. (1988), *Language and Representation: A Socio-Naturalistic Approach to Human Development*. London: Harvester-Wheatsheaf.

Sinnott, J., and T. L. Williamson (1999), Can macaques perceive place of articulation from formant transition information? *Journal of the Acoustic Society of America*, 106: 929–37.

SLI Consortium (2002), A genomewide scan identifies two novel loci involved in Specific Language Impairment. *American Journal of Human Genetics* 70: 384–98.

Smith, A. (1786), *An inquiry into the nature and causes of the wealth of nations*, 5th edn. 3 vols. London: A. Strahan, and T. Cadell.

Smith, A. D. M. (2001), Establishing communication systems without explicit meaning transmission. In J. Kelemen and P. Sosik (eds.), *Advances in Artificial Life: Proceedings of the European Conference on Artificial Life*. Heidelberg: Springer, 381–90.

Smith, F., and G. A. Miller (eds.) (1966), *The Genesis of Language: A Psycholinguistic Approach*. Cambridge, Mass.: MIT Press.

Smith, J. M. B. (2001), Did early hominids cross sea gaps on natural rafts? In I. Metcalfe, J. M. B. Smith, M. Morwood, and I. Davidson (eds.), *Faunal and Floral Migration and Evolution in SE Asia-Australia*. Lisse, Netherlands: Swets & Zeitlinger, 409–16.

Smith W. J. (1977), *The Behaviour of Communicating*. Cambridge, Mass.: Harvard University Press.

Snowling, M., et al. (2001), Educational attainments of school leavers with a pre school history of speech-language impairments. *International Journal of Language and Communication Disorders* 36: 173–83.

Soffer, O., J. M. Adovasio, J. S. Illingworth, H. A. Amirkhanov, N. D. Praslov, and M. Street (2000), Palaeolithic perishables made permanent. *Antiquity* 74: 812–21.

Sommers, M. S., D. B. Moody, C. A. Prosen, and W. C. Stebbins (1992), Formant frequency discrimination by Japanese macaques (*Macaca fuscata*). *Journal of the Acoustical Society of America*, 91: 3499–3510.

Sonntag, C. F. (1921), The comparative anatomy of the Koala (*Phascolarctos cinereus*) and Vulpine Phalanger (*Trichosurus vulpecula*). *Proceedings of the Zoological Society of London* 39: 547–77.

Sowell, T. (1997), *Late-Talking Children*. New York: Basic Books.

Sperber, D., and G. Origgi (2000), Evolution, communication and the proper function of language. In P. Carruthers and A. Chamberlain (eds.), *Evolution and the Human Mind*. Cambridge: Cambridge University Press.

Staddon, J. (1988), Learning as inference. In R. Bolles and M. Beecher (eds.), *Evolution and Learning*. Hillside, NJ: Erlbaum.

Stanford, C. B. (1996), The hunting ecology of wild chimpanzees. *American Anthropologist* 98: 96–113.

Stebbins, W. C. (1983), *The Acoustic Sense of Animals*. Cambridge, Mass.: Harvard University Press.

Steels, L. (1997), Self-organizing vocabularies. In C. G. Langton and K. Shimohara (eds.), *Proceedings of the 5th International Workshop on Artificial Life: Synthesis and Simulation of Living Systems, ALIFE-96*. Cambridge, Mass.: MIT Press, 179–84.

——and F. Kaplan (2002), Bootstrapping grounded word semantics. In Briscoe, 53–73.

————A. McIntyre, and J. Van Looveren (2002), Crucial factors in the origins of word-meaning. In Wray, 252–71.

Stevens, K. N. (1972), Quantal nature of speech. In E. E. David, Jr., and P. B. Denes (eds.), *Human Communication: A Unified View*. New York: McGraw-Hill, 51–66.

——(1989), On the quantal nature of speech. *Journal of Phonetics* 17: 3–45.

——and A. S. House (1955), Development of a quantitative description of vowel articulation. *Journal of the Acoustical Society of America* 27: 484–93.

Stiner, M. C. (1998), Comment on Marean and Kim. *Current Anthropology*, 39 (suppl.): S98–S103.

——N. D. Munro, and T. A. Surovell (2000), The tortoise and the hare: small-game use, the broad-spectrum revolution, and paleolithic demography. *Current Anthropology* 41(1): 39–73.

Stoel-Gammon, C. (1985), Phonetic inventories, 15–24 months: a longitudinal study. *Journal of Speech and Hearing Research* 18: 505–12.

Stringer, C., and C. Gamble (1993), *In Search of the Neanderthals*. London: Thames & Hudson.

Stromswold, K. (2001), The heritability of language: a review and metaanalysis of twin and adoption studies. *Language* 27: 647–723.

Studdert-Kennedy, M. (1983), On learning to speak. *Human Neurobiology* 2: 191–5.

——(1998a), Introduction [to part II]: the emergence of phonology. In Hurford et al., 169–76.

——(1998b), The particulate origins of language generativity: from syllable to gesture. In Hurford et al., 202–20.

——(2000), Evolutionary implications of the particulate principle: imitation and the dissociation of phonetic form from semantic function. In Knight et al., 161–76.

——(2002), Mirror neurons, vocal imitation, and the evolution of particulate speech. In M. Stamenov and V. Gallese (eds.), *Mirror Neurons and the Evolution of the Brain and Language*. Amsterdam: Benjamins, 207–27.

——and E. W. Goodell (1995), Gestures, features and segments in early child speech. In B. de Gelder and J. Morais (eds.), *Speech and Reading*. Hove, UK: Erlbaum, 65–88.

Stuss, D. T., and D. F. Benson (1986), *The Frontal Lobes*. New York: Raven.

Sudendorff, T., and A. Whiten (2001), Mental evolution and development: evidence for secondary representation in children, great apes and other animals. *Psychological Bulletin* 127: 629–50.

Suthers, R. A., D. J. Hartley, and J. J. Wenstrup (1988), The acoustic role of tracheal chambers and nasal cavities in the production of sonar pulses by the horseshoe bat, *Rhilophus hildebrandti*. *Journal of Comparative Physiology, A* 162: 799–813.

Svoboda, J. (2000), The depositional context of the Early Upper Paleolithic human fossils from the Koneprusy (Zlaty kun) and Mladec Caves, Czech Republic. *Journal of Human Evolution* 38(4): 523–36.

Swisher, C. C. III, W. J. Rink, S. C. Antón, H. P. Schwarcz, G. H. Curtis, A. Suprijo, and Widiasmoro (1996), Latest *Homo erectus* of Java: potential contemporaneity with *Homo sapiens* in Southeast Asia. *Science* 274: 1870–74.

Szathmary, E. (2001), Origin of the human language faculty: the language amoeba hypothesis. In J. Trabant and S. Ward (eds.), *New Essays on the Origin of Language*. Berlin: de Gruyter.

Tanner, J. E., and R. W. Byrne (1996), Representation of action through iconic gesture in a captive lowland gorilla. *Current Anthropology* 37: 162–73.

Taylor, A. E., J. A. Saint-Cyr, and A. E. Lang (1990), Memory and learning in early Parkinson's disease: evidence for a 'frontal lobe syndrome'. *Brain and Cognition* 13: 211–32.

Teal, T., and C. Taylor (1999), Compression and adaptation. In D. Floreano, J. D. Nicoud, and F. Mondada (eds.), *Advances in Artificial Life*. Berlin: Springer.

Terrace, H., S. Chen, and A. Newman (1995), Serial learning with a wild card by pigeons: effect of list length. *Journal of Comparative Psychology* 109: 162–72.

Thelen, E. (1984), Learning to walk: ecological demands and phylogenetic constraints. In L. Lipsitt (ed.), *Advances in Infancy Research*, vol. 3. Norwood NJ: Ablex, 213–50.

Thieme, H. (1997), Lower Palaeolithic hunting spears from Germany. *Nature* 385: 807–10.

Thompson, D. (1961), *On Growth and Form*, abridged edn., ed. J. T. Bonner. Cambridge: Cambridge University Press.

Tinbergen, N. (1951), *The Study of Instinct*. New York: Oxford University Press.

——(1952), 'Derived' activities: their causation, biological significance, origin and emancipation during evolution. *Quarterly Review of Biology* 27: 1–32.

Titze, I. R. (1994), *Principles of Voice Production*. Englewood Cliffs, NJ: Prentice-Hall.

Tobias, P. V. (1987), The brain of *Homo habilis*. *Journal of Human Evolution* 16: 741–61.

Tomasello, M. (1992), *First Verbs: A Case Study in Early Grammatical Development*. Cambridge: Cambridge University Press.

——(1995), Language is not an instinct. *Cognitive Development* 10: 131–56.

——(ed.) (1998a), *The New Psychology of Language: Cognitive and Functional Approaches*. Mahwah, NJ: Erlbaum.

——(1998b), Reference: intending that others jointly attend. *Pragmatics and Cognition* 6: 219–34.

——(1999), *The Cultural Origins of Human Cognition*. Cambridge, Mass.: Harvard University Press.

——(2000a), The item-based nature of children's early syntactic development. *Trends in Cognitive Sciences* 4(4): 156–63.

——(2000b), Do young children have adult syntactic competence? *Cognition* 74: 209–53.

——Akhtar, N., K. Dodson, and L. Rekau (1997), Differential productivity in young children's use of nouns and verbs. *Journal of Child Language* 24: 373–87.

——and M. Barton (1994), Learning words in non-ostensive contexts. *Developmental Psychology* 30: 639–50.

——and P. Brooks (1998), Young children's earliest transitive and intransitive constructions. *Cognitive Linguistics* 9: 379–95.

————(1999), Early syntactic development. In M. Barrett (ed.), *The Development of Language*. London: UCL Press.

——and J. Call (1997), *Primate Cognition*. Oxford: Oxford University Press.

————and A. Gluckman (1997), The comprehension of novel communicative signs by apes and human children. *Child Development* 68: 1067–81.

————K. Nagell, K. Olguin, and M. Carpenter (1994), The learning and use of gestural signals by young chimpanzees: A trans-generational study. *Primates* 35: 137–54.

————J. Warren, G. T. Frost, M. Carpenter, and K. Nagell (1997), The ontogeny of chimpanzee gestural signals: a comparison across groups and generations. *Evolution of Communication* 1: 223–59.

——B. George, A. Kruger, J. Farrar, and E. Evans (1985), The development of gestural communication in young chimpanzees. *Journal of Human Evolution* 14: 175–86.

——D. Gust, and G. T. Frost (1989), The development of gestural communication in young chimpanzees: a follow up. *Primates* 30: 35–50.

Tomasello, M. and K. Zuberbühler (2002), Primate vocal and gestural communication. In M. Bekoff, C. Allen, and G. Burghardt (eds.), *The Cognitive Animal: Empirical and Theoretical Perspectives on Animal Cognition.* Cambridge, Mass.: MIT Press.

Tomblin, J. B. (in press), Genetics and language. In *Encyclopedia of the Human Genome.* London: Nature Publishing Group.

Tonkes, B. (2002), On the origins of linguistic structure: computational models of the evolution of language. Ph.D. thesis, University of Queensland.

——and J. Wiles (2002), Methodological issues in simulating the emergence of language. In Wray, 226–51.

Tooby, J., and I. DeVore (1987), The reconstruction of hominid evolution through strategic modeling. In W. G. Kinzey (ed.), *The Evolution of Human Behavior: Primate Models.* Albany, NY: SUNY Press.

Toth, N. (1985), Archeological evidence for preferential right-handedness in the lower and middle Pleistocene, and its possible implications. *Journal of Human Evolution* 14: 607–14.

Trask, L. (1996), *Historical Linguistics: An Introduction.* New York: St Martin's Press.

Traugott, E. C. (1994), Grammaticalization and lexicalization. In R. E. Asher and J. M. Y. Simpson (eds.), *The Encyclopedia of Language and Linguistics.* Oxford: Pergamon, 1481–6.

——and B. Heine (eds.) (1991a), *Approaches to Grammaticalization,* vol. 1: *Focus on Theoretical and Methodological Issues.* Amsterdam: Benjamins.

————(eds.) (1991b), *Approaches to Grammaticalization,* vol. 2: *Focus on Types of Grammatical Markers.* Amsterdam: Benjamins.

Trout, J. D. (2000), The biological basis of speech: what to infer from talking to the animals. *Psychological Review* 108: 523–49.

Trudgill, P. (1992), Dialect typology and social structure. In E. H. Jahr (ed.), *Language Contact.* New York: Mouton de Gruyter, 195–211.

——(1999), *The Dialects of England,* 2nd edn. Oxford: Blackwell.

Turing A. M. (1936), On computable numbers with an application to the *Entscheidungsproblem. Proceedings of the London Mathematical Society* 2(42): 230–65.

——(1950), Computing machinery and intelligence. *Mind* 49: 433–60.

Turkel, W. (2002), The learning guided evolution of natural language. In Briscoe 235–54.

Underhill, P. A., P. D. Shen, A. A. Lin, L. Jin, G. Passarino, W. H. Yang, et al. (2000) Y chromosome sequence variation and the history of human populations *Nature Genetics* 26: 358–61.

Valiant, L. G. (1984), A theory of the learnable. *Communications of the ACM* 27 1134–42.

Vallortigara, G., L. J. Rogers, and A. Bisazza (1999), Possible evolutionary origins of cognitive brain function. *Brain Research Reviews* 30: 164–75.

van der Lely, H. K. J., and V. Christian (1998), Lexical word formation in specifically language impaired children. MS, Department of Psychology, Birkbeck College

University of London.

——S. Rosen, and A. McClelland (1998), Evidence for a grammar-specific deficit in children. *Current Biology 8*, 1253–8.

——and L. Stollwerck (1996), A grammatical specific language impairment in children: an autosomal dominant inheritance? *Brain and Language* 52: 484–504.

'an Everbroeck, E. (1999), Language type frequency and learnability: a connectionist appraisal. In *Proceedings of the 21st Annual Cognitive Science Society Conference*. Mahwah, NJ: Erlbaum, 755–60.

'an Valin, R. D., and R. J. Lapolla (1997), *Syntax: Structure, Meaning, and Function*. Cambridge: Cambridge University Press.

'apnik, V. N. (1988), *Statistical Learning Theory*. New York: Wiley.

——and A. Y. Chervonenkis (1971), On the uniform convergence of relative frequencies of events to their probabilities. *Theory of Probability and its Applications* 17: 264–80.

————(1981), The necessary and sufficient conditions for the uniform convergence of averages to their expected values. *Teoriya veroyatnostei i Ee Primeneniya* 26: 543–64.

'argha-Khadem, F., K. Watkins, R. Passingham, and P. Fletcher (1995), Cognitive and praxic deficits in a large family with a genetically transmitted speeech and language disorder. *Proceedings of the National Academy of Sciences* 92: 930–33.

——C. J. Price, J. Ashbruner, K. J. Alcock, A. Connelly, R. S. J. Frackowiak, K. J. Friston, M. E. Pembrey, M. Mishkin, D. G. Gadian, and R. E. Passingham (1998), Neural basis of an inherited speech and language disorder. *Proceedings of the National Academy of Sciences* 95: 12695–12700.

'ihman, M. M., and S. Velleman (1989), Phonological reorganization: a case study. *Language and Speech* 32: 149–70.

'illa, P. (1981), Matières premières et provinces culturelles dans l'Acheuléen français. *Quaternaria* 23: 19–35.

'leck, E. (1970), Etude comparative onto-phylogénétique de l'enfant du Pech-de-L'Aze par rapport à d'autres enfants neanderthaliens. In D. Feremback (ed.), *L'enfant Pech-de-L'Aze*. Paris: Masson, 149–86.

'oegelin, C. F., and F. M. Voegelin (1977), *Classification and Index of the World's Languages*. New York: Elsevier.

Vaddell, P. J. and D. Penny (1996), Evolutionary trees of apes and humans from DNA sequences. In Lock and Peters (eds.), 53–73.

Vaddington, C. (1942), Canalization of development and the inheritance of acquired characters. *Nature* 150: 563–5.

Valker, S. (1983), *Animal Thought*. London: Routledge & Kegan Paul.

Vallman, J. (1992), *Aping Language*. New York: Cambridge University Press.

Vang, W. S. Y. (1998), Language and the evolution of modern humans. In K. Omoto and P. V. Tobias (eds.), *The Origins and Past of Modern Humans*. Singapore: World Scientific, 247–62.

Vanner, E., and L. Gleitman (1982), Introduction. In Wanner and Gleitman (eds.),

*Language Acquisition: The State of the Art.* Cambridge, Mass.: MIT Press, 3–

Ward, R., and C. Stringer (1997), A molecular handle on the Neanderthals. *N* 388: 225–6.

Watkins, K. E., N. F. Dronkers, and F. Vargha-Khadem (2002), Behavioural ana of an inherited speech and language disorder: comparison with acquired sia. *Brain* 125: 452–64.

Webb, K. E. (1977), An evolutionary aspect of social structure and a verb *American Anthropologist* 79: 42–9.

Weiner, J. (1994), *The Beak of the Finch.* New York: Vintage.

Weiskrantz, L. (ed.). (1988), *Thought Without Language.* New York: Oxford Ur sity Press.

Weissengrüber, G. E., G. Forstenpointner, G. Peters, A. Kübber-Heiss, and W. T. *i* (2002), Hyoid apparatus and pharynx in the lion (*Panthera leo*), jaguar *thera onca*), tiger (*Panthera tigris*), cheetah (*Acinonyx jubatus*), and domest *(Felis silvestris f. catus). Journal of Anatomy* 201: 195–209.

Wendt, W. E. (1976), 'Art mobilier' from the Apollo 11 Cave, South West A *South African Archaeological Bulletin* 31: 5–11.

Werner, H., and B. Kaplan (1963), *Symbol Formation: An Organismic-Develop tal Approach to Language and the Expression of Thought.* New York: Wiley.

Wernicke, C. (1874), *Der Aphasische Symptomencomplex.* Breslau: Cohn & Wei

West, G.B., J. H. Brown, and B. J. Enquist (1997), A general model for the orig allometric scaling laws in biology. *Science* 276: 122–6.

Wexler, K., and P. Culicover (1980), *Formal Principles of Language Acquis* Cambridge, Mass.: MIT Press.

White, R. (1989), Production complexity and standardization in early Aur cian bead and pendant manufacture. In P. A. Mellars and C. Stringer (eds.) *Human Revolution.* Princeton, NJ: Princeton University Press, 366–90.

Whiten, A. (2000), Primate culture and social learning. *Cognitive Science 2* 477–508.

Wilkins, W. K., and J. Wakefield (1995), Brain evolution and neurolinguistic pr ditions. *Behavioral and Brain Sciences* 18: 161–226.

Williams, G. C. (1966), *Adaptation and Natural Selection: A Critique of Current Evolutionary Thought.* Princeton, NJ: Princeton University Press.

Wind, J., E. G. Pulleyblank, E. de Grolier, and B. H. Bichakjian (eds.) (1989), *St* in *Language Origins,* vol. 1. Amsterdam: Benjamins.

Worden, R. (1995), A speed limit for evolution. *Journal of Theoretical Biology* 137–52.

——(1998), The evolution of language from social intelligence. In Hurford 148–68.

——(2002), Linguistic structure and the evolution of words. In Briscoe, 75–11

Wray, A. (1998), Protolanguage as a holistic system for social interaction. *Lang and Communication* 18: 47–67.

——(2000a), Holistic utterances in protolanguage: the link from primates to humans. In Knight et al., 285–302.

——(ed.) (2002b), *The Transition to Language*. Oxford: Oxford University Press.

Wright, A. A., and J. J. Rivera (1997), Memory of auditory lists by rhesus monkey (*Macaca mulatta*). *Journal of Experimental Psychology: Animal Behavior Processes* 23: 441–9.

Wurz, S. (1999), The Howiesons Poort backed artefacts from Klasies River. *South African Archaeological Bulletin* 54(169): 38–50.

Xuerob, J. H., B. E. Tomlinson, D. Irving, R. H. Perry, G. Blessed, and E. K. Perry (1990), Cortical and subcortical pathology in Parkinson's disease: relationship to Parkinsonian dementia. In M. B. Streifler, A. D. Korezyn, J. Melamed, and M. B. H. Youdim (eds.), *Parkinson's Disease: Anatomy, Pathology and Therapy*. New York: Raven Press, 35–9.

Yamauchi, H. (2000), Evolution of the LAD and the Baldwin Effect. MA dissertation, University of Edinburgh.

——(2001), The difficulty of the Baldwinian account of linguistic innateness. In J. Keleman and P. Sosik (eds.), Advances in Artificial Life: *Proceedings of the 6th European Conference on Artificial Life*. Heidelberg: Springer.

Yellen, J. E., A. S. Brooks, E. Cornelissen, M. J. Mehlman, and K. Stewart (1995), A Middle Stone Age worked bone industry from Katanda, Upper Semliki Valley, Zaire, *Science* 268: 553–6.

Zipf, G. K. (1935), *The Psycho-biology of Language: An Introduction to Dynamic Philology*. Boston, Mass.: Houghton Mifflin.

Zsiga, E. C. (1995), An acoustic and electropalatographic study of lexical and postlexical palatalization in American English. In B. Connell and A. Arvaniti (eds.), *Phonology and Phonetic Evidence*. Cambridge: Cambridge University Press, 282–302.

Zuberbühler, K. (2002), A syntactic rule in forest monkey communication. *Animal Behaviour* 63: 293–9.

Zukow-Goldring, P., M. A. Arbib, and E. Oztop (2003), Language and the mirror system: a perception/action based approach to communicative development (to appear).

# Index

The index covers *Notes on Contributors* and Chapters 1–17 and should be used in conjunction with the list of contents. Author index entries are not exhaustive, referring to those mainly discussed or referred to in the text and cited in the further reading sections.